COMPUTATIONAL CULTURAL NEUROSCIENCE

This book provides novel insights into the study of empirical computational approaches in the field of cultural neuroscience. It discusses and analyses topics such as cultural intelligence, cultural machine learning, cultural brain dynamics and cultural security.

This comprehensive text engages with computational principles to guide the research on the influence of cultural environments on human genetics. It explores the theoretical and methodological approaches involved in computational neuroscience. The author elucidates how cultural processes intersect with the structural organization of the nervous system, contributing to the study of computational principles and neural information-processing mechanisms at the cultural level. Research in this subject area can help provide better understanding of the role of computation in cultural neuroscience, stimulating further research into practice and policy.

Computational Cultural Neuroscience: An Introduction is the ideal resource for academics, researchers and students of psychology, neuroscience, computer science or philosophy, who are interested in cultural neuroscience.

Joan Y. Chiao is the Director of the International Cultural Neuroscience Consortium and the Director of the International Consortium for Computational Cultural Neuroscience. She received her Ph.D. from Harvard University and B.S. from Stanford University. Her research on computational cultural neuroscience examines the computational principles of cultural processes in the structure and function of the nervous system.

Essays in Cultural Neuroscience
Series Editor: Joan Y. Chiao

Essays in Cultural Neuroscience showcases scholarly work over a wide range of areas taking a cultural neuroscientific approach. The series is designed to highlight foundations of the fields for interdisciplinary discovery that contribute to our understanding of the neurobiological bases of culture and the mutual construction of culture and neurobiology across evolutionary and developmental timescales. Themes and topics in cultural neuroscience may include a range of cross-disciplinary perspectives, including ethical, scientific and philosophical issues related to the study of the neurobiological bases of cultural processes. Books are designed to examine the systematic study of theoretical principles, methodological approaches and empirical paradigms in the fields, including developmental and evolutionary perspectives. This is fascinating reading for students, researchers, and academics across a range of disciplines including psychology, neuroscience, cultural studies, sociology and the social sciences more generally.

Philosophy of Computational Cultural Neuroscience
Joan Y. Chiao

Computational Cultural Neuroscience
An Introduction
Joan Y. Chiao

For more information about this series, please visit: www.routledge.com/Essays-in-Cultural-Neuroscience/book-series/ECN

COMPUTATIONAL CULTURAL NEUROSCIENCE

An Introduction

Joan Y. Chiao

Routledge
Taylor & Francis Group

NEW YORK AND LONDON

Designed cover image: © Getty Images\bestbrk

First published 2025
by Routledge
605 Third Avenue, New York, NY 10158

and by Routledge
4 Park Square, Milton Park, Abingdon, Oxon, OX14 4RN

Routledge is an imprint of the Taylor & Francis Group, an informa business

British Library Cataloguing-in-Publication Data
A catalogue record for this book is available from the British Library

ISBN: 9781032470177 (hbk)
ISBN: 9781032470160 (pbk)
ISBN: 9781003384236 (ebk)

DOI: 10.4324/9781003384236

Typeset in Galliard
by Deanta Global Publishing Services, Chennai, India

CONTENTS

11 Cultural Intelligence 367

12 Computational Cultural Neuroscience and Public Policy 399

Conclusion 436

ILLUSTRATIONS

Figures

Tables

PREFACE

The goal of the book *Computational Cultural Neuroscience – An Introduction* is to provide an overview of the study of computational approaches to cultural neuroscience. Computational cultural neuroscience is a field of study that investigates the computational foundations of culture and neuroscience. Research on computational cultural neuroscience generates systematic studies of the computational basis of the cultural basis of brain and behavior. Since antiquity, scholars and philosophers alike have long shown interest in the nature of the mind and its physical basis. The development of empirical approaches to the study of science and technology of the 18th century revealed novel theoretical, methodological and empirical approaches to the scientific study of culture and neuroscience and the foundations of computation as a basis of the physical states of culture and the brain. The scientific and technological advancement of the study of computation in parallel with the burgeoning of empirical progress on the understanding of the nature of culture and neuroscience encompasses a wide expanse of theoretical approaches and empirical studies from which to investigate the computational basis of the cultural brain.

The book entails review of the causation of computation and culture on the physical states of the nervous system and how culture influences the computational basis of the mind, brain and behavior. Evidence-based approaches to the study of computational cultural neuroscience contribute to the innovation of practical applications of artificial intelligence, bioengineering and translational medicine among others. The study of computational cultural neuroscience provides a foundation for the understanding of the fundamentals of computational cultural neuroscience as well as its practical applications in health, medicine, public health and related fields of study of technology, law, public policy among others. The book incorporates review of interdisciplinary themes on technology, society and public policy as well as biology, artificial intelligence and cultural studies at the intersection of the scholarly studies of humanities and basic sciences of broad interest.

ACKNOWLEDGEMENTS

The goal of the book is to provide a review of fundamentals of computational cultural neuroscience. The book reviews theoretical, methodological and empirical approaches to the study of computational approaches to cultural neuroscience. The advancement of the field of computational cultural neuroscience allows for novel insights into the study of computation and culture, culture and neuroscience, and the computational neuroscience of culture and behavior. The study of computational principles guides the study of the culture and biology of mind, brain and behavior. Research on computational cultural neuroscience informs broader issues of computation and culture and its translation into practice and policy.

The book is aimed at undergraduates, graduate students, researchers and faculty interested who are teaching on psychology, neuroscience, philosophy, computer science or related interdisciplinary fields of study.

The author is grateful to the early mentorship of Nalini Ambady, Jennifer Eberhardt, John Gabrieli, Alexandra Golby, Hazel Markus and Robert Zajonc of Stanford University, Matthew Lieberman of University of California Los Angeles, Ken Nakayama and Dan Schacter of Harvard University and Tetsuya Iidaka of Nagoya University for their tutelage. She is also grateful to Peter Godfrey-Smith, Daphne Kohler, Eric Roberts, Kenneth Tayler and Tom Wasow of Stanford University for their dedication and encouragement of interdisciplinary studies.

The author thanks Pamela Collins, Beverly Pringle, Su Yeon Lee-Tauler, Makeda Williams, Tamara Lewis-Johnson and Ishmael Amarreh of the National Institutes of Health. She is grateful to Tokiko Harada, Yoko Mano, Hidetsugu Komeda and Norihiro Sadato of National Institutes of Physiological Sciences, Jack van Honk of Utrecht University for their intellectual and scientific commitment to international collaboration and international cooperation. She is grateful to Michio Nomura, Zhang Li, Genna Bebko, Vani Mathur and Bobby Cheon of the Laboratory for Social Affective and Cultural Neuroscience for their intellectual and scientific contributions and international cooperation on the innovation of science and technology in psychology, neuroscience and related fields of study.

INTRODUCTION

The book *Computational Cultural Neuroscience: An Introduction* investigates computational approaches to the study of cultural neuroscience. Research on computational cultural neuroscience generates novel knowledge of the computational foundations of the cultural basis of the mind, brain and behavior. The systematic study of computational cultural neuroscience entails the theoretical, methodological and empirical foundations of the systematic study of cultural influences on the neural basis of behavior. The interdisciplinary study of computational cultural neuroscience contributes to the scientific and technological innovation on the role of computation on the cultural and neural basis of mind and behavior. The systematic investigation of computational principles that underlie cultural neuroscience contributes to the understanding of the computational and biology of culture and behavior.

Fundamental research on computational cultural neuroscience investigates the lawlike principles that underlie the computational basis of cultural patterns of the brain and behavior in the world. The systematic investigation of computational cultural neuroscience entails the postulation of the theoretical foundations of computational principles and their physical instantiation in the biology and culture of behavior. Theoretical foundations of the field of study include the formal modeling of the causation of the basic mechanisms and processes that contribute to the computational basis of the biology and culture of behavior across levels of analysis. The theoretical advancement of computational cultural neuroscience contributes to the understanding of the generalization of the lawlike principles of computation as an organizing principle of the biological and cultural

DOI: 10.4324/9781003384236-1

basis of behavior. The theoretical foundations of computational cultural neuroscience have practical applications to the innovation of scientific and technological development of artificial intelligence, bioengineering and translational medicine, among others.

The theoretical development of computational cultural neuroscience includes the formalization of causal models that posit on the functional pathways of cultural information processing of the brain and its output. The formal theories of the field of study provide a conceptual basis for the postulation on the directionality and magnitude of causation of information processing of mechanisms and processes of the cultural brain and its relation to behavior. The scientific theories of computational cultural neuroscience advance mechanistic understanding of the causation of functional mechanisms of biological organisms that are responsible for the cultural and biological patterning of behavior. The identification of the functional mechanism of biological organisms that contribute to the cultural patterns of brain and behavior that facilitate understanding of the functionalism of biological organisms.

The fundamental principles of computation as an organizing principle of the cultural mind, brain and behavior serve as a foundation for the theoretical understanding of the computational basis of the cultural brain and the cultural basis of mental and biological computation of biological organisms. The fundamental basis of computation provides conceptual approaches to the understanding of biological organisms as computing machines. The biological computation of organisms details the role of computational principles on the physical states of minds and machines. The cultural mind as a biological computing machine illustrates the plausibility of biological organisms as cultural machines. The study of the computation of the cultural brain contributes to the understanding of the computational basis of culture and its role in the functional basis of brain and behavior.

Methodological approaches to the study of computational cultural neuroscience consider the role of causation on the identification of the functional pathways of the brain and behavior that are fundamental to the cultural neurocomputation of biological organisms in the world. The development of methodological approaches to the study of the neurocomputation of the brain that is culturally and ethnically diverse contributes to the understanding of the fundamental organization of the nervous system of diverse populations, cultures and settings. The methodological development of tools and technologies for the study of the neurocomputation of the brain and its functional relation to culture and behavior broadens the understanding of the biology and culture of behavior and the wide variation of human populations and cultures.

Empirical foundations of computational cultural neuroscience illustrate the systematic approaches to the testing of scientific theories on the computation of the cultural brain. Empirical investigation of the computation of the cultural brain shows the contribution of observation and inference on the empiricism of the field of study. The systematic study of computational approaches to the cultural brain provides a breadth of empirical approaches to the identification of the functional mechanisms of cultural organisms. The empirical study of computational cultural neuroscience contributes to the evidence-based approaches on the understanding of computation and culture and the computational level of analysis of the cultural brain.

The study of computational cultural neuroscience is integral to the interdisciplinary study of computer science, culture and neuroscience that informs practical applications on health, medicine, public health and related fields of technology, public policy among others. Evidence-based approaches to research on computational cultural neuroscience inform the innovation of scientific and technological approaches on culture and computation and the biology and culture of behavior at the computational level. The building of evidence-based knowledge contributes to the breadth of evidence base that informs on the development of medical applications of health systems in medicine. The scientific and technological innovation of computational cultural neuroscience is of importance to the health and health equity of diverse populations and cultures. The multidisciplinary study of computational cultural neuroscience illustrates the breadth of evidence-based research on culture and health and its practical application on culture, health and society.

Theoretical Foundations

Theoretical foundations of computational cultural neuroscience serve as a fundamental basis for the integrative study of the computational basis of the biology and culture of behavior. Scientific theories of computation and culture contribute to the conceptualization of the fundamental principles that characterize the organization of the nervous system and its neurocomputation of culture and behavior. The conceptual development of scientific theories includes the formulation of conceptual language for the identification of the functional mechanisms and processes that are of importance to the computational study of culture, brain and behavior. The computational level of analysis of the nervous system encompasses a wide breadth of functional mechanisms and computational algorithms for the production of complex behavior. The characterization of the functional mechanisms of the computational cultural brain contributes to the conceptual understanding of the fundamental principles of the nervous system and its role on the cultural computing of biological organisms.

Chapter 1 describes the foundations of computational cultural neuroscience. The chapter reviews how culture affects neural information processing at the computational level of the brain and presents a rationale of computational approaches for the study of the biology and culture of behavior.

Chapters 2 to 4 encompass a broad overview of the cultural computation of mind and machines. Chapter 2 discusses the role of cultural computation as a fundamental basis of biological organisms and the cultural computing of the biological machine. Chapter 2 investigates how the computation of biological organisms produces the cultural information of functional mechanisms and processes. Chapter 3 reviews the cultural mental computation as a functional process of the cultural mind. The chapter discusses the cultural computation of the mind as a computing machine. The chapter reviews cultural mental computation as a fundamental basis of the functional processes of the cultural mind. Chapter 4 introduces cultural machine computation as a conceptual basis for the understanding of the computation of culture as a mechanistic process. The chapter reviews a mechanistic understanding of cultural computation and the role of machine computation on cultural information processing.

Chapter 5 provides a review of the cultural neurocomputation of the brain. Cultural neurocomputation entails the understanding of cultural information processing of neural mechanisms of the nervous system of biological organisms. Chapter 5 discusses the role of cultural neurocomputation on the generation and production of cultural information processing as the functional processes of neurons and populations of neurons. Chapter 5 reviews the role of neurocomputation as a functional basis of the neuronal processing of cultural information of the brain.

Chapter 6 reviews the foundations of cultural neural networks. The chapter provides an overview of the fundamental basis of cultural neural networks and the identification of neural networks of cultural information processing. Chapter 6 investigates the functional basis of cultural neural networks and the role of functional mechanisms on the neural information processing of culture. Chapter 6 includes a review of the role of interconnectivity of the brain regions on the patterns of cultural neural networks.

Chapter 7 introduces the fundamentals of cultural brain dynamics. It provides a review of the fundamental principles of cultural brain dynamics and the contribution of dynamical systems and information theory to the understanding of the autonomous dynamical processes of the cultural brain and behavior. The chapter discusses the importance of cultural brain dynamics for the understanding of cultural neural networks and the functional relations of culture, brain and behavior.

Chapter 8 details the fundamentals of cultural machine learning. Cultural machine learning encompasses the investigation of computational discovery on the characterization of the patterns of large data sets. The chapter reviews the importance of the formulation of cultural models on the characterization of the prediction and inference of the cultural patterns of neural networks.

Chapter 9 describes the foundations of the cultural connectome. The cultural connectome describes the interconnectivity of brain regions and its role in the computation of culture and behavior. Chapter 9 reviews the role of structural and functional connectivity of the brain on the identification of the computational basis of culture and behavior of the brain.

Chapter 10 examines the practical applications of computational cultural neuroscience for innovation of culture and technology. The chapter discusses the impact of evidence-based research on computational cultural neuroscience on the research and development of culture and technology. Chapter 10 reviews the importance of the evidence base of basic research on the scientific and technological innovation of culture and the development of culture and technology for the educational and social enrichment of society and the public.

Chapter 11 reviews the foundations of cultural intelligence. The chapter discusses the fundamental basis of cultural intelligence and the role of artificial technology on cultural intelligence. It discusses the impact of basic research on the development of artificial cultural intelligence and technological applications that are built on artificial cultural intelligence for the practical benefit of society and the public.

Chapter 12 discusses the foundations of policy-based approaches to computational cultural neuroscience. The chapter reviews the advancement of basic research on culture and technology and its practical benefits on the development and implementation of policy on culture and health. Chapter 12 reviews the fundamentals of the scientific and technological innovation on computational cultural neuroscience and its practical applications.

The chapters review fundamental themes of computational cultural neuroscience and its practical importance on the related fields of health, medicine, public policy and related areas of study. The chapters of the book provide an introduction to the computational level of the cultural brain and the understanding of the computational foundations of culture and neuroscience. The scientific and educational understanding of computational cultural neuroscience contributes to the societal enrichment of the public on cultural computation and its practical applications in health, medicine, public policy among others. The advancement of research on computational cultural neuroscience contributes to the innovation of science and technology and broadens the conceptual scope of interdisciplinary research on humanities and basic sciences.

Implications

The development of evidence-based research on computational cultural neuroscience provides a wide range of benefits to the innovation of science and technology on culture and health and its practical applications for the benefit of society and the public. The development of tools and technologies of computational cultural neuroscience provides unprecedented approaches to the empirical study of the cultural computation of biological organisms of the natural world and the bioengineering of the cultural computation of artificial and real systems. The advancement of evidence-based approaches on computational cultural neuroscience illuminates the development of tools and technologies for the improvement of the practical applications of health systems in medicine and the development of culture and technology for the benefit of society and the public.

The policy-based research on computational cultural neuroscience is of importance to the understanding of how research on computational cultural neuroscience informs policy development and implementation on culture and technology. The policy development on culture and technology is of importance to the understanding of the policies and programs that contribute to the breadth and scope of the practical applications of culture and technology for the benefit of society and the public. The development of policy-based approaches on computational cultural neuroscience facilitate the formulation of policies and programs on culture and technology that provide scientific and educational enrichment on the impact of culture and technology for the benefit of the public.

1

COMPUTATIONAL CULTURAL NEUROSCIENCE

Computational approaches to cultural neuroscience

Introduction

History of computation

From the early 19th to 20th centuries, scholarly interest in computation arose from the early notion of the computation of machines as automated production. From the analytic engine and the Turing machine to the modern computer and artificial intelligence, the elaboration of scholarly inquiry on computation and its physical manifestation has brought forth an amalgamation of conceptual designs of machines capable of automated function. The early insights into the computing machine led to the development of the notions of the fundamental units of information and its representation as physical states. The development of machines as causal systems with sets of physical states has brought forth a range of machine designs that provide a foundation for the construction of machines or physical devices with capabilities of functional performance. The design of computing machines has brought forth an evidential basis for the understanding of computation and inferential reasoning in the natural world.

The early insights of the design of the computing machine as physical devices serve as a fundamental principle of computation. The systematic development of computing machines as physical devices with capability for automated functional performance illustrated the feasibility of the construction of physical systems for machine computation. The conceptualization of machine computation as consistent with multiple representations ("analog", "digital") illustrates the notion of computing machines

DOI: 10.4324/9781003384236-2

as physical systems with distinct sets of physical states that show a range of functional performance. The theory confirmation of scientific hypotheses on computation regarding the physical realism of machine computing led to the design of the modern computer as one of the physical devices capable of automated machine production.

The earliest notions of machine computation as akin to biological computation brought forth philosophical interest in the functional equivalence of minds and machines. The scholarly inquiry into the fundamentals of computation is a pursuit of the understanding of the mind and its physical realizers. The philosophical inquiry into the mind brought forth early insights into the fundamental principles of computation and its role in calculation. The development of the conceptual landscape of the mind as a computing machine brought forth elaboration of the natural realism of logical reasoning and its physical basis. The foundations of computation illustrate the plausibility of the mind as a computing machine with multiple physical realizers.

The scientific study of computational cultural neuroscience illustrates the reliance on empiricism in the conceptual and practical understanding of the computation of biological organisms and the natural world and the role of computation on the biology and culture of behavior. The conceptualization of the cultural computation of biological organisms provides a mechanistic understanding of the functional mechanisms of biological organisms that govern culture and behavior. The characterization of the lawlike principles of the computational level of the cultural brain are foundational to an understanding of the cultural patterns of the brain and behavior and its causal and functional significance in the natural world. The identification of the cultural patterns of biological organisms and the natural world illustrates the functional plausibility and physical instantiation of culture as biological computation. The understanding of the computation of cultural life patterns contributes to the bioengineering of the cultural computation of real and artificial systems.

Computational cultural neuroscience

Computational cultural neuroscience is the study of the computation of culture and the nervous system. The study of computational cultural neuroscience investigates computational principles that characterize the biology and culture of behavior. The identification of the computational principles of the nervous system provides a conceptual foundation for the study of the physiological basis of culture and behavior. International, cross-cultural research on the computational level of the brain provides a comparative approach to the understanding of the computational basis of cultural information-processing mechanisms that underlie the consciousness and higher-level function of behavior. The international comparison of the computational level of the cultural brain contributes to the understanding of the neuro-computation of functional processes and mechanisms that underlie culture and behavior. The characterization of the functional mechanisms of culture and behavior informs on the understanding of the fundamental basis of the wide variation of behavior of populations and cultures.

Research on computational cultural neuroscience includes a wide range of thematic approaches to the study of the computational level of the cultural brain. Themes of computational cultural neuroscience comprise the systematic study of the computational approaches to cultural neuroscience, neurocomputational analysis of culture and behavior, cultural neural networks, cultural brain dynamics, cultural connectome and the cultural machine learning of behavior among others. The wide range of thematic approaches to the understanding of the cultural computation of biological organisms contributes to the characterization of the cultural information-processing mechanisms of the brain and its relation to higher-level function and behavior. The research on computational cultural neuroscience is important to the study of the brain basis of culture and behavior, the brain–behavior relationships of culture and the individual and cultural variation of behavior of human populations.

The study of computational cultural neuroscience informs on the considerations regarding the philosophical inquiry of the mind. The scholarly inquiry of computational cultural neuroscience entails postulation on the relation of the cultural mind to the cultural brain, the function of the cultural mind as a cultural computer, and the causality and actuality of the cultural mind as a form of cultural computation. The detailed characterization of the computational level of the cultural brain encompasses an understanding of the determinism of the cultural mind as a cultural brain, the functional mapping of cultural states and the multiple realizability of the cultural mind as a computing machine. The scholarly inquiry of computational cultural neuroscience broadens the scope of scholarly inquiry on culture, mind and machines with the classical and contemporary approaches of multidisciplinary study.

The "Computational Cultural Brain Hypothesis"

The "computational cultural brain hypothesis" postulates on the encoding of cultural information within patterns of functional brain activity of the brain. The "computational cultural brain hypothesis" asserts that the computational level of cortical organization is comprised of the functional components of the core capacities that are essential to the functional performance of culture and behavior. From cellular and molecular mechanisms to cortical neural networks, the computational mechanisms of the cultural brain are foundational to the representation and processing of cultural information and the functional activity of cultural and behavioral performance. The study of the neurocomputation of functional mechanisms of the brain is important for an understanding of the encoding and decoding of cultural information as the representational content of the physiological activity of the brain. The understanding of the brain basis of culture and behavior contributes to the understanding of the computational mechanisms that contribute to the functional organization of the nervous system and its functional consequences for multilevel adaptation.

The study of the "computational cultural brain hypothesis" involves understanding how neurons and networks of neurons encode and decode cultural information at the level of computation of the brain and its cortical organization. The systematic investigation of the "computational cultural brain hypothesis" ascertains on the distinct types of neuronal response that are responsible for the representation of cultural information and its transformation into cultural output. The cultural computation of the brain demonstrates the importance of the computation of neuronal mechanisms on the encoding and decoding of sensory data of the cultural environment. The detailed characterization of the neurocomputation of the functional mechanisms of the cultural brain contributes to a broader understanding of the multiple realizability of culture in the world.

The "computational cultural brain hypothesis" entails understanding of the structural and dynamical aspects of cortical brain organization and its relation to culture and behavior. The structural and functional principles of cortical organization facilitate the information processing of functional mechanisms across levels of processing. The identification of the computational principles that are of importance to the processing of cultural information within cortical organization shows the role of neurocomputation on the representation and interpretation of cultural patterns of brain activity. The computational level of cortical organization broadens the scope of understanding of the role of cultural information processing mechanisms on the adaptability and causality of culture and behavior.

Fundamentals of computational cultural neuroscience

The fundamentals of computational cultural neuroscience comprise the understanding of the organizing principles of the cultural brain. The structural and functional principles of the brain organization contribute to the understanding of the computational level of cultural brain organization. The structural principles of brain organization consist of the multilevel processing of the nervous system and the interrelations of the representation of information across levels of processing. The multilevel processing of the nervous system is comprised of the computational level and the algorithmic level of the information processing of functional mechanisms across cortical brain organization. The levels of processing of cortical organization describe the functional role of the computation of neural mechanisms and the representation of information processing across levels of analysis. From neurons to networks of neurons, the computation of information-processing mechanisms illustrates the role of the physiological basis of biological organisms as a means of the functional basis of cultural and behavioral adaptation. Across levels of analysis, the representation of information-processing mechanisms of cortical brain organization contributes to the transformation of sensory input into motor output of the biological organism that facilitates multilevel adaptation.

The structural principles of brain organization consist of the characterization of the functional pathways that show dedicated information processing that is content-specific. The functional pathways of specialized information-processing mechanisms illustrate the functional specificity of cortical brain regions and its role in the structural and functional organization of the brain. The content-specificity of cortical brain regions demonstrates the domain specificity of information processing mechanisms and the functional role of specialized processing on the automaticity of inference and behavior. The structural basis of cortical brain organization shows the functional mapping of cortical brain regions to dedicated processing that is content-specific. The functional specificity of cortical pathways shows the importance of dedicated, specialized information processing on cortical brain organization.

The cultural patterns of cortical brain organization illustrate the responsivity of the functional pathways of sensory and cognitive systems to information processing that it is culture-specific. The cultural specificity of the information processing of the functional pathways of higher-level function demonstrates the cultural patterns of neurocomputation that are important to culture and behavior. The information processing of functional mechanisms that is culture-specific shows the role of cultural patterns on the functional pathways of consciousness and higher-level function. The detailed characterization of the cultural patterns of functional pathways illustrates the importance of culture on the information-processing mechanisms of cortical brain organization.

The structural and functional connectivity of brain organization facilitates an understanding of the functional principles of cultural information-processing mechanisms and the mechanistic basis of the cultural computation of brain organization. The interconnectivity of the cultural brain comprises the neural networks of culture and behavior and its functional basis. The cultural patterns of neural networks consist of the structural and functional connectivity of functional pathways that show co-occurrence with culture and behavior. The characterization of the interconnectivity of the cultural brain organization is foundational to the understanding of the structural and functional connectivity of cultural neural networks and their relation to behavior.

The dynamical organization of the cultural brain describes the autonomous brain dynamics that contribute to culture and behavior. The dynamical patterning of cultural brain activity describes the changes of autonomous brain dynamics that are characteristic of the cultural patterns of brain and behavior. The identification of cultural brain dynamics provides an electrophysiological basis for the understanding of the dynamical aspects of the functional relationships of brain and behavior across cultures and populations. The fluctuations of electrophysiological activity of brain dynamics describes the role of physiological activity as a functional basis of culture and behavior. The characterization of cultural brain dynamics describes the role of culture and complexity on the functional patterns of brain and behavior and its causal–functional significance.

The description of the cultural connectome broadens the understanding of the role of connectivity on the neurocomputation of culture and behavior. The interconnectivity of brain organization illustrates the level of structural and functional connectivity of interconnected brain regions and its causal influence on culture and behavior. The directionality and magnitude of interconnectivity of brain organization is important to the characterization of the cultural patterns of brain activity and behavior. The study of the cultural connectome contributes to the characterization of the role of connectivity on cultural neurocomputation and the functional role of interconnectivity on brain-behavior relationships.

The study of the cultural patterns of the brain and behavior provides large-scale data as informatics that are of importance to the prediction and explanation of the cultural patterns of brain activity and behavior. The machine learning of culture and behavior is foundational to the prediction and inference of cultural patterns of large data sets. The prediction and inference of cultural patterns is fundamental to the learning of culture and behavior. The computational discovery of the cultural patterns of large data sets facilitates the extraction of cultural patterns that are predictive of the inferential learning of culture and behavior. The development of cultural machine learning approaches to the study of culture and behavior is fundamental to the understanding of the complexity of large data sets and the prediction and inference of cultural patterns from information.

The development of learning machine algorithms of culture for the prediction and extraction of the cultural patterns of the brain that predict mental inference is fundamental to the understanding of the neural information processing of the cultural brain. The characterization of cultural learning algorithms for the identification of the cultural patterns of functional brain activity contributes to the understanding of the functional relations of brain activity and mental inference. The cultural patterns of functional brain activity that predict mental inference demonstrate the functional use of learning machine algorithms for the prediction and inference of culture and behavior.

Cultural computation

Research on computational cultural neuroscience is important to the study of cultural computation. The study of cultural computation is fundamental to understanding the computational foundation of the culture of real and artificial systems. The computational foundation of culture consists of the characterization of culture as the information processing of real and artificial systems. Cultural computation describes the computational processing of cultural information of biological and physical systems and the physical states of culture as a mechanism of computation. The description of culture as a computing machine illustrates the importance of physical states as a characterization of culture as information processing (Table 1.1). The cultural computation of biological organisms demonstrates the functional plausibility of cultural information processing of the functional mechanisms of biological organism as a mechanistic basis of computation. The physical states of the biological organism serve as a description of the states of change of the biological computing machine and its functional role. The cultural computation of biological organisms describes the changes of physical states that comprise the processing of cultural information. The formulation of the cultural computation of biological organisms as computational models facilitates the design of algorithms that facilitate the cultural computation of real and artificial systems.

The computational level of the cultural mind describes the functional basis of the mental content of culture. Cultural mental computation consists of the computational and functional basis of cultural mental content. The informational content of the cultural mind comprises cultural information that contributes to the learning and experience of the biological organism. The cultural computation of the mind consists of an understanding of the functionalism of culture and the mind. The cultural and mental computation of functional operations comprise a basis for the understanding of learning and experience that are important to cultural knowledge. The informational content of mental states as states of knowledge entails the functional basis of the mental computation that comprises the cultural mind. Cultural mental computation as informational content and states of knowledge details the importance of the distinction between representation and information processing on the cultural and mental computation of biological organisms.

Cultural machine computation consists of the functional operations that comprise the information processing of the cultural mind. The computation

TABLE 1.1 History of Computation

Computing Machine	Type of Computation	Physical States	Unit of Computation
Mind	Mental computation	Functional mechanisms	Cognitive representation
Brain	Neurocomputation	Biophysical mechanisms	Neural representation
Machine	Analog computation	Mechanical states	Analog representation
Computer	Digital computation	Electromechanical to Electronic states	Digital representation ("classic bit")
Quantum Computer	Quantum computation	Quantum states	Digital representation ("qubit")
Supercomputer	Supercomputer computation		

of the cultural mind entails the understanding of the mind as a computing machine. The cultural computation of the mind comprises the computing machine of the mind and its functional operations. The functional operations of the cultural computing machine – the cultural mind – illustrate the structural organization of functional tasks and the specific operations or algorithms that facilitate the information processing of culture and behavior. Cultural machine computation implies the functional processes that demonstrate the fundamental processes that show the mental capacities of biological organisms. The cultural machine computation of biological organisms entails the performance of functional operations that facilitate the functional and mechanistic basis of cultural computation.

The identification of the computational level of culture as real and artificial systems facilitates the naturalism of cultural information and its physical basis. The neurocomputation of biological mechanisms as a physical realizer provides plausibility of a computational basis for the representation of cultural information at the level of information processing of the nervous system. From neurons to networks of neurons, cultural neurocomputation is comprised of the distinct types of representations and information processing that contribute to the multilevel adaptation of culture and behavior. The cultural computation of neurons is comprised of the signal communication of cellular and molecular mechanisms that contribute to the representation and information processing of functional and cultural adaptation. The encoding and decoding of culture as patterns of neuronal mechanisms illustrates the representation and transformation of cultural information-processing mechanisms across spatiotemporal scales. The synaptic plasticity of cellular and molecular mechanisms comprises a physiological basis for the learning and experience of culture. The description of the physical instantiation of cultural information as patterns of the functional activity of cortical brain regions illuminates the importance of a mechanistic basis of culture in the life patterns of biological organisms. Across levels of processing, the neurocomputational analysis of cultural computation entails the characterization of the representational content and functional processes that comprise a mechanistic basis of culture computation.

Research on cultural neurocomputation investigates the neurocomputation of cultural information processing mechanisms. The cultural neurocomputation of cortical brain organization demonstrates the cultural computation of the physiological basis of behavior across levels of processing. The fluctuations of physiological mechanisms in response to cultural information processing detail the dynamics of the cultural brain. The cultural neurocomputation of cortical brain organization consists of the functional specificity and functional integration of cortical pathways that show multiple pathways of cultural information processing that contribute to the constraint satisfaction of the computational function of the cultural brain.

The cultural neurocomputation of cortical brain organization is comprised of functional pathways of sensory and cognitive systems that contribute to the dedicated and content-specific information processing of neural mechanisms. The functional pathways of sensory and cognitive systems of the brain show the functional specialization of cortical organization and its functional role as processing mechanisms of cultural information. The representation of cultural information of sensory and cognitive systems shows the role of early processing streams on the cultural specificity of information processing. The representation and transformation of sensory information and cognitive representation into cultural learning and experience illustrates the functional significance of higher-level function on culture and behavior.

The functional integration of the cortical pathways of brain organization demonstrates the role of structural and functional connectivity on the information processing of multilevel adaptation. The functional integration of cultural neural networks describes the structural and functional connectivity of functional pathways of interconnected brain regions and their function on the information-processing mechanisms of culture and behavior. The functional pathways of interconnected brain regions comprise cortical pathways of culture that contribute to the semantic representation of the abstract knowledge of culture. The functional integration of interconnected brain regions facilitates the cognition and higher-level function of culture and behavior. The cultural neurocomputation of functional pathways demonstrates the role of connectivity on the cultural information processing of functional mechanisms.

The cultural patterns of biological organisms in the natural world provide a foundation for the design of artificial systems that demonstrate the capabilities of cultural computation. The development of artificial systems that show the capability of cultural computation provides a formal model of cultural computation. The formulation of the computational modeling of culture contributes to the conceptual basis of the physical realizer of culture and allows for the testing of scientific theories on the formal modeling of the cultural patterns of real and artificial systems. The computational modeling of culture in biological organisms allows for the identification of the causality of the cultural patterns and the functional mechanisms of biological organisms as a component of the cultural life patterns of the natural world. The computational modeling of the biological basis of culture demonstrates the formulation of a conceptual basis of the cultural patterns of biological mechanisms as a functional basis of information processing. The detailed characterization of the cultural patterns of the information processing of biological organisms and its functional basis contributes to the study of the computational biology of culture and behavior.

The computational principles of artificial systems serve as a foundation for the physical realization of cultural computation as synthetic devices or computers. The principles of computation entail the functional equivalence of cultural computation across distinct physical realizers. From biological to digital computation, cultural computing abounds in the functional performance of information-processing mechanisms and the capabilities of cultural computation of distinct kinds of physical implementation. The functional basis of the information processing of computational mechanisms contributes to the autonomous information flow of cultural computation. The computational principles of cultural computation demonstrate the multiple realizability of culture across distinct kinds of physical realizers.

Cultural mental computation

Research on cultural mental computation is important for the understanding of the cultural computation of the mind. The cultural computation of the mind consists of the mental content or mental property of the mind as a physical realizer of culture. The mental content of the cultural mind as a functional or physical property in the natural world illustrates a functional and physical basis of the cultural computation of the mind. The cultural computation of the mind demonstrates that the physical realization of culture unfolds in the consciousness and higher-level function of behavior. The understanding of the cultural mind as states of consciousness and cognition comprises the functional basis of the cultural mind in the natural world. The characterization of the cultural computation of the mind is important for the understanding of the range of physical realizers of culture as mental content and states of consciousness of biological organisms in the natural world.

Across species, the study of the cultural computation of biological organisms demonstrates the range of physical realizers of culture that serves as a mechanistic basis of functional and cultural adaptation. From simple to complex organisms, the comparative study of the social and emotional capacities shows that the functional basis of cultural perception is a process-specific mechanism that is of importance to the functional and cultural adaptation of biological organisms. The functional basis of cultural perception illustrates the appearance of adaptive machinery as a physical realizer of cultural information processing. The presence of functional machinery for the information processing of cultural capacities demonstrates the importance of functional mechanisms on the cultural computation of biological organisms.

The study of cultural mental computation illustrates the role of sensory and cognitive systems in the higher-level function of consciousness and culture of complex organisms. The processing of cultural information as representational content, the semantic processing of cultural meaning from symbolic understanding and the capacities of rule-based cognition in the control of cultural information processing demonstrate the importance of higher-level function as a foundation of the cultural computation of the mind. The integration of sensory and cognitive systems and distinct types of representation of cultural content illustrate the importance of information representation in cultural processing. The functional capacities of sensory and cognitive systems detail the role of higher-level function in the states of consciousness that comprise cultural information processing of complex organisms.

Cultural mental computation is a component of the abstract knowledge that is essential for understanding the intentions and actions of others. The shared understanding of the functional and social significance of cultural objects and actions is fundamental to the understanding of the mind in interaction with the cultural environment. The semantic representation of language and action understanding contributes to the interpretation of the meaning of abstract concepts, which contributes to the understanding of the functional or social significance of the cultural environment. The shared understanding of the interpretation of abstract meaning is a process of cognition and higher-level function that is fundamental to abstract cultural knowledge. The building of abstract cultural knowledge facilitates the shared representation and interpretation of the objects and actions in the cultural environment. Abstract cultural knowledge, as a functional component of the cultural mind, demonstrates the functional importance of semantic processes and memory in the building of cultural capacities.

Cultural mental computation is fundamental to consciousness and higher-level function. The cultural mental computation of biological organisms illustrates the stream of consciousness that comprises cognition and higher-level processes such as intentionality and action understanding. The consciousness and cognition of culture detail the role of higher-order reasoning in the inferential processes of social cognition. The inferential reasoning on the consciousness and cognition of agency and animacy serves as a foundation for the understanding of other minds. The inferential processing of the perception of agency as a dimension of mental experience is of importance to the understanding of others. The inference of the perception of agency as a mental dimension ("God") demonstrates the representation of the intentionality or goal states of other minds. The understanding of the motivational and intentional states of others facilitates inferences of intentionality and action understanding.

The perception of animacy or biological motion consists of the detection of apparent or implied motion. The apparent or implied movement of biological motion serves as a perception of the motivation and intentionality of the mind. The perception of animacy implies the biological motion of others and the inferential reasoning on biological movement. The detection of animacy ("living things") serves as an attributional basis of the category-specificity of social and biological knowledge. The perception and detection of animacy contribute to the understanding of the motivational and intentional states of goal-directed action of the self and others.

The scaffolding of the perception of other minds entails the understanding of the agency of other minds. The perception of mental experience as comprised of intentional and motivational states imbues the experience of mental life with the understanding of the intentionality and goals of the cultural environment. The intentional goals and motivational states of the agency of other minds elicit the social inferences that are important to the motivational and goal states of the cultural environment. The inferential reasoning on the motivational and goal states of the cultural environment contributes to the understanding of other minds as intentional agents of motivated action. The social reasoning on the motivation and intentionality of moral thought and action is foundational to the understanding of other minds as intentional actors.

The understanding of the intentionality and action of self and others is fundamental to the social and moral cognition of other minds. The inferential reasoning on the motivational and goal states of the self is a functional basis for self-concept and the knowledge states of the self. The representation of the motivated and intentional goal states of self and others facilitates the attainment of goals and actions. The understanding of the self and others demonstrates the importance of states of knowledge as a functional basis for the inferential reasoning on self and others. The social knowledge of self and others consists of the states of consciousness and higher-level functions that facilitate the understanding of the goals and actions of the social world.

The inferential reasoning on the supernatural world comprises the states of knowledge that are consistent with supernatural agency. The understanding of other minds ("supernatural beings") consists of the social inferences on the motivational and intentional goals of supernatural actors in a counterintuitive world. The understanding of the moral thought and action of supernatural agency entails the inferential reasoning on the benevolence and malevolence of other minds as supernatural actors. The moral motivation and action of supernatural actors details the expansion of supernatural agency into the naturalistic realism of the cultural environment. The understanding of the motivational and social significance of the moral action of supernatural agency elicits the higher-level processes that serve as a functional basis of the motivation and intentionality of supernatural actors in a natural world.

The detection of the agency of supernatural actors elicits understanding of the inferential reasoning on the moral thought and action of others.

The understanding of the moral thought and action of supernatural agency facilitates the representation of counterintuitive beliefs and the inferential reasoning on the moral action of others. The belief states of supernatural agency entail the understanding of the counterintuitive states of belief and the world. The deliberation of moral thought and action of supernatural agency consists of the consideration of the wide realm of beliefs in novel states of truth conditions. The consideration of moral thought and action encompasses the motivation and intentions of supernatural actors and the role of supernatural agency on the resolution of the counterintuitive beliefs as a natural realism of the world.

Cultural mental computation entails the understanding of naturalism and supernaturalism of the world. The characterization of the functional components of cultural mental computation reveals a set of functional capacities that are essential to the understanding of the mind and its interaction with the cultural environment. The functional capacities of the cultural mind encompass the mental content of the world that is cultural and the representation of cultural information as mental content. Culture as mental representation facilitates the understanding of its causality and functionality across multiple physical realizers. The mental representation of culture as informational content of the cultural environment facilitates the understanding of the causal–functional role of the cultural mind in the natural world.

The cultural computation of the mind entails the understanding of the mental property of the mind and its functional basis. The core capacities of the mind comprise the functional properties of mental processes, its representational content and functional consequences. The cultural computation of the functional processes of the mind describes the functional tasks of cultural computation as a functional component of the mental processes and core capacities of functional adaptation. The core capacities of biological organisms for cultural computation illustrate the importance of functional machinery on the physical implementation or realization of cultural computation. The functional implementation of cultural computation across multiple physical realizers reveals the plausibility of the computational level of the cultural environment in the natural world. The fundamentals of cultural mental computation contributes to the characterization of the mental and physical property of culture and its functional significance.

Cultural machine computation

The study of cultural machine computation is the understanding of culture and its physical instantiation as a computing machine. The notion of the mind as a computing machine connotes on the mind as capable of functional performance. The mind as a computing machine implies that the functional performance of the mind is sufficient and necessary as a computational mechanism. The computation of the mind resembles that of a computing machine capable of functional operations of simple to complex machines (Figure 1.1). The notion of cultural machine computation harks on the philosophical assumptions of machine functionalism and the understanding of the mind as a causal power.

The cultural mind as a cultural computing machine entails the notion of the cultural mind as capable of cultural performance. The cultural mind consists of functional components that demonstrate cultural capacities. The cultural computation of the mind as a machine demonstrates the functional capacities of the mind as cultural capacities. Cultural computation as machine computation implies the functional capacities of mental output

FIGURE 1.1 History of computation – Analytical Engine

that is cultural. The cultural capacities of the mind are comprised of functions that operate on cultural input and output. The cultural machine computation of the mind details the functional components of the mind that demonstrate cultural capacities with real-world advantages. The cultural capacity of the mind to perform cultural tasks illustrates the causal–functional role of the cultural computing machine as capable of cultural performance.

The mind as a cultural computing machine assumes the casual functionalism of the cultural mind as a cultural machine. The cultural machine computation of the mind demonstrates the functional operations of cultural performance as part of the complexity of causal events. The cultural input–output of the machine computation of the mind details the causality of cultural events and the complexity of its interaction with the causal structure of the world. The cultural computation of the mind is important for the understanding of the causal significance of cultural events that demonstrate mental causation. The cultural significance of mental events elicits the causal implications of cultural causation that arise from mental or functional performance.

The machine computation of the cultural mind elicits cultural performance as mental events. The cultural performance of the mind as cultural input–output manifests complexity in the causal realism of the natural world. The mental causation of cultural performance encompasses a wide expanse of mental causality that imbues the causal structure of the world with cultural events of real-world significance. The cultural causation of functional or mental performance shows the causal significance of culture as real-world events that inhabit the causal structure of real and possible worlds. The cultural mind as machine computation illustrates the causal realism of cultural performance and the functional significance of cultural causation.

The cultural performance of the cultural computing machine details the determinism of cultural computation. The conceptualization of the cultural computing machine as comprised of functional operations on cultural input and output details a computational basis of machine determinism. The functional mapping of cultural input–output based on algorithmic rules demonstrates the functional capabilities of cultural machine computing for cultural performance. The cultural performance of mental computation from rule-based algorithms illustrates the production of cultural output that is consistent with rule-based function. The functional performance of rule-based cultural algorithms shows the functional role of algorithmic rules on the determination of cultural input and cultural output and its functional performance.

The detailed characterization of the functionalism of cultural performance illustrates how machine computation contributes to the determinism of cultural performance. The description of a functional mapping of cultural input and cultural output demonstrates a functional basis for the determination of cultural performance. The simple computation of cultural performance based on input–output mapping demonstrates a functional basis for cultural computation. The complex computation of cultural performance based on rule-based algorithms details the mapping of cultural input and cultural output that varies and shows the causal role of cultural rules or algorithms in the machine performance of culture. The determinism of cultural machine computation implies that simple to complex computation is sufficient for the functional equivalence of cultural machine performance.

The fundamentals of cultural machine computation are foundational to the design of machines that show cultural capabilities. The design of cultural machines that demonstrate performance capabilities that are cultural demonstrates the realization of cultural function in the natural world. Cultural capabilities consider the breadth of causal realism of cultural design and the scope of causal responsibility of cultural machines in a natural world. The cultural capabilities of machine performance elicit natural and artificial realism that presuppose the causal complexities of naturalism. The causal capabilities of cultural machine performance detail the causal realism of cultural machine computation and its functional significance on the causality of the world that is responsible for real-world outcomes.

The design of artificial systems that show cultural capabilities demonstrates the practical impact of cultural machine computation on the implementation of cultural performance capabilities. The development of artificial cultural systems that show functional capabilities illustrates the role of machine computation on the intelligent design and performance of culture. The functional use of multiple realizers on the design of artificial cultural systems ("cultural automaton") demonstrates the flexibility of cultural machine computation and its implementation. The design of artificial systems allows for the function of capabilities that advance the design and performance of cultural computation. The artificial design of cultural computation allows for the development and implementation of performance systems with functional capabilities for the advancement of the real-world advantages of cultural computation.

The study of cultural computation as real and artificial systems allows for the understanding of the distinct types of computation on culture. The computation of biological computing machines – such as the cultural brains of complex organisms – demonstrates the core capacities of biological organisms for the cultural computation of the natural world. The cultural computation of biological organisms illustrates the core capacities of cultural species and organisms for the functional computation of cultural task performance. The enactment of cultural performance of cultural species and organisms details the functional capacities of cultural computation as computational mechanisms that are fundamental to the adaptability of biological organisms to the cultural environment. The functional equivalence of cultural performance across species demonstrates the computational capabilities of biological organisms for functional and cultural adaptation.

The complexity of biological organisms illustrates computational principles of cultural computation. Complex biological organisms demonstrate computational machinery that facilitates the functional performance of tasks that are autonomous and self-organizing. The cultural computation of complex biological organisms shows that the cultural performance of functional capacities is autonomous and self-organizing for the biological organism as a cultural species. The biological organism as a cultural species automatically displays the adaptability and complexity that is sufficient and necessary for the functional performance of cultural capacities. The autonomous self-organization of biological organisms as cultural species illustrates the adaptability and complexity of the cultural mind and brain for the performance of cultural computation.

The multiple realization of cultural machine computation from simple to complex machines suggests the considerations of functional equivalence and understanding of its role on cultural machine performance. From simple to complex machines, cultural computation abounds with the functions and algorithms that are essential to the performance of cultural tasks. The functional equivalence of multiple realizers of culture entails the consideration of the scope of the causality and functionality of the physical implementation of cultural computation. From real to artificial systems, cultural computation illustrates the plausibility of the computation of cultural performance as a causal mechanism in the structure of the world. The detailed understanding of cultural computation and its multiple realizability broadens perspectives on the computational basis of culture and its natural realism in the world.

Cultural neurocomputation

Research on cultural neurocomputation investigates the computational level of the cultural brain as a physical implementation of the cultural computation of biological organisms as a cultural species. The computational level of neuronal mechanisms illustrates the functional processes that are of importance to the performance of functional tasks. The cultural neurocomputation of cortical mechanisms across levels of processing details the specific functional mechanisms that contribute to the information processing of behavior. Cultural neurocomputation is the representation and processing of the functional mechanisms of cortical organization and its relation to culture and behavior. The detailed characterization of the functional mechanisms of cortical organization is of importance to multi-level adaptation.

Across levels of processing, cortical organization is comprised of distinct types of mechanisms and processes that are essential to culture and behavior. The cultural information processing of functional mechanisms facilitates the representation and interpretation of cultural information and its transformation into patterns of behavior. The cortical representation of the perceptual and cognitive processing of culture illustrates the functional specialization of dedicated cortical mechanisms that is content-specific. The domain specificity of functional specialized mechanisms demonstrates the automatic and innate specificity of information-processing mechanisms. The functional specialization of cortical mechanisms to content-specific information illustrates the role of automaticity and innateness in the physical implementation of multilevel adaptation.

The functional specialization of cortical mechanisms abounds with the categorical specificity of domain-specific knowledge. The categorical specificity of social and physical knowledge contributes to the specialized processing of category-specific information that is of importance to functional and behavioral adaptation. The functional responsivity of cortical mechanisms to category-specific knowledge demonstrates the domain-specificity of the functional processes of cortical organization. From faces and biological motion to objects and scenes, the perceptual and cognitive processing of category-specific knowledge illustrates the specificity of neuronal mechanisms in response to the cognitive representation of the environment. The sensory and cognitive processing of the environment demonstrates the cortical processing of domain-specific knowledge.

The cognitive representation of social information such as faces illustrates the domain specificity of category-specific mechanisms of cortical organization. The functional specialization of cortical mechanisms that are dedicated to the processing of social information demonstrates the domain-specificity of the perceptual and cognitive processing of faces. The functional specificity of cortical organization shows the functional involvement of occipitotemporal cortices and the ventral visual pathway as cortical mechanisms ("face cells") dedicated to the processing of social information. The functional processes of social information detail the functional specialization of the cortical mechanisms for the adaptive responsivity to faces.

The cognitive representation of domain-specific knowledge entails the functional responsivity to objects and scenes of the environment. The functional specialization of cortical organization that is dedicated to the response of neuronal mechanisms to objects and scenes illustrates the category-specificity of domain-specific knowledge. The functional response of cortical mechanisms that shows the response specificity to objects and scenes of the ventral visual pathway illustrates the dedicated, specialized machinery ("place cells") for the perceptual and cognitive processing of objects and the environment. The functional activity of cortical mechanisms of the lateral occipital cortex and entorhinal cortex located within the occipitotemporal cortices of the ventral visual pathway illustrates the domain-specific content of specialized mechanisms. The functional specialization of the ventral visual pathway for the processing of cognitive information demonstrates the content-specific mechanisms of perceptual and cognitive systems.

The functional processes of cognition and higher-level function further show the functional specialization of the processing of social information. The functional specialization of cortical pathways dedicated to social cognition such as the temporal lobe and frontal cortex illustrates the cortical mechanisms that are specialized for the processing of social cognition. The neuronal mechanisms of the superior temporal sulcus show the dedicated, specialized functional machinery that is responsible for the perception of animacy and social cognition. The functional specificity of the superior temporal sulcus shows the specificity of neuronal mechanisms to the perception of animacy and biological motion. The perception of apparent motion of faces and bodies elicits inferences of social cognition and intentionality. The responsivity of the temporal lobe to the processing of animacy and biological motion facilitates the perception and cognition of social information.

The functional response of cortical mechanisms of the temporal lobe shows dedicated, specialized mechanisms for the processing of social cognition. The neuronal mechanisms of the superior temporal sulcus show the responsivity of cortical mechanisms to the perception of social and relational status. The perception of the facial and bodily motion of gestures and expressions facilitates social inferences on the motivation and intentionality of others. The inferential reasoning on the motivation and intentional action of others from perceptual cues demonstrates the role of the apparent motion of faces and bodies in the inferential reasoning of others. The functional specialization of neuronal mechanisms dedicated to the processing of social status and relational information illustrates the dedicated, specialized machinery of cortical organization that is responsible for social cognition.

The cortical representation of cognitive information of the ventral striatum demonstrates the encoding of social status information and its role in decision-making and higher-level processes. The cognitive representation of social information of faces and bodies elicits the perceptual and motivational salience of motivational objects that show value representation of the cultural environment. The perceptual and motivational salience of objects of the cultural environment details the motivational value of the processing of social information. The encoding of the motivational value or significance of the objects of the cultural environment details the functional specificity of the limbo-thalamic circuitry for the processing of the motivational and social significance of the cognitive representation of the environment.

The cultural tuning of the social and cognitive processing of the environment illustrates the experience-dependent activity of cortical mechanisms. The malleability of cortical mechanisms to the cultural environment demonstrates the responsivity of cortical mechanisms to the sensory input of culture-specific information. The perceptual tuning of cortical mechanisms shows the responsivity of neuronal mechanisms to the cultural environment. The functional specificity of cortical response to the cultural environment illustrates the tuning of perceptual and cognitive mechanisms that is culture-specific. The functional pathways of perceptual and cognitive mechanisms and its cultural responsivity detail the role of experience-dependent activity on the malleability of cortical–processing mechanisms.

The cultural specificity of the perceptual and cognitive processing of higher-level function demonstrates the role of the identity and expression of culture. The responsivity of neuronal mechanisms to the identity and expression of culture details the functional specificity of responses of cellular and molecular mechanisms to the cultural environment. The identity and expression of culture from facial and bodily representation details the characteristics that are of perceptual and motivational significance to the cultural environment. The encoding and decoding of facial and bodily expression illustrates the cognitive representation of cultural information and the responsivity of neuronal mechanisms to functional and cultural adaptation.

The cultural specificity of cortical mechanisms details the perceptual and motivational salience of cultural objects in the cultural environment. The perceptual and motivational salience of cultural objects illustrates the valuation of the motivational and functional significance of cultural objects in the environment. The information processing of functional mechanisms of cultural objects facilitates the motivational valuation of cultural objects and the anticipation of functional outcomes. The functional response of cortical mechanisms to the motivational valuation of cultural objects elicits the patterning of neuronal response that is of importance to the anticipatory expectation of decision-making. The encoding of the anticipation and functional significance of cultural objects in limbo-thalamic circuitry illustrates the functional significance of motivational valuation on the encoding and decoding of the cultural objects of the environment.

The encoding of cultural information as the response patterns of neuronal mechanisms details the representation of the cultural environment. The facial and bodily expression of cultural identity and expression illustrates the cultural patterns of response of neuronal mechanisms as representations of the cultural environment. The cultural patterns of cortical neurons demonstrate the encoding specificity of functional mechanisms that is expression- and identity-specific to culture. The functional specificity of the neuronal responsivity to the faces and biological motion of others in the cultural environment details the characterization of the motivational and social significance of cultural identity and expression. The cognitive representation of social and cultural capacities illustrates the importance of functional processes and mechanisms for cultural and behavioral adaptation.

The social and cognitive processing of the social information in the cultural environment shows the importance of functional integration on the cortical organization of functional machinery. The functional integration of cortical pathways details the role of connectivity of the functional response to the cultural environment. The structural and functional connectivity of cultural neural networks shows the relaying of cultural information of interconnected brain regions and its functional significance on behavior. The inhibition and control of cultural information demonstrates the importance of corticolimbic circuitry on cognitive and behavioral regulation of culture. The cognitive and behavioral regulation of cultural identity and expression illustrates the role of rule-based cognition on the processing of core capacities and multilevel adaptation.

Functional specialization and functional integration as computational principles illustrate the importance of the level of computation on the processing of the cortical organization of the nervous system. The detailed characterization of the computational principles of cultural computation within cortical organization demonstrates an organizing principle of culture in the nervous system of biological organisms and, in a broader sense, as a functional adaptation of cultural organisms and species. The functional pathways of cultural computation within neurons and networks of neurons demonstrate the computational principles of cortical organization that are descriptive of a mechanistic basis of cultural representation. The study of the functional and mechanistic basis of cultural computation illustrates the importance of the mind and brain as physical realizers of culture in the natural world and beyond.

The functional pathways of culture and behavior comprise the information processing mechanisms of the cortical pathways of simple to complex behavior. Cultural information processing mechanisms describe the functional pathways that are important to the core capacities of functional and behavioral adaptation. The core capacities of culture and behavior consist of the fundamental processes of cognition, emotion and motivation and the social and cultural capacities that are important to behavioral adaptation. Understanding of the cultural influences on the functional pathways of cognition and behavior contributes to the understanding of computational mechanisms and their functional significance. The cultural influences on functional pathways detail the top-down and bottom-up processing that affect the control or regulation of the information processing of computational mechanisms.

The cortical organization of functional pathways details the information-processing mechanisms that are important to behavioral and cultural adaptation. The identification of neurons and networks of neurons dedicated to the processing of cultural information details the computational mechanisms that are responsible for the cultural representation of biological organisms and its role in functional adaptation. The detailed characterization of specific functional pathways for core capacities of cognition, emotion and motivation illustrates the role of culture on the information-processing streams of core capacities. The cultural differences in the functional patterns of specialized pathways of core capacities demonstrate the role of individual and the cultural variation on brain–behavior relationships. The cultural differences of functional patterns of neural networks show the functional differentiation of cortical activity that is responsible for the cultural differences of the brain basis of behavior. The identification of the computational mechanisms that show the cultural differences of representation and response of neurons and networks of neurons illustrates the importance of culture on the computational level of neural machinery.

The cultural processes of cortical networks show the automaticity of the information processing of cultural neural networks. The cultural patterns of cortical brain networks demonstrate the automatic processing of core capacities that contribute to the adaptability and causality of the functional mechanisms of cortical organization. The automatic responsivity of limbic brain regions ("amygdala neurons") shows the automatic and reflexive response of cortical brain regions to the detection of the identity or expression of culture. The neurocomputation of the subcortical brain regions that comprise limbic circuitry illustrates the automatic responsivity of subcortical mechanisms to the detection of culture-specific information. The cortical information processing of subcortical brain regions shows the automaticity of subcortical brain regions as a computational basis of cultural information processing.

The cultural patterns of the functional activity of cortical networks illustrate the functional mechanisms of cultural information processing across levels of processing. Cultural patterns of cortical neural networks ("corticolimbic circuitry") demonstrate the functional properties of the functional response to cognition and behavior. The cultural differences in functional brain activity show the responsivity of cortical neural networks to the task-based or effortful processing of culture and behavior. The cortical basis of task-based processing demonstrates the fluctuations of functional activity based on the levels of cognition and behavior. The cultural patterns of functional brain activity demonstrate the task-based processing of culture on cognition and behavior. The cultural patterning of cortical brain networks that is task-specific details the functional mechanisms and information processing that is specific to the performance of cultural tasks. The cultural processes of cortical brain networks illustrate the responsivity of functional mechanisms to cultural tasks and performance.

The cultural processing of task-based brain activity shows the causal role of effortful processing on the functional performance of cortical neural networks and behavior. The states of cognition and behavior that comprise effortful processing contribute to the controlled processing of functional brain activity. Cultural processing that is effortful or task-based implies the functional significance of states of cognitive or conscious activity on the controlled processing of cortical neural networks and behavior. The cultural patterns of functional activity that are congruent with states of conscious or mental activity illustrate the role of brain–behavior relationships on levels of performance of culture and behavior.

The cultural differences of rule-based cognition demonstrate the cultural patterning of cognition that contributes to the functional brain activity of cultural neural networks. Cultural processing as rule-based cognition illustrates the cultural differences in patterns of functional brain activity that coincide with levels of cultural and behavioral performance. The cultural variability in the cortical processing of functional tasks demonstrates the role of algorithmic-based cultural tasks on functional performance. The cultural patterns of rule-based cognition illustrate the cultural processing of effortful or task-based brain activity. The cultural patterning of rule-based cognition is of importance to the understanding of the functional performance of the cultural mind and brain as a cultural machine.

Cultural response in cortical neural networks also describes the patterning of functional brain activity that is intrinsic to cortical function and behavior. The intrinsic brain response of cortical brain networks shows the default mode function of cortical brain regions ("default mode network") to resting state activity and behavior. The cultural patterning of cortical mechanisms entails the cultural differences in the levels of resting state activity of cortical brain regions and behavior. Cultural differences in levels of resting state activity of cortical neural networks imply the variability in tonic levels of cortical brain activity that is specific to culture. Cultural patterns of intrinsic brain activity of cortical brain networks show that tonic levels of functional activity are sufficient to demonstrate the autonomous responsivity of biological organisms as a cultural species. Cultural differentiation of functional brain activity shows the role of cultural influences on the intrinsic and extrinsic activity of cortical brain regions across levels of processing.

Cultural brain dynamics

Cultural brain dynamics describe the computational principles of large-scale organization and interactivity as components of cortical neural networks. The large-scale organization of the cortical brain as a computational principle integrates structural and dynamical facets of cortical organization. The structural and functional connectivity of cortical brain networks illustrate the role of interconnection on the structural and dynamical principles of cortical brain organization. The large-scale organization of the cortical brain contributes to the understanding of the functional architecture of the brain. Cultural brain dynamics further describes the interactivity of information processing across interconnected brain regions of cortical neural networks. The structural and functional connectivity of cortical brain networks details the role of interconnection on the flow of cultural information processing of functional pathways. The identification of the functional architecture of the brain is of importance to the understanding of the structural and dynamical facets of cortical organization.

The cortical neural networks of brain organization demonstrate the functional integration of cortical brain activity and its relation to behavior. The cortical neural networks of functional processes illustrate the functional specificity of cortical brain organization for the information processing of culture and behavior. The detailed characterization of the functional specificity of cortical brain organization is of importance to the understanding of the brain basis of culture and its role in multilevel adaptation. The functional architecture of cortical organization shows the role of functional specificity in the processing of core capacities. The functional integration of cortical brain networks contributes to the understanding of the role of interconnectivity on the functional capacities of core processes.

The identification of cortical neural networks that are responsible for the processing of culture is of importance to the understanding of the functional architecture of the brain. The detailed characterization of the cortical neural networks of culture entails the identification of specific brain regions and their interconnections and the functional correspondence of brain–behavior relationships with culture. The study of the functional architecture of the cultural brain contributes to the understanding of the functional mechanisms of specific brain regions and the functional role of cortical brain circuits on behavior.

The functional connectivity of the cortical networks of culture describes the interconnection of cortical brain regions that is responsible for cultural information processing and its functional relation to behavior. The interconnection of frontal and limbic circuitry ("corticolimbic circuitry") describes a cortical neural network that is of importance to the relay of information processing of culture and behavior. The interconnectivity of corticolimbic circuitry describes a neural network that shows cultural patterns of functional brain activity. The cultural patterns of functional brain activity of cortical neural networks illustrate the responsivity of networks of neurons that is culture-specific and the relaying of cultural information processing across interconnected brain regions as nodes of neural networks.

The directionality and magnitude of cortical neural networks show the functional connectivity that is of importance to the cultural processing of behavior. The top-down and bottom-up processing of the frontal cortex and its interconnected brain regions demonstrate the role of controlled processing in the inhibition or regulation of cognitive information processing of culture. The cognitive representation of information in the frontal cortex demonstrates the selection and maintenance of social and cultural knowledge. The reliance on cognition as controlled processing facilitates the inhibitory processing of information flow within the frontal cortex and interconnected brain regions. The controlled processing of the frontal cortex shows the feedforward and inhibitory processing of cortical neural networks that are responsible for the control or regulation of culture and behavior.

The cultural brain dynamics of neural networks entail the cultural information processing of cortical brain regions. Cultural entrainment of endogenous brain activity describes the autonomous response of cortical brain regions that comprise active cultural representation that is maintained at rest. The endogenous brain activity of cortical brain regions during resting state activity details the autonomous brain dynamics of cultural neural networks. The autonomous brain dynamics of resting state activity is comprised of the fluctuations of patterns of cortical brain activity that demonstrate the autonomous and self-organizing principle of cultural brain dynamics. The spontaneous fluctuations of cultural brain dynamics illustrate the changes of cultural brain activity that arise based on the intrinsic properties of cultural neural networks. The entrainment of cultural processing of cortical brain regions describes active and itinerant cultural representations of functional brain activity that are spontaneous and intrinsic to the cultural neural network.

The study of cultural brain dynamics is of importance for the characterization of the entrainment of culture in the functional brain activity of cortical neural networks. The identification of the autonomous brain dynamics of resting state activity describes the intrinsic properties of cultural neural networks. The understanding of the dynamical processes of the cultural brain at rest illustrates the autonomous and self-organizing dynamics of cortical brain networks. The functional role of autonomous brain dynamics in cultural entrainment supports a putative mechanism for the intrinsic properties of cultural processes. The notion of the intrinsic properties of cultural brain dynamics demonstrates the importance of spontaneous and autonomous resting state activity on the cultural entrainment of cortical brain function.

The global dynamics of cultural neural networks describe the network dynamics of neural networks. The network dynamics of neural networks is comprised of the global states of activation across networks of neural networks ("connectome"). The characterization of the functional properties of the global dynamics of cultural neural networks illustrates the extrinsic connectivity of cortical networks. The extrinsic connectivity of cortical neural networks demonstrates global dynamics that describe the synaptic density or structural connectivity of networks of neural networks. The global dynamics of cultural neural networks consist of the structural connectivity of networks of neural networks that are involved in cultural processing. The detailed characterization of global dynamics facilitates the understanding of the extrinsic properties of cultural brain regions that is based on connectivity.

The dynamics of cortical neural networks describe the changes of causality of information processing that are of importance to the functionality of interconnected brain regions. The dynamical processing of cultural information of cortical neural networks illustrates the role of causality and functionality in the information processing of cultural capacities. The causal dynamics of cortical processing reflect the directionality of information flow across interconnected brain regions that are responsible for the representation and interpretation of mental states of activity and behavior. The causal dynamics of cortical neural networks illustrate the directionality and causality of interconnection of cortical brain regions that are essential for the information processing of core capacities.

Cultural neural networks

Cultural neural networks consist of the cortical brain networks that are responsible for the cultural information processing of behavior. Research on cultural neural networks entails the detailed characterization of the specific cultural brain regions and their interconnection. The study of the brain dynamics of cultural neural networks facilitates the understanding of the processing of cultural information as the interconnection of brain regions. The processing of cultural information of interconnected brain regions consists of the aggregation of the neuronal population response as activation patterns of brain activity. The functional connectivity of cultural neural networks contributes to the dynamical processing of cultural information across interconnected brain regions of cortical brain networks.

The functional integration of cortical brain regions facilitates the interconnectivity of cultural brain regions. The identification of core brain regions that comprise cultural neural networks contributes to the characterization of the structural and functional connectivity of cortical brain dynamics. The differential patterns of functional brain activity of cultural neural networks illustrate the patterns of activation that coincide with cultural mental activity and behavior. The interconnection of cortical brain regions facilitates the bidirectional information processing of cultural information across multiple brain regions. The integrative processing of interconnected brain regions illustrates the role of feedforward and inhibitory processing on the brain dynamics of cultural neural networks.

The functional brain activity of cultural neural networks describes the activity of interconnected brain regions and its causal interactions. The population-level activity of networks of neurons comprises an aggregation of neuronal population response and its transformation into cultural patterns of functional brain activity. The population-level neuronal activity describes the encoding of the representation and interpretation of cultural information into cultural output. The causal interaction of interconnected brain regions details the functional role of specific brain regions as nodes of cultural information processing. The causation of interconnection contributes to the functional specificity of interconnected brain regions and the interrelation of the cultural patterns of functional brain activity and behavior.

The characterization of the cortical brain regions that comprise cultural neural networks consists of the understanding of the cultural processing that coincides with the interconnection of cortical brain regions and its interrelation to behavior. The study of the structural connectivity of cultural neural networks provides a foundation for understanding the strength or magnitude of interconnection of the functional pathways that comprise the cognition and behavior of culture. Interconnectivity of functional pathways arises from the interconnection of networks of neurons across nodes of cortical brain regions. The interconnection of cortical brain regions illustrates the magnitude of information flow across distinct brain regions and the transformation of its representational content into patterns of motor output.

The functional connectivity of functional pathways that comprise cultural neural networks demonstrates the flow of cultural information across interconnected brain regions. The flow of cultural information across distinct brain regions details the bidirectionality of processing that is of importance to the patterning of cultural input and cultural output. The processing of cultural information flow across interconnected brain regions demonstrates the directionality of the information flow of cultural neural networks from sense data to motor output. The bidirectional processing of cultural information implies a functional basis for the understanding of the automaticity and controlled processing of cortical mechanisms and their relation to culture and behavior. The characterization of the causality of the cultural information flow of cortical brain regions contributes to the understanding of the causal directionality of functional pathways that are important to culture and behavior.

The information processing that arises from feedforward processing describes the automaticity of cultural patterns of functional activity across cortical brain regions. The automatic processing of cultural information of cortical brain regions describes the functional specificity of cultural neural networks. Feedforward processing ("excitatory pyramidal cells") implies the mechanistic transformation of cultural representation into cultural motor output. The automaticity of cultural processing illustrates the dedicated and specialized processing of specific brain regions that comprise cultural neural networks. The feedforward processing of cultural neural networks illustrates the automatic response of the functional mechanisms of cortical brain regions to the cultural environment. The feedforward processing streams of functional pathways demonstrate the automatic processing of cultural brain regions.

Feedforward processing of cultural neural networks implies the patterning of functional activation that shows the causality of cultural processing from streams of functional pathways. The characterization of the functional pathways of cultural processing details the role of the feedforward processing of cultural information flow on the causality and functionality of cultural neural networks. The feedforward processing of cultural information flow contributes to the understanding of the spatiotemporal dynamics of cultural neural networks. The causal directionality of information flow from early to late processing streams across cortical brain regions implies the spatiotemporal dynamics of functional pathways.

Other types of cultural information processing mechanisms of cultural neural networks entail the role of inhibitory processes on the functional patterns of brain activity. The interconnection of cortical brain regions illustrates the feedforward and inhibitory processing of cultural information flow. The feedback or inhibitory processing ("inhibitory interneurons") of cultural information flow across multiple brain regions demonstrates the functional role of cortical brain regions as information relays of inhibitory processing in the functional pathways of culture and behavior. The inhibitory population of neuronal processing shows the change of the causality and directionality of cultural information flow as a functional component of cortical pathways. The inhibitory processing of the cultural neural network illustrates the controlled processing of information flow that contributes to changes in cognition and behavior.

The inhibitory processing of functional pathways demonstrates the role of the control or regulation of cognition and behavior on the cultural information processes of behavior. The controlled processing of cultural information elicits top-down and bottom-up processing of cortical brain regions that are responsible for control or regulatory function. The control or regulation of cognition and behavior changes patterns of cortical brain activity and mental function. The controlled processing of cognition and behavior contributes to the interpretive layer of information processing on the representational content of culture and behavior. The interpretive layer of information processing facilitates the evaluation of types of cultural representations and their functional significance. The cultural patterning of inhibition and controlled processing demonstrates the selection and maintenance of cultural representational content that is essential for multilevel adaptation. The core capacities of cognition and regulatory function facilitate the changes of cognition and behavior, showing cognitive and cultural adaptation.

The study of the functional connectivity of cultural neural networks provides a foundation for the characterization of the bidirectional processing of cultural information. The characterization of functional connectivity of cultural neural networks involves the identification of the functional role of specific cortical brain regions and the distinct types of cultural processing that arise across the information processing stream. The identification of the functional role of the interconnectivity of cultural neural networks demonstrates the causal significance of the functional pathway. The detailed characterization of the distinct types of cultural processes that arise from the patterning of the functional activation of the neural networks of culture illustrates the functional basis of the interconnectivity of cultural brain regions.

The cultural patterns of cortical brain networks entail the identification of cortical brain regions that are responsive to the brain dynamics of cortical networks. The interconnectivity of cortical brain networks shows the patterns of activation that are of importance as states ("attractor states") of cultural brain dynamics. The detailed characterization of the interconnected brain regions that comprise functional patterns of brain activity describes points of activation states that show multiple constraint satisfaction. The functional patterns of cultural information processing of cultural neural networks show the activation patterns that comprise the fixed point or stable states of the functional patterns of cultural brain dynamics. The attractor states of cultural brain dynamics illustrate the harmonization of the states of brain activity into a cultural pattern of functional brain activation that is important to cultural information processing.

Theoretical foundations of computational cultural neuroscience

Theoretical foundations of computational cultural neuroscience consist of the understanding of the fundamental principles of computation and their role in the study of the brain basis of culture. Scientific theories of computation abound in the understanding of the lawlike principles that govern the computational level of the cultural brain. The formulation of the conceptual landscape of computational cultural neuroscience is comprised of the understanding of the role of computational principles as an organizing principle of the cultural brain. The conceptual theories of the field of study contribute to the formalization of scientific concepts and their causal relations.

Theory building on computational cultural neuroscience arises from the understanding of the structural and functional principles of cortical brain organization and their role in the information processing of cultural capacities. The structural and functional principles of cortical brain organization facilitate the detailed characterization of the functional pathways of neuronal mechanisms and their relation to cultural computation. The structural and functional organization of the cortical brain describes the functional specialization and functional integration of cortical mechanisms and their role in multilevel adaptation. The structural and functional principles of cortical organization serve as a foundation for the study of the computational level of the cultural brain.

The *functional specialization* of cortical brain regions describes a computational principle of brain organization. Across sensory and cognitive systems, the functional specialization of cortical pathways describes the dedicated, specialized functional machinery that is responsible for the processing of core capacities. The functional specificity of specialized functional machinery illustrates the automatic, specialized information processing of core capacities of cognition and behavior. The functional specialization of cortical pathways of perception and cognition illustrates the importance of this computational principle on the representation and transformation of sense data into motor output.

The computational principle of functional specialization contributes to the theory building on the specialized functional machinery of the cultural brain. Theory building on the functional specialization of the brain entails the identification of specific types of computational mechanisms that are dedicated and specialized to cultural processing. Across levels of processing, computational mechanisms demonstrate the functional specificity of neuronal response to the stimuli of the cultural environment. Scientific theories of the computational cultural brain arise from the conceptual development of the plausibility of computational mechanisms that show dedicated and specialized processing of culture. From neurons to networks of neurons, the study of culture-specific processing mechanisms formulates a conceptual basis from which to build scientific theories of the functional specialization of cortical organization.

The development of the conceptual language of computational cultural neuroscience illustrates the breadth and scope of scientific theories of culture-specific processing in the cortical organization of the brain. The study of the cultural brain across levels of processing demonstrates the implementation of functional mechanisms that show dedicated information processing that is essential to functional and cultural adaptation. From "culture neurons" to "cultural neural networks", cortical brain organization abounds of the computational mechanisms that show dedicated, specialized processing of cultural information that is content-specific. Scientific theories of "culture cells" – neurons that show culture-specific response – demonstrate the theory building on functional specialization that contributes to the broader understanding of the computational level of cortical brain organization. The theory building on the cultural specificity of cortical processing facilitates the design of empirical research to investigate the functional specificity of the cultural brain across levels of processing.

The *functional integration* of cortical brain regions demonstrates the importance of interconnectivity on the brain dynamics of information-processing mechanisms. Functional integration as a computational principle describes the functional properties of cortical brain networks and their interconnections. The structural and functional connectivity of cortical brain networks describes the functional architecture of the brain and its relation to core capacities of mental activity and behavior. The functional integration of cortical pathways details the interconnectivity of interconnected brain regions and their role in the processing of cultural information. The interconnection of cortical brain regions show the importance of the flow of information processing on core capacities.

Theory building on the functional integration of cortical brain organization provides a foundation for the understanding of the interconnectivity of the cultural brain. The conceptual notion of cultural neural networks entails the understanding of the structural and functional properties of the cultural brain. The characterization of the structural and functional properties of cultural brain dynamics facilitates the investigation of the interconnectivity of cortical brain regions that show culture-specific processing. The development of scientific theories that explain the functional significance of the structural and functional properties of the cultural brain contributes to the theoretical foundations of computational cultural neuroscience.

Scientific theories of "cultural neural networks" – neural networks that show culture-specific processing – illustrates the theory building on functional integration that contributes to the understanding of the computational level of cortical organization. The testing of the scientific theories on cultural neural networks comprises the design of computational models that predict and explain culture-specific processing of networks of neurons. Computational modeling of cultural neural networks seeks to identify and predict the cultural processing of interconnected brain regions that comprise cultural neural networks. The computational study of cultural neural networks facilitates the characterization of the types of information processing that comprise culture-specific responses at the computational level of functional brain organization.

The *large-scale organization* of the cortical brain is an important computational principle that describes the structural and functional connectivity of the cultural neural networks of the cortical brain organization. Functional patterns of brain activity of cortical neural networks are of importance to the understanding of the large-scale organization of the brain. Cortical brain activity describes the patterns of functional activity that are important to cultural information processing. The large-scale organization of the cortical brain entails the characterization of multiple neural networks that are responsible for core capacities. The distribution of information-processing mechanisms as functional pathways of multiple neural networks illustrates the large-scale organization of the brain.

The *interactivity* of cortical brain organization is a computational principle of brain organization that is of importance for the characterization of cultural brain dynamics. The interactivity of cortical brain regions describes the directionality and causality of information processing across interconnected brain regions. The feedforward and feedback processing of information flow that comprises the functional pathways of cortical brain regions describes the interaction of interconnected brain regions during the processing of core capacities. The feedforward and inhibitory processing of information across interconnected brain regions demonstrates the importance of interactivity on the cortical processing of neural networks.

The interactivity of cultural brain regions details the role of interconnection in the information processing of culture and behavior. The interactivity of the interconnected brain regions of cultural neural networks describes the bidirectional processing of cultural information that comprises cortical neural networks. The functional patterns of cultural brain activity identify the interconnected brain regions that show bidirectional processing of cultural information. The identification of the interconnected brain regions of cultural information processing that comprise cultural neural networks is foundational to understanding the interactivity of the cultural brain. The understanding of the interactivity of the cultural brain contributes to the study of the computational level of the cultural brain.

The computational principle of "*multiple constraint satisfaction*" details the functional basis for the patterns of functional activity that arise across cortical brain regions. The information processing of cultural and functional mechanisms demonstrates the multitude of cultural processes that are physically instantiated within networks of neurons that comprise cultural neural networks. The local dynamics of cultural neural networks describes the aggregate population-level neuronal activity of networks of neurons that show the distinct types of response of bidirectional processing that is responsible for cultural processes. The feedforward and feedback processing of local networks of neurons show the population-level activity of neurons that comprise the local dynamics of cultural neural networks. At the level of networks of neurons, the cortical brain regions show the patterns of functional response of cultural neural networks as cortical brain activity that demonstrate the satisfaction of internal and external constraints of the environment. The population-level activity of cortical brain regions demonstrates the optimized response of networks of neurons to the cultural environment.

The global dynamics of cultural neural networks details the extrinsic properties of networks of neural networks that are important to cultural processing. The extrinsic connectivity of networks of neural networks describes the dynamical processing of cultural information flow that arises based on the structural connectivity of networks of neurons. The global dynamics of extrinsic connectivity elicit dynamical processing that is specific to the extrinsic properties of structural connectivity. The characteristics of structural connectivity such as synaptic density and interconnection contribute to the characterization of the extrinsic properties of cultural brain dynamics. The understanding of the global dynamics of cultural neural networks is essential for the formulation of the intrinsic and extrinsic properties of multilevel cultural processing from simple neurons to networks of neural networks.

Methodological foundations of computational cultural neuroscience

The methodological foundations of computational cultural neuroscience build on the observation and measurement of cortical brain activity and the study of its relation to cultural processes and behavior. The observational study of the processes and mechanisms of culture relies on the measurement of the fluctuations of cortical brain activity and its relation to mental activity and behavior. The multilevel methods of computational cultural neuroscience provides an integral approach to the systematic study of the computational level of the cultural information processes of the brain. The development of the methodological foundations of computational cultural neuroscience contributes to the tools and techniques for understanding of the brain basis of culture and behavior.

One important method for the study of computational cultural neuroscience is single-cell recording or the direct measurement of the response of neurons to the environment. Research on single-cell recording of neurons provides a foundation for the development of methodological tools and techniques that facilitate the observation and measurement of the neuronal response of culture. Single-cell recording facilitates the direct measurement of the neuronal response of cells to experimental stimuli.

The use of single-cell recording allows for the measurement of the patterns of neuronal response as distinct types of representation of the cultural environment. The direct recording of single-cell recording techniques provides direct evidence of neuronal mechanisms that are specialized or dedicated to specific types of cultural information processing ("culture neurons"). The study of the patterns of neuronal response that comprise cultural representation contributes to the understanding of the computational level of neurons as a cellular and molecular basis of cortical brain activity.

Single-cell recording research is important for the identification of the functional role of cellular and molecular mechanisms of cortical brain organization. The identification of specific cells that show dedicated or specialized processing of cultural information is of importance to the understanding of the computational level of cultural processes. The design of single-cell recording research to investigate neurons and the functional response of neuronal mechanisms to the cultural environment contributes to the observation and measurement of the neural activity of the cultural brain. The computational modeling of the response patterns of neuronal mechanisms to the cultural environment illustrates the importance of distinct types of representations that comprise cultural neurocomputation.

Another important method of computational cultural neuroscience is electrophysiology or the study of the electrophysiological basis of behavior. Electrophysiology is a measurement tool that is important to the observation of cortical brain activity and its relation to the mental activity and behavior of culture. Electrophysiology research includes the measurement of the fluctuations of brain activity and its functional correspondence to mental function and behavior. The direct recording of neuronal activity from the scalp illustrates the observation and measurement of the functional activity of cortical brain mechanisms within milliseconds of stimulus onset. The development of methodological tools and techniques of electrophysiological research contributes to the means of measuring the cultural patterns of electrophysiological activity and its relation to mental activity and behavior. The use of electrophysiology in the study of the neural and mental activity of behavior demonstrates the feasibility of the measurement tool as an assay of the computational level of the cultural brain.

Electrophysiological research consists of the use of large-scale data sets that detail the study of cortical brain activity and its relation to behavior. The study of electrophysiological research facilitates the compilation of large-scale data sets for the use of computational tools and techniques to test theories of causal interaction of cultural brain dynamics. The computational modeling of the causal interaction of cortical brain regions in response to cultural brain activity and behavioral performance illustrates the use of computational approaches in the study of the electrophysiological basis of culture and behavior. Formal computational models are of importance to the understanding of the causal interaction of cortical brain regions and the cultural patterning of the interaction of cortical brain regions.

The understanding of the causal interaction of cultural brain dynamics contributes to the identification of the cultural patterns of brain activity. The study of cultural brain dynamics with electrophysiological methods encompasses the study of the causal interaction of cortical brain regions and the spatiotemporal dynamics of cultural brain processing. The study of the spatiotemporal dynamics of cultural brain processing provides detailed characterization of the electrophysiological basis of cultural processing and the causal interaction of cortical brain regions that show functional correspondence with cultural processes of behavior. The identification of the spatiotemporal dynamics of cultural brain processing is of importance for the characterization of the causal interaction of the cultural brain.

The spatiotemporal precision of electrophysiological methods is essential as an integral part of the measurement tool that is of importance to the study of computational cultural neuroscience. The measurement of the spatiotemporal properties of cultural brain dynamics is fundamental to the understanding of the causal interaction of the cultural brain. Electrophysiological tools and techniques are beneficial as a measurement tool for the observation and measurement of the spatiotemporal dynamics of cultural brain processing. The neurocomputational analysis of cultural brain dynamics with electrophysiological measurement contributes to the understanding of the causal interactivity of the cultural brain.

Event-related potential studies (ERP) are an important methodological approach to the study of the spatiotemporal dynamics of cultural brain activity and its relation to behavior. The design of ERP studies facilitates the use of electrophysiological techniques for the purpose of the systematic investigation of the fluctuations of electrophysiological activity that is time-locked to specific events. The ERP methodology is important for the identification of specific patterns of electrophysiological activity ('electrophysiological waveforms') that characterize specific types of cultural processing. The directionality and magnitude of the changes of amplitude of specific waveforms detail the functional role of the event-related response of cortical brain regions to the cultural environment. The study of ERP waveforms illustrates the functional role of specific cortical brain regions in response to cultural processing.

The ERP methodology is appropriate for the study of the patterning of electrophysiological activity that is culture-specific. The design of ERP studies of cultural processing is important for the identification of specific brain regions that are involved in cultural processing and the spatiotemporal dynamics of electrophysiological activity that are linked to mental activity and behavior that is culture-specific. The ERP methodology is beneficial for the identification of culture-specific waveforms that show specialized processing and response to the stimuli of the cultural environment. The ERP waveforms that show culture-specific processing contribute to the understanding of the fluctuations of electrophysiological activity that are linked to the cultural processing of the brain and behavior. The characterization of the spatiotemporal characteristics of culture-specific waveforms is important for the understanding of the computational level of cultural brain dynamics. The design of ERP methodological tools and techniques for the identification of culture-specific waveforms contributes to the methodological approaches of computational cultural neuroscience.

A third methodological approach to the study of computational cultural neuroscience is functional magnetic resonance imaging (FMRI). Functional neuroimaging is an important neuroscience tool and technique for the indirect measurement of the functional patterns of brain activity and behavior. The use of functional neuroimaging as a measurement tool provides indirect observation and measurement of the cerebral blood flow of the brain with spatiotemporal resolution (mm^3). The method of functional neuroimaging can be utilized for the study of the functional patterns of cultural brain activity. Functional neuroimaging research provides large-scale datasets of brain imaging data that facilitate the testing of scientific theories on the causality and functionality of the structural and dynamical facets of cultural brain activity with computational tools. The study of cultural brain activity with functional neuroimaging illustrates the use of neuroscience tools and techniques for the systematic investigation of the structural and functional properties of cortical organization.

Functional neuroimaging studies of culture contribute to the neuroscience tools and techniques for the indirect observation and measurement of the cortical brain activity of culture and behavior. The study of the functional patterns of cultural brain activity facilitates the identification of the specific brain regions that show functional correspondence with cultural brain activity and behavior. The identification of specific brain regions that are responsible for the cortical processing of cultural information relies on the design of functional tasks that are specific to the study of culture and behavior. The design of functional neuroimaging tasks for the study of culture and behavior demonstrates the feasibility of the use of functional neuroimaging for the systematic investigation of cultural processing.

Research on functional neuroimaging contributes to large-scale data sets on cortical brain activity and behavior. The use of large-scale datasets is important the study of cultural brain activity and behavior. Computational approaches are beneficial for the formalization of models that perform data extraction to predict and explain cultural patterns of functional brain activity. The development of large-scale data sets of functional neuroimaging contributes to the computational approaches for the study of the prediction and explanation of cultural patterns of brain activity. The design of neuroimaging tools and techniques contributes to the computational approaches for the identification of cortical brain regions responsible for cultural processing.

The development of the study paradigms of functional neuroimaging for the study of culture contributes to the methodological tools and techniques that facilitate the measurement and observation of the cultural patterns of functional brain activity of core brain regions. The design of functional neuroimaging paradigms for the study of the resting-state and task-based activity of the cultural brain facilitates the identification of the core brain regions that are dedicated or specialized for cultural information processing. The identification of the core brain regions that are dedicated to cultural information processing contributes to the understanding of the functional properties of the cultural brain. The identification of core brain regions that comprise cultural processing is important for the characterization of the functional specificity of the cultural brain.

Computational tools and techniques of functional neuroimaging are essential to the systematic investigation of the types of representations that comprise the functional patterns of cultural brain activity of core brain regions. Computational modeling of the spatial patterning of functional brain activity contributes to the understanding of the distributed representation of cultural information across cortical brain regions. Computational tools and techniques such as multivoxel pattern analysis ("MVPA") facilitate the identification of the representational content of cultural information across distinct brain regions of cortical function. The prediction of the spatial patterning of cultural information across multiple brain regions illustrates the distributed representation of cultural processing. The characterization of the distributed representation of cultural processing across distinct brain regions demonstrates the types of representation that comprise the functional brain activity of core brain regions.

Studies of functional neuroimaging also contribute to the identification of the core brain regions that comprise cultural neural networks. Functional neuroimaging tools and techniques are beneficial for the characterization of the cultural patterns of functional brain activity. The use of functional neuroimaging tools and techniques for the study of the cultural patterns of functional brain activity illustrates the importance of computational methods in the characterization of cultural neural networks.

Computational methods contribute to the characterization of the structural and functional connectivity of cultural neural networks. The use of computational tools and techniques is important to the study of the structural and functional connectivity of cultural neural networks. Computational tools are beneficial to the characterization of the structural and functional connectivity of interconnected brain regions that are responsible for cultural processing. The identification of the directionality and magnitude of structural and functional connectivity of cultural neural networks is important for the characterization of the functional role of the functional patterns of cultural neural networks. The identification of the core brain regions of cultural neural networks and their interconnectivity provides a foundation for the study of the directionality and causality of the information processing of cultural processes and functional mechanisms.

Computational modeling provides a means for the systematic investigation of the causal dynamics of cultural neural networks. Computational models of cultural neuroscience include the systematic investigation of the causal interaction of the functional brain activity of culture and behavior. Computational models are comprised of the formalization of causal pathways that predict the relative weight of cortical brain regions as a component of the causal interaction of functional mechanisms. The dynamic causal modeling ("DCM") of the directionality and causality of cultural neural networks facilitates the characterization of the types of information processing that contribute to the neurocomputation of culture and behavior. Computational models facilitate the identification of the core brain regions of cultural neural networks and the interactivity of cortical brain regions that comprise the interconnectivity of cultural neural networks and their functional significance.

Bayesian approaches are of importance to the computational study of the patterns of large data-sets. Bayesian approaches contribute to the computational approaches to the study of the cultural brain and its function. Bayesian modeling of cultural brain activity may facilitate the characterization of the patterns of cultural brain activity that predict cultural knowledge. Other Bayesian approaches may be appropriate for the understanding of the prediction and explanation of cultural brain activity from large data-sets. Bayesian approaches to the study of cultural brain activity are essential to the identification of cultural brain regions that show patterns of cultural brain activity predictors of culture ("culture-based predictors").

Multilevel methods to the study of computational cultural neuroscience provide a multitude of basic neuroscience tools and techniques for the study of the computational level of the cultural brain. From single-cell recording to functional magnetic resonance imaging, research on computational cultural neuroscience benefits from a breadth of methodological approaches that provide an integral approach to the study of the spatiotemporal characteristics of cultural brain activity. The use of the multilevel, multimethod approach to the study of computational cultural brain mechanisms illustrates the complementarity of tools that are beneficial to the identification and characterization of cultural brain activity across levels of processing. Methodological foundations of computational cultural neuroscience illustrate the wide range of neuroscience tools and techniques for the systematic investigation of the neurocomputation of culture across levels of cortical organization.

Empirical foundations of computational cultural neuroscience

The study of computational cultural neuroscience comprises the empirical investigation of cultural neurocomputation across levels of analysis. Empirical research on computational cultural neuroscience investigates the interconnection of brain regions that comprise cultural neural networks and their dynamical interaction. The functional mechanisms of the core capacities of cultural information processing describe a mechanistic basis for the neurocomputation of culture. The systematic investigation of computational cultural neuroscience illustrates the role of the neurocomputational machinery of core capacities in the information processing of culture and behavior.

The empirical study of cultural neurocomputation arises from the testing of scientific theories on the neurocomputational study of culture and behavior. The conceptual landscape of scientific theories of computational cultural neuroscience provides a conceptual basis for the systematic investigation of the computational level of cultural brain activity. The study of the cultural neurocomputation of neurons and networks of neurons illustrates the distinct types of representations that comprise the information processing of functional and cultural adaptation. The design of the empirical studies on cultural neurocomputation encompasses a wide range of empirical approaches.

Formal empiricism on computational cultural neuroscience entails multilevel, multimethod approaches to the empirical investigation of the neurocomputational mechanisms of culture. First, the identification of the core brain regions that comprise cultural brain circuitry is essential to the understanding of the cultural processes of cortical brain organization. Second, the characterization of the functional mechanisms or dedicated, specialized mechanisms responsible for cultural processing is foundational to the understanding of the computational basis of cultural processes and mechanisms. Third, the computational modeling of the dynamical aspects of the cultural brain is important for the understanding of the intrinsic and extrinsic properties of cultural brain circuitry. Fourth, the study of the cultural entrainment of cortical brain organization is foundational to the understanding of the spontaneous and autonomous brain dynamics that are essential to the cultural processing of culture and behavior.

Empirical research on computational cultural neuroscience provides systematic studies on the understanding of the cultural neurocomputation of cortical brain circuitry. The capacities of biological organisms for cultural neurocomputation across distinct functional mechanisms demonstrate the adaptability of cortical brain organization for functional and cultural adaptation. Empirical studies on cultural neurocomputation detail the specific processes and mechanisms that comprise the information processing of cultural mechanisms. The identification of the specific types of processes and mechanisms of cultural neurocomputation is fundamental to the study of computational cultural neuroscience.

Cultural affective neurocomputation

Research on cultural affective neurocomputation is the neurocomputational study of culture and emotion. The cultural neurocomputation of cortical brain circuitry entails the understanding of the representation and interpretation of cultural information processing and its functional output. The cultural processes and mechanisms of emotional brain circuitry detail the functional role of culture on the information processing of brain circuits of emotion. The cultural representation of functional mechanisms that are located within emotional brain circuitry facilitates understanding of the functional and computational basis of culture and emotion.

Empirical research on cultural affective neurocomputation describes the neurocomputational study of culture and emotion. The cultural processing of emotion entails the functional basis of the encoding and decoding of emotional information in the cultural context. The processing of culture-based emotion entails the functional response of dedicated cortical brain regions that are specialized for the cultural processing of emotional information. The functional and computational basis of culture and emotion consists of the characterization of the interactivity of cultural and emotional processing, in particular, the cultural processing of emotional information that is content- and task-specific. The understanding of the neurocomputational mechanisms of culture and emotion is foundational to the understanding of the cultural processing of emotional information.

The neuroanatomical structures of emotional brain circuitry describe cortical brain regions that are responsible for the automatic and controlled processing of emotion. Limbic brain circuitry such as the amygdala, hippocampus, thalamus and interconnected brain regions describe cortical brain areas that are responsible for the automatic and reflexive response of neurons to emotion. The identification of limbic brain structures as emotional brain circuitry facilitates the understanding of the types of neuronal representation that comprise the processing of emotional information. The detailed characterization of limbic brain structures shows the importance of the structural components of emotion brain circuits on the functional mechanisms of emotion.

Limbic brain structures are comprised of functional brain circuits that consist of cellular and molecular mechanisms for the dedicated and specialized processing of culture and emotion. Functional brain circuits such as the amygdala show the cortical processes of functional mechanisms responsible for the processing of culture and emotion. The neurotransmission of the amygdala demonstrates the distinct types of cells that comprise the amygdala and its synaptic transmission. The neurotransmission of functional mechanisms shows the specificity of neuronal response responsible for cultural and emotional processing. The identification of the structural and functional properties of limbic brain structures provides a foundation for the empirical investigation of the neurocomputation of culture and emotion.

The cultural affective neurocomputation of cortical brain circuitry describes the cultural processing of emotional information within discrete brain regions of cortical brain circuits. The identification of the core brain regions that comprise the cultural processing of emotion details the functional specialization of cortical brain areas for the dedicated and specialized processing of culture and emotion. Limbic brain structures such as the amygdala show patterns of functional response that contribute to the automatic and reflexive processing of culture and emotion. The functional responsivity of the amygdala to the processing of emotion within cultural context illustrates a functional and mechanistic basis for the culture-based processing of emotion.

The distinct types of representation of the neurons located within the amygdala demonstrate a cellular and molecular basis for cultural and emotional processing. The distinct patterns of response of amygdala neurons demonstrate the automatic and reflexive processing of subcortical mechanisms to the processing of cultural and emotional information. The response of amygdala neurons shows automatic detection of the emotional information in the cultural environment. The functional involvement of the amygdala and interconnected brain regions in the emotional learning and experience of the cultural environment contributes to the understanding of the computational level of subcortical brain circuitry.

Corticolimbic circuitry demonstrates a functional basis for the cultural processing of emotion. Corticolimbic circuitry describes the interconnection of frontal and limbic brain structures as a cortical brain circuit that is responsible for the control or regulation of cultural and emotional processing. The cortical brain regions of frontal cortex and the amygdala comprise the neuroanatomical basis of corticolimbic circuitry. The structural and functional connectivity of corticolimbic circuitry illustrates the role of structural and functional properties on the processing of cultural and emotional information.

The functional brain activity of corticolimbic circuitry demonstrates the response of networks of neurons that comprise cortical brain regions responsible for cultural and emotional processing. The cultural processing of emotional information describes the distinct types of cultural representations that are located within frontal cortex and limbic circuitry. The cultural representation of emotion in functional mechanisms details the functional properties of corticolimbic circuitry that are responsible for the processing of cultural and emotional information. The cultural processing of emotional information relies on the functional involvement of corticolimbic brain structures.

The automatic and controlled processing of corticolimbic circuitry to the identity and expression of culture and emotion illustrates a functional basis for cultural affective neurocomputation. The encoding and decoding of the identity and expression of culture and emotion from facial and bodily expression detail the cultural processing of emotional information from sensory data. The cultural patterns of functional brain activity of corticolimbic circuitry illustrate the responsivity of core brain regions to the cultural processing of emotional expressivity. The functional responsivity of corticolimbic brain circuitry to the cultural expression of emotion illustrates the automatic and controlled processing of culture and emotion.

The functional patterns of cultural brain activity of corticolimbic circuitry demonstrate the information processing of functional mechanisms that is culture-specific. The cultural specificity of functional brain activity to emotional expression demonstrates the role of cultural processing on emotion that is culture-specific. The cultural representation of emotion located within corticolimbic brain structures illustrates the functional role of cultural processing on emotion. The generation and maintenance of cultural representations of emotion located within frontal cortex, the amygdala and interconnected brain regions show the importance of functional mechanisms on the selection, maintenance and response of cultural processes to emotion. The functional role of emotional expression on cultural processing illustrates the importance of emotion on the processing of cultural information.

The cultural processing of emotion describes the functional components of cultural tasks that are emotion-specific. The cultural specificity of functional brain activity to facial expression that is emotion-specific ("fear expression") demonstrates the functional basis of the cultural processing of emotional information. The contextualizing influence of culture on emotion illustrates the role of the context layer as an interpretive layer of information processing. The abstract cultural representation of emotion is comprised of multiple interpretations of complex sensory signals that comprise cultural expressions of emotional significance.

The study of cultural affective neurocomputation entails the understanding of the neurocomputation of the cultural processing of emotion. The neurocomputational study of culture and emotion comprises the identification of core brain regions that are responsible for the cultural processing of emotion. The cultural processing of emotion illustrates the representation and interpretation of cultural and emotional processing into motor output. The characterization of cortical brain circuitry dedicated to cultural processing contributes to the understanding of the computational level of cortical brain organization.

Cultural cognitive neurocomputation

Research on cultural cognitive neurocomputation is the study of the cultural processing of cognitive information at the level of computation. The study of the neurocomputation of culture and cognition consists of the understanding of the functional processes and mechanisms that are responsible for the cultural processing of cognition. The neurocomputational study of culture and cognition is comprised of the functional and mechanistic basis of the cultural and cognitive processes of multilevel adaptation. Cultural processes contribute to the cognitive understanding of the environment and its physical realizers.

The study of cultural cognitive neurocomputation is dedicated to the systematic investigation of the cultural processing of cognition and behavior at the level of computation. The functional components of cultural and cognitive processes comprise a set of tasks that show the computational basis of functional and cultural adaptation. The functional basis of cultural and cognitive processing details the mechanistic basis of specialized and dedicated machinery. The characterization of the neurocomputational mechanisms of culture and cognition contributes to the identification of the functional mechanisms that are important for the understanding of the cultural environment.

Empirical research on the cultural processing of cognitive information seeks to characterize the core cognitive capacities that are responsive to cultural contexts. The functional pathways of early cognitive processing illustrate the functional mechanisms that are dedicated to the cognitive representation of the cultural environment. From faces and objects to scenes and biological motion, the functional pathways of sensory and cognitive processes ("what" and "where" pathways) detail the core brain regions that show category-specific responses to the cultural environment. The early perceptual processing of social and cultural information is comprised of the distinct types of cognitive representations that have perceptual or motivational salience in the cultural environment. The functional basis of cognitive neurocomputation consists of the category-specific responses of cortical machinery for the cultural processing of cognition and behavior.

The functional specialization of cortical pathways facilitates the encoding and decoding of the cognitive representations of social and cultural information. The encoding of sensory and perceptual information into cognitive representations elicits the transformation of sense data into motor output. The cultural specificity of cognitive representation details the encoding and decoding of the social and cultural information from early perceptual cues. The cognitive representations of social and cultural information consist of the representational content that is culture-specific. The representation and interpretation of cognitive representations that is culture-specific facilitates the patterning of motor output within the cultural context.

The functional pathway of early perceptual processing details the specialized response of cortical mechanisms to social and cognitive information. The cortical mechanisms located within occipitotemporal cortices show culture-specific response to social ("faces") and cognitive information ("objects"). The specialized mechanisms for the cultural processing of social and cognitive information detail the differential patterns of face and object processing that is culture-specific. The cultural differences in the cortical mechanisms of face and object perception detail the processing of social and cognitive information that show functional specificity to the cultural context. The cultural specificity of face and object processing mechanisms serves as a functional basis of the cultural processing of cognitive representation of the environment.

The cultural specificity of cognitive processing demonstrates the distinct response of functional mechanisms to the cognition and higher-level processes of behavior. The cultural differences in cognitive processing detail the representation and interpretation of cognitive information of functional significance that is culture-specific. The cultural patterning of cognitive representation shows the specific responsivity of functional processes and mechanisms that illustrate the detection of the perceptual and motivational salience of cultural processing that is of importance. The cultural processing of cognitive information contributes to the detection of the perceptual and motivational salience of goals and actions. The understanding of cognition and higher-level processes relies on the cognitive representation of goals and actions that is responsive to the cultural environment.

The encoding and decoding of the cognitive representation of culture within neurons and networks of neurons illustrate functional mechanisms of culture and cognition. Neurons located within occipitotemporal cortices show the functional specificity of response to distinct types of representations of simple to complex stimuli. The neuronal response of cortical brain regions details the local dynamics of cultural processing that is specific to cognition and behavior. The bidirectional processing of cortical mechanisms localized within occipitotemporal cortices and adjacent regions show the population-level activity of neuronal mechanisms that demonstrate response to the cultural processing of cognitive information.

The neurocomputation of cultural and cognitive processing encompasses the understanding of the interpretation or meaning of cognitive representations and their functional significance. The cognitive processing of culture is comprised of the semantic processes that facilitate the interpretation of the shared meaning or significance of conceptual representation. The active representation and maintenance of semantic representation of abstract concepts located within parietal cortex is based on cognition and language and facilitates the interpretative layer of cultural and cognitive processing. The semantic processing of cultural information entails the interpretation of the cognitive representation of the cultural environment. The semantic understanding of abstract conceptual knowledge contributes to the neurocomputation of culture and cognition.

The cultural processing of cognitive information entails the generation, selection and maintenance of active representations of the cultural environment. The prefrontal cortex located within the frontal lobe is responsible for the representation and interpretation of cognitive processes. The cognitive representation of social and emotional information within the prefrontal cortex demonstrates the role of active representations as a conceptual basis of social and cultural knowledge. The dorsal and ventral portions of the medial prefrontal cortex show the functional subdivisions of the cognitive processing of social and cultural information. The cognitive representation of social and cultural information within the prefrontal cortex shows the functional significance of abstract knowledge on social and cultural adaptation.

The prefrontal cortex is also responsible for the control and regulatory function of cognitive inhibition. The neurocomputational study of the cultural processing of cognitive inhibition details the characterization of the bidirectional processing of networks of neurons. The functional brain activity of lateral prefrontal cortex and its interconnected brain regions shows the core brain regions dedicated to the cultural processing of inhibitory control. The functional involvement of lateral prefrontal cortex in inhibitory control and its interconnectivity to cognitive neural networks demonstrates a functional mechanism for the control and regulation of the cultural processing of cognition. The controlled processing of inhibitory control illustrates a mechanistic basis for the inhibition or regulation of cultural processes.

The cultural patterns of cognitive brain activity illustrate the causality and directionality of the functional basis of cognition and higher-level processes of brain function. Cultural patterning of cortical brain activity of neural networks demonstrates the causal directionality of bidirectional processing that is responsible for cognition and higher-level function. The feedforward and feedback processing of cortical brain regions responsible for cognition detail the top-down and bottom-up processing of cognition that contributes to higher-level processes. The cultural processing of cognitive information relies on the bidirectional processing of cortical brain regions that comprise cognitive neural networks of culture.

The cultural brain dynamics of cognitive neural networks detail the causal role of structural and functional properties of interconnectivity on the global dynamics of neuronal mechanisms. The brain dynamics of cultural cognitive neural networks demonstrate structural properties that are of functional importance to the dynamical processing of culture and cognition. The structural connectivity of cultural cognitive neural networks shows the depth of interconnection of functional pathways that are dedicated to the information processing of culture and cognition. The global dynamics of cultural cognitive neural networks show the intrinsic properties of cultural and cognitive processes that are based on the structural aspects of networks of neural networks.

The functional basis of cultural cognitive neurocomputation demonstrates the cultural processing of cognitive information at the level of computation. The neurocomputational study of culture and cognition details the role of culture on the cognition and higher-level processes of brain function. The identification of the core brain regions that describe the cultural cognitive neural networks contributes to the understanding of the functional basis of the cultural processing of cognition and behavior. The characterization of the local and global dynamics of cultural cognitive neural networks facilitates the understanding of the population-level activity of neurons to networks of neural networks and their functional significance.

Research on cultural cognitive neurocomputation is important for the understanding of the culture and cognition of cortical brain organization at the level of computation. The study of the neurocomputation of culture and cognition demonstrates the mechanistic basis of the cultural processing of cognitive information across levels of processing. The identification of the cultural neurocomputation of cognitive processes and mechanisms illustrates the representation and transformation of sense data into motor output. The functional basis of the core brain regions of cultural capacities details the types of processing at the computational level.

Cultural social neurocomputation

Research on cultural social neurocomputation consists of the study of the cultural neurocomputation of social information. The study of cultural social neurocomputation investigates the cultural processing of social information of neuronal mechanisms across levels of processing. From neurons to networks of neurons, the identification of the distinct types of representations of neuronal mechanisms illustrates the representational content important for multilevel adaptation. The cultural representation of social information of cortical mechanisms details the causal–functional role of functional machinery on the representation and interpretation of the social and cultural environment.

Cultural social neurocomputation describes the cortical mechanisms that comprise the cultural processing of social information and its functional significance. The functional basis of cultural processing illustrates the role of cortical brain mechanisms in the representation and interpretation of the social and cultural information of the environment. The encoding and decoding of social information into the functional mechanisms of cortical brain circuits detail the representation and transformation of sense data into motor output that is culture-specific. The cultural representation of social knowledge illustrates the role of perceptual and cognitive processing in the generation and maintenance of representations of the social and cultural environment.

Functional mechanisms of the cultural processing of social information consist of core brain regions that show the functional specificity of neurons and networks of neurons to social and cultural processing. The core brain regions of social brain circuitry illustrate the dedicated and specialized machinery for social processing. The functional machinery of cortical brain mechanisms demonstrates the response of core brain regions to the social processing of the cultural environment. From faces and biological motion to agency and intentionality, core social capacities show the functional role of core processes in the understanding of other minds.

The core brain regions of social brain circuitry detail the specialized and dedicated machinery of social and cultural processing. The core brain regions located within occipitotemporal, parietal and frontal cortices show the functional response of neuronal mechanisms that are specialized for the social processing of the cultural environment. The social processing of functional mechanisms illustrates the automatic response of cortical brain regions to the detection of social information. From faces to biological motion, the occipitotemporal lobes show neuronal mechanisms that are dedicated to the processing of social information. The automatic response of neuronal mechanisms to category-specific social knowledge illustrates the functional basis of the social processing of the cultural environment.

The cultural patterns of the cortical brain activity of core brain regions show the processing of social information that is culture-specific. The cultural specificity of functional brain activity located within the temporal lobe shows the dedicated, specialized response of cortical mechanisms to the detection of the perceptual and motivational salience of social cues. The functional brain activity of cortical brain regions such as superior temporal sulcus located within the upper and lower bank of the temporal gyrus shows the cultural specificity of functional response to mental state understanding that is specific to culture. Cultural patterns of functional brain activity demonstrate the functional specificity of social brain circuitry to the inferential reasoning on the motivation and intentionality of goals and actions in the cultural environment. The cultural specificity of functional responsivity of core brain regions facilitates social cognition within the cultural context.

The neurocomputational study of the cultural representation of social knowledge facilitates understanding of the functional role of cortical mechanisms. The prefrontal cortex located within the frontal lobe is important to the generation, selection and maintenance of active representations of information. The medial prefrontal cortex is important to the representation and maintenance of cognitive representation of social knowledge. The functional brain activity of the medial prefrontal cortex shows a response to social processing that is culture-specific. The cultural patterns of functional brain activity of the medial prefrontal cortex demonstrate the cultural processing of social information. The cultural specificity of cortical brain activity of social processing facilitates the understanding of the self and others.

The functional role of prefrontal cortex in the control or regulation of social capacities details a core brain region that is important to the controlled processing of social and cultural regulation. The prefrontal cortex is involved in inhibitory control or regulatory function of social information processing within cultural context. The controlled processing of social information demonstrates the inhibition or regulatory function of prefrontal cortex that is responsible for sociocultural processing. The control or regulation of social and emotional capacities demonstrates the role of top-down processing on the cultural representation of social and emotional information. The functional specificity of core brain regions demonstrates the network of brain regions that are responsible for the social processing of the cultural environment.

The culture-based social processing of core brain regions underlies the cognition and higher-level processes of social and cultural adaptation. The local dynamics of cortical neurons that comprise social brain regions show the population-level activity of neurons and their role in the cultural processing of social information. The neuronal response of social brain regions to social and emotional cues shows the information processing of functional mechanisms that is culture-specific. The feedforward and feedback processing of social information of cortical brain regions illustrates the brain dynamics of localized brain circuits for culture-specific processing. The characterization of local brain dynamics contributes to the understanding of the neurocomputation of social and cultural adaptation.

The global dynamics of cultural social neural networks describe the brain dynamics of the cultural processing of social information. Global brain dynamics of cultural neural networks detail the structural and functional properties of connectivity that are intrinsic to cultural and social processing. The global dynamics of the connectivity of cultural neural networks show the role of depth of interconnection and structural connectivity as intrinsic properties of cultural and social processing. The study of the global dynamics of cultural social neural networks contributes to the neurocomputational study of cultural and social processing.

Research on the computational approaches to the study of cultural social neurocomputation enables the understanding of the cultural and social processes and mechanisms of cortical brain organization at the level of computation. Across levels of processing, the cultural processing of social information relies on functional mechanisms that show distinct types of representation and interpretation of the social and cultural environment. From neurons to networks of neuronal networks, cultural neurocomputation entails the characterization of the specific mechanisms that contribute to the representation of the social and cultural environment. The study of the cultural patterning of social brain circuitry is foundational to the broader understanding of the cultural computation of the social environment.

The brain basis of culture and behavior

Research on computational cultural neuroscience contributes to the understanding of the brain basis of culture and behavior. The conceptual development of the scientific language of the field of study demonstrates the plausibility of the cultural mind as a biological computing machine. The physical realization of the cultural mind as a biological computing machine raises a range of notions of the core capacities of the cultural brain and its functional role in adaptation. The study of the brain basis of culture and behavior is fundamental to the understanding of the biology and culture of behavior. The systematic investigation of the brain basis of culture and behavior contributes to the characterization of the functional mechanisms of cortical brain organization at the level of computation.

The identification of cortical brain regions that comprise cultural brain function is foundational to the understanding of the brain–behavior relationships of culture. The detailed characterization of the cultural patterns of functional brain activity that are important to behavior contributes to the understanding of cultural capacities and its functional basis. The functional processes of cultural capacities contribute to multilevel adaptation. The study of the cultural processing of cortical brain circuits details the functional mechanisms that predict cultural patterns of behavior. The cultural patterning of cortical brain function and behavior facilitates the understanding of the cultural differences in cognition and brain function.

The scholarly interest in the brain basis of culture and behavior yields insight into the biological instantiation of cultural computation and its functional relation. The cultural computation of biological organisms as social and cultural species shows the adaptability of functional adaptation. The cultural computation of cortical brain function demonstrates the functional processes and mechanisms of the nervous system that serve as a mechanistic basis of culture and behavior. The detailed understanding of the biophysical mechanisms of neurocomputation contributes to the satisfactory explanation on the origin and nature of cultural computation in real and artificial systems.

Discussions

Research on computational cultural neuroscience is fundamental to the understanding of the computational level of the cultural brain and its structure and function. The computational level of cortical brain organization details the mechanistic basis of functional processes that are essential to the biological basis of culture and behavior. The systematic study of computational cultural neuroscience provides a conceptual foundation for the understanding of the neurocomputational mechanisms of biological organisms as cultural species. The empirical studies of computational cultural neuroscience contribute to the scientific and educational enrichment on the understanding of the brain basis of culture and its broader implications for culture and society.

The building of scientific knowledge on computational cultural neuroscience is an advancement of the science and technology of culture and health. The scientific and technological innovation of the computational tools and technologies contributes to the research capabilities of health and medicine. The development of the scientific infrastructure on computational cultural neuroscience is important to the broadening of scholarly understanding of the computational level of cortical brain organization and cultural computation in the natural world.

The building of scientific knowledge on cultural computation provides a foundation for the bioengineering or construction of artificial or synthetic devices that are capable of cultural computation. The innovation of artificial intelligence and modern computing shows the feasibility of the design of physical devices that have functional capabilities of automated computation and intelligence. The scientific and technological advancement of the functional capabilities of artificial intelligence and modern computing toward the automated production of culture allows for the scientific and educational innovation of programs and resources that promote the social and cultural enrichment of society.

Conclusion

The study of computational cultural neuroscience is foundational to the investigation of the computational level of cortical brain organization and their role in the higher-level processes of brain function. The cultural capacities of biological organisms represent one of the most complex functional adaptations that contribute to ecological niche construction. Research on computational cultural neuroscience builds knowledge on the functional mechanisms of biological organisms that are important for multilevel adaptation. The biological plausibility of the cortical brain as a mechanistic basis of cultural computation broadens the understanding of culture as a causal system and the causal realizers of culture in the natural world.

Practical applications of computational cultural neuroscience

The study of computational cultural neuroscience informs on the evidence-based and policy-based approaches of culture in health and medicine and related fields of study. The scientific study of computational cultural neuroscience builds the evidence base of scientific knowledge on the computational tools and technologies that contribute to the understanding of the brains of diverse populations and cultures. The empirical study of computational cultural neuroscience is beneficial to the innovation of science and technology and its practical applications. Research on computational cultural neuroscience informs policy-based approaches to the development and implementation of policies and programs of culture and technology and its practical impact on society and the public.

Public health

Evidence-based research on computational cultural neuroscience builds scientific knowledge on the computational mechanisms of culture and brain function. The generation of novel knowledge on computational cultural neuroscience provides a foundation for an understanding of the computational level of cultural brain function. The design of computational tools and technologies is a scientific and technological innovation that is beneficial to the advancement of research and development on computational cultural neuroscience. The promotion of culture and mental health benefits from the building of scientific knowledge on computational science and informatics and its practical applications on culture and mental health. The building of evidence based on culture and computation contributes to the scientific and educational enrichment of culture and mental health.

Policy-based research on computational cultural neuroscience is beneficial to the design of prevention and intervention strategies that contribute to the promotion of culture and mental health. The innovation of computational tools and technologies constitutes scientific and technological advancement that is beneficial to the public interest. Prevention strategies contribute to the design of public health programs that are important to the promotion of culture and mental health. The development of policy-based intervention is foundational to the design of programs and policies that lead to the improvement of societal and health outcomes in the public interest. The development and implementation of policy-based research contributes to the policies and programs that advance the highest standards of health and health equity of diverse populations and cultures.

Medicine

Research on computational cultural neuroscience contributes to the evidence-based resources for the prevention and treatment of disease in medicine. The development of evidence-based resources serves as an evidence base and information source for the design of prevention and intervention in health and medicine. The design of computational tools and technologies broadens opportunities for the discovery of cures and the prevention and treatment of complex diseases. Computational tools and informatics are important to the design of medical capabilities that inform on the treatment and rehabilitation of complex diseases. Prevention medicine benefits from the innovation of computational tools and technologies that inform the medical capabilities of health systems. The innovation of computational tools and informatics leads to the advancement of the prevention and treatment of complex diseases.

Public policy

The consideration of public interests on culture and health benefit from the integral discourse on technology and health and its impact on society. The development of evidence-based resources on computational cultural neuroscience as a field of study contributes to the scientific infrastructure and database informatics that inform on the computational tools and technologies of culture and mental health. The innovation of computational tools and technologies is beneficial to the design of prevention and intervention strategies in health, medicine and applied fields of bioengineering and computer science that are important to the research and development of culture and health. The development of policy-based approaches on computational cultural neuroscience contributes to the design of programs and policies that inform on the societal benefits of culture, technology and health.

International cooperation on the innovation of science and technology in computational cultural neuroscience informs the development of international policy and standards on health in the public interest. The development of international policy and standards on health contributes to the understanding of the cultural and societal considerations that impact mental health and health promotion. The promotion of mental health benefits from the consideration of the cultural and societal issues that are of importance to the population. The design of international guidance and standards informs the development of international policy on health that leads to the improvement of societal and health outcomes.

Policy development and implementation on culture and mental health promotion benefit from the scientific and technological innovation in computational cultural neuroscience. The design of policy development in the areas of health, trade and defense among others, provides guidance on international and national standards that are important to implementation of policies and programs. The development of national policies and programs on culture and health benefits from scientific and educational resources that inform on the benefits of culture, technology and health to public policy.

2

CULTURAL COMPUTATION

Introduction

Early philosophical thought posits the scholarly and scientific interest on the nature of cultural life and the cultural world. From Lao Tzu to Locke, the philosophical postulation of cultural life and the cultural world as societal discourse arises from the early contemplation of the spiritual realm as the social and cultural thought of civilization and society. The early scholarly interest on social and cultural thought as being and existence illustrates the importance of culture as a means of the understanding of the mind and world and its meaning. The concept of the ideal being and ways of life arises from the metaphysics and epistemology of the material and immaterial world that comprises social and cultural thought. The metaphysics and epistemology of spirituality and religion imbues pontification on cultural life and the cultural world with the understanding of ideal and material forms of being and existence of the world. From ancient civilizations to contemporary societies, culture as systems of shared meaning contributes to the understanding of fundamental ways of cultural life in the world.

The study of cultural computation arises from scholarly and scientific interest in the understanding of the fundamentals of culture at the level of computation. The foundations of cultural computation was built on the philosophical postulation of culture as an ideal and material notion. The idealism and materialism of culture encompass the conceptual formulation of the idealism and materialism of culture. The scholarly inquiry of culture as ideals and matter arouses interest in the conceptual foundation of culture. Culture as ideas and material realism with a functional basis entails

DOI: 10.4324/9781003384236-3

the understanding of the origin and nature of culture. The notion of cultural ideas with material realism connotates on the function of culture in the natural world. The scientific interest in culture as a physical realism reflects on the naturalism of culture as ideas and materialism. The physical realization of material culture implies a functional and mechanistic basis of culture.

The naturalism of cultural ideas and material culture entails the understanding of the physical realization of culture in the natural world. The naturalistic understanding of culture as a material realism implies the multiple realizability of ideal and material forms of culture as physical systems. Culture as a causal system of physical states of the natural world implies the notion of the cultural ideas and material culture as physical realizers. The understanding of the functional or causal role of culture in the natural world imbues shared meaning and significance in the understanding of culture as mental and physical events. Cultural systems as mental and physical events describe states of the natural world with a functional or causal significance. The characterization of the cultural systems within spatiotemporal dimensions facilitates understanding of culture as a causal system.

The conceptual foundations of culture as a byproduct of the adaptation of biological organisms arise conceptual notions of the multiple realizability of culture in the natural world. The scientific empiricism of the 18th and 19th centuries brought forth interest in the naturalism of biology. The scientific study of biology as systems of life placed emphasis on the functional and mechanistic understanding of life systems. The development of the scientific theories of evolutionary biology furthered the conceptual landscape on the fundamentals of life and its mechanisms. Theories of evolutionary biology serve as a foundation for the understanding of the laws and principles of living systems. The characterization of the lawlike principles of evolutionary biology demonstrates the importance of selection and adaptation as fundamental principles of the origin of species.

The adaptability of biological organisms as cultural species presents a compendium of interests and considerations on the adaptation of culture. The study of cultural computation comprises the scholarly inquiry of the physical realization of culture as a life system. The biological adaptation of cultural species illustrates the causality and functionality of culture and its function in the natural world. The understanding of the multilevel adaptation of biological organisms entails the understanding of the functional machinery of cultural mind at the level of computation. The characterization of the functional basis of the cultural mind within structural and functional organization of the nervous system illustrates the cultural adaptation of biological organisms across levels of processing.

The goal of the chapter is to provide a review on the fundamentals of cultural computation. Research on cultural computation provides a foundation for the scientific discovery of the culture and computation of the natural world. The study of cultural computation encompasses a wide range of thematic topics including cultural mental computation, cultural machine computation and cultural neurocomputation. The scholarly study of cultural computation serves as a foundation for the understanding of a range of topics from the cultural mind as a computing machine to the biological basis of culture and brain function and its interrelations. The practical applications of cultural computation for related fields of study are discussed.

Cultural Computation

Research on cultural computation investigates the fundamentals of cultural processing of computational mechanisms and its physical realization in the world. Cultural computation entails the conceptual understanding of cultural processing at the computational level. Culture as computation refers to the physical states of cultural systems and its physical state transitions. The concept of cultural systems as physical systems encompasses the conceptualization of culture as physical systems that show the functional capabilities of machine performance. The notion of culture as physical systems implies the idea of physical systems that are capable of shared or symbolic meaning. The concept of cultural machines or machines with cultural capacities illustrates a physical instantiation of a functional basis of culture. Cultural machines with functional capacities for shared or symbolic understanding illustrate a physical basis of culture and its function.

Early notions of culture as shared systems of meaning refer to a wide realm of ideas and materialities of culture in the natural world. The breadth of culture and physicalism entails the understanding of the multiple physical realizers of culture in the world. Across multiple physical realizers, culture is comprised of the set of physical states and its state transitions that show functional capability for shared or symbolic meaning. The cultural mind demonstrates how the functional capacities of cognition and language contribute to the shared understanding of others. The core capacities of cognition and language demonstrate how cultural processes of semantic understanding facilitate the representation and interpretation of shared meaning systems. The core capacities of the cultural mind demonstrate the functional basis of cultural processing in the world.

Cultural mental computation details the cultural computation of the mind and its function. The study of cultural mental computation ascribes mental states as the functional basis of cultural computation. The mental states of cognition and language demonstrate the functional components of the cultural processing of the mind. The cognition and language of semantic understanding details the functional role of processing mechanisms on the shared understanding of symbolic systems and its functional significance. The mental state understanding of others as a functional capacity further describes the core capacities of emotion and social cognition as important in the shared understanding and interpretation of mental states and their social and cultural significance.

The cultural computation of machines provides a conceptual foundation of the understanding of the computational level of culture. Cultural machine computation describes how the mind as computing machine serves as a functional basis of cultural computation. The conceptual language of cultural machine computation broadens the understanding of the mind as a computing machine to the role of the cultural mind in the automated production of cultural performance. The cultural mind as a byproduct of cultural performance implies the functional significance of computation on the tacit habituation of cultural practices. The functional capacities of the cultural mind serve as a functional basis of the learning and experience of culture and behavioral performance.

The study of culture at the level of computation is comprised of the understanding of the machine states or mechanistic basis of culture and its automated performance. The notion of cultural computation broadens the notion of culture from a system of physical states to a computing machine or system of machine states and its transitions. On the one hand, the cultural mind as a computing machine with core capacities of shared or symbolic understanding illustrates a physical realizer of cultural computation as a functional adaptation of biological organisms in the natural world. The cultural mind shows how the core capacities of the mind such as cognition and language serve as a functional basis of cultural competence. On the other hand, the cultural computer as a physical device shows how the functional capabilities of machine computation are sufficient for the automated production of cultural performance. The cultural computer demonstrates the functional capabilities of machine computation for the automated production of language and culture. From the cultural mind to the cultural computer, the study of cultural computation abounds with interest on the multiple realizability of culture across physical realizers.

Culture as a byproduct of the adaptation of biological organisms delves into the manifestation of culture as a functional mechanism in the natural world. Culture as a functional adaptation of biological organisms as cultural species highlights the role of core capacities on the functional performance of multilevel adaptation. The cultural processing of biological organisms illustrates the cultural capacities of functional machinery at the level of computation. The cultural capacities of the brain demonstrate the functional machinery of biological organisms for functional and cultural adaptation. The cultural brain shows how the structural and functional organization of the nervous system serves as a biological basis of culture and adaptation.

The study of the cultural processes of cortical brain organization contributes to the understanding of a functional and mechanistic basis of culture that is characteristic of biological organisms. The cultural computation of the brain structure and function details the role of cortical brain organization on the cultural processing of biological organisms across levels of processing. The computational level of the cultural brain entails the understanding of the functional components that comprise the core capacities of functional and cultural adaptation. The functional components of cultural capacities illustrate the role of functional tasks on the performance of cultural processes. The systematic study of the cultural brain contributes to a broad understanding of the cultural computation of the brain as a physical realizer of culture and its function in the natural world. The cultural neurocomputation of cortical brain organization describes a mechanistic basis of cultural computation.

The cultural neurocomputation of functional mechanisms details the biophysical mechanisms that comprise the cultural processing of cortical brain organization and behavior. The characterization of the cultural processing of cortical brain mechanisms details the causal flow of information that comprises the functional capacities of cultural adaptation. The cultural processing of information flow across cortical brain regions demonstrates a functional basis for the cultural computation of neural machinery. The study of cultural neurocomputation arises from the understanding of the functional architecture of culture and cortical brain organization.

The cultural computation of minds and machines demonstrates a functional and mechanistic basis of cultural performance. From simple to complex machines, cultural computation illustrates the functional capabilities of mental and machine computation for cultural production. The understanding of cultural computation across multiple physical realizers details on the generalizability of the causality and functionality of culture as a causal system. The demonstration of the plausibility of cultural computation of biological organisms as cultural species provides a foundation for the design of cultural computation of real and artificial systems. Practical applications of cultural computation include the innovation of computational tools and technologies of culture such as the construction of physical devices and machines that show artificial cultural intelligence or the design of computational tools and technologies for the study of cultural computation in the natural world.

Cultural Mental Computation

The cultural mental computation of the mind in the world describes the cultural content of mental computation. The understanding of the cultural computation of the mind relies on the core capacities of mental computation and its functional performance in the world. The study of cultural mental computation consists of the characterization of the role of mental function on the generation of culture in the mind and in the world. Cultural mental computation also entails the description of the cultural content of mental states and its causal role in the functional performance of culture in the mind and in the world. The cultural computation of the mind encompasses the functional components of core capacities that are necessary and sufficient for cultural adaptation.

The core capacities of the cultural mind demonstrate the functional role of cultural mental computation. The computation of mental function is comprised of a wide range of core capacities that are fundamental to the understanding of culture in the mind and in the world. From sensation and cognition to emotion and consciousness, the core capacities of the mind bestow functional roles on the mental function of thought and action. The interaction of the mind with the cultural environment evokes patterns and regularities of mental function that are concordant of cultural thought and action. The functional capacity of the cultural mind for the cultural patterning of thought and mind illustrates the causal role of environmental influence on mental function.

The cultural processing of mental function includes the core capacities of functional processes of the mind and its relation to behavior. The functional components of core capacities describe the cultural processing of functional tasks as a functional basis of cultural mental computation. The cultural processing of functional tasks elaborates on the role of mental function in the problem-solving of cultural tasks and performance. The wide range of cultural tasks that comprise cultural-mental computation demonstrates the functionality and complexity of the cultural mind. The recognition of the identity and expression of culture is an example of a cultural task that relies on the functional capacities of the mind. The shared understanding of the meaning or significance of events is another example of cultural task that is based on the core capacities of the mind. The inferential reasoning on the intentionality and motivation of others constitutes a functional basis of the cultural capacities of the mind. The functional components of mental function are essential to the wide range of cultural capacities that are important to multilevel adaptation.

The mental function of cultural capacities displays levels of complexity that are emblematic of cognition and higher-level processes of behavior. The cultural capacities of cognition and language describe the mental function of cultural tasks that arise from early to late processing streams. The cognition and language of culture illustrate how core capacities subserve the functional processes of cultural adaptation. The cultural processing of cognition and language demonstrates the functional basis of cultural and behavioral performance of the mind. The cultural capacities of cognition and higher-level processes are fundamental as instances of functional adaptation.

The concept of the cultural mind as a computing machine implies the performance of mental function in the production of cultural content. The notion of the cultural mind as a cultural computing machine elaborates on the notion of the mind and its causal role in the generation of culture. The core capacities of mental function depict functional components that subserve the performance of task-based cultural processing. Cultural processing entails the functional components that are essential to the task-based processing of mental function. The cultural processing of the mind illustrates how task-based processing is sufficient as a functional component of culture and behavior.

The cultural mind also shows functional capacities for cultural thought and action that are based on the spontaneous activity of mental function. The spontaneous activity of mental function implies that the automaticity of cognition and higher-level processes of consciousness. The cultural patterns of consciousness and cognition detail the streams of conscious thought that are elicited from the spontaneous activity of mental function. The understanding of the functional role of spontaneous mental thought as a functional component of the cultural mind contributes to the characterization of the functional basis of cultural mental computation.

Cultural Machine Computation

Cultural machine computation describes the computing machine and its functional performance. The study of cultural machine computation entails the understanding of how machine computation and its functional performance facilitates the automated production of culture. Research of machine computation has long shown interest in the design of computing machines that have functional capabilities of mental function. Early computing machines sought to demonstrate the capabilities of computing machines to perform simple to complex mental function such as calculation (Figure 2.1). Modern computers show how the functionality and capabilities of physical devices demonstrate performance that shows functional equivalence with that of mental computation.

Scholarly interest in cultural machine computation arises from the notion of culture in the mind and in the world as a function of culture and machine computation. The early notions of the history of computing sought to determine a design of a computing machine with capabilities for the higher-level processes of mental function. The earliest analog computers were designed to show the functional capabilities of symbolic production ("numerical calculation", "alphanumeric symbols"). The early insight of analog computing to transform mechanical states into analog

FIGURE 2.1   Photo of ENIAC

representations of numerical relations was an innovation of machine computation. Other analog computers aimed to show how the physical realism of the computing machine impacts the fundamental parameters of machine computation. Early efforts aimed to regulate the efficiency, reliability and complexity of symbolic production from the analog representation of computing machines.

Modern computing with electromechanical machines demonstrates how culture and machine computation arises from computing machines with functional capabilities. The computing machine shows how digital representation contributes to the efficiency and reliability of simple to complex symbolic production. The transformation of electromechanical states into digital representation of modern computers details how machine computation shows performance that is of functional equivalence to mentality and intelligence. From simple calculation to complex language production, digital computers show the determinism of functional capabilities that are comparable to mental function and intelligence. The design of functional components of computing machines with memory and processing function led to the improvements of the efficiency and reliability of computational functions.

TABLE 2.1 Cultural Computation

Types of Cultural Computation	List of Examples
Cultural mental computation	Ruled-based cognition
Cultural neurocomputation	Amygdala neurons
Cultural machine computation	Symbolic production
Cultural computation	Automated language production
Cultural quantum computation	2-D to 4-D performance
Cultural supercomputation	Simulated performance

Research on cultural machine computation shows how the history of computing has sought to demonstrate the physical realism of mental function. From early analog computing to modern computers, the history of computing illustrates the innovation of physical devices that perform machine computation with functional capabilities that are comparable to thought and intelligence. Functional capabilities that are comparable across multiple physical realizers of the computing machine demonstrate the formidable impetus of innovation on the establishment of standards on machine computation in science and technology. The history of computing details how the standard of functional performance of modern computing shows the adaptability and complexity of functional capabilities that are fundamental to culture and computation.

Contemporary innovation of computing has introduced machine computation that is capable of the adaptability and complexity of computing that is conditional. The machine computation of quantum computing that incorporates conditions of functional performance illustrates functional capabilities that adapt to conditional states. The use of digital representation with quantum computing shows how quantum states show the functional capability for performance of automated production that is conditional. The innovation of quantum computing advances the notion that machine computation arises from the automated production of complex functions. The functional capabilities of machine computation for automated production that occurs under conditions demonstrates how the quantum computing machine shows levels of adaptability of mentality and intelligence in functional performance that is comparable in real-world advantages.

The notion of the computing machine is integral to the philosophical postulation on the dualism and materialism of the mind and machine. The early philosophical thought on the minds and machines has long sought to ascertain the ideal and natural realism of the mind and machine and the interrelations. The demonstration of the multiple realizability of computing machines advances the philosophical thought on mind and machines. The unified and parsimonious explanation of the computation of minds and machines as patterns and regularities of complex systems illustrates the material relations of mind and machines that are consistent with natural realism. The contribution of the history of computing to the philosophical traditions illuminates the breadth of conceptual landscape that encompasses the development of machine computation as a source of cultural causation in the world.

Cultural Neurocomputation

The study of cultural neurocomputation refers to the characterization of the multilevel mechanisms of cortical brain organization that subserve cultural capacities. Cultural capacities of biological organisms demonstrate the wide range of functional processes and mechanisms that show the cultural processing of mind and behavior. The cultural neurocomputation of cortical brain organization details how functional processes and mechanisms of culture contribute to the functional role of cultural adaptation. Cultural capacities are essential to the demonstration of the core function of cultural thought and action that is of importance to multilevel adaptation. Research on cultural neurocomputation is fundamental to the identification of causal mechanisms that comprise the functional processes of cultural capacities. The detailed characterization of the structural and functional properties of cortical brain organization are fundamental to the functional processes and mechanisms of cultural adaptation.

The systematic investigation of cultural neurocomputation entails the study of empirical paradigms that comprise the functional components of cultural capacities and their underlying mechanisms. The study design of empirical paradigms illustrates the functional operationalization of functional tasks and components that comprise cultural tasks as functional capacities. The functional operationalization of empirical paradigms facilitates the observation of cultural tasks that are consistent across empirical investigation. The functional design of task-based cognition implies the role of task based processing in the demonstration of the functional components of cultural capacities.

The cultural neurocomputation that occurs across multilevel mechanisms demonstrates the cultural processing that arises based on levels of processing. Across levels of processing, cortical brain organization details the multilevel mechanism that subserves the functional capacities of cultural adaptation. From neurons to networks of neural networks, the neurocomputational study of cultural processing details how the aggregation of response of functional mechanisms reveals the representation and transformation of information processing as input–output relations. The empirical study of the structural and dynamical principles of cortical brain organization contributes to the detailed characterization of multilevel mechanisms that underlie cultural neurocomputation.

The structural and dynamical principles of cortical brain organization entail the fundamental principles of computation that govern the mechanisms of complex systems. The functional mechanisms of the nervous system demonstrate the computational level of mechanistic processing that is consistent with the structural and dynamical principles of cortical brain function. The structural principles of cortical brain organization describe how the structural connectivity of the cortical brain regions contribute to the relaying of information processing that is necessary and sufficient for cultural capacities. The structural connectivity of cortical brain networks shows how the strength of interconnectivity is of importance to the functional processing of cultural capacities.

The dynamical principles of cortical brain function show the autonomous self-organization of cortical brain function that is based on the entrainment of cortical brain activity for cultural capacities. The functional activity of cortical brain mechanisms demonstrates autonomous brain dynamics that illustrate the self-organizing principle of brain function. The entrainment of cortical mechanisms to the functional activity of cortical brain function that is spontaneous and active demonstrates how the patterning of functional activity contributes to the functional basis of cultural capacities. The active representations of cultural processing within cortical brain function elicits cultural patterns of functional brain activity that are spontaneous and self-organizing. The cultural entrainment of cortical brain function shows how cultural brain dynamics are sufficient as the functional machinery of cultural capacities.

Culture, Mind and Machines

The conceptual foundations of cultural computation inform on the understanding of culture, mind and machines and its societal impact. Culture as systems of shared meaning of groups of people is of importance to the understanding of cultural knowledge. Cultural knowledge consists of the knowledge states of cultural systems that describe the societal or cultural significance of events in the natural world. The cultural capacities of biological organisms contribute to the naturalism of cultural systems of knowledge. The characterization of the cultural capacities facilitates understanding of the functional components that comprise cultural computation. Understanding the fundamental principles of computation provides a foundation for the understanding of the natural realism of culture.

The notion of the mind as a computing machine is foundational to the understanding of the functional significance of culture, minds and machines. The mind as a computing machine suggests that machine computation is capable of the automated production of mental content; similarly, the mind is comprised of functional capacities that are essential to the representation and interpretation of the environment. The functional capacities of the mind demonstrate the causal role of mental function of thought and reasoning. The core capacities of the mind detail the cognition and higher-level function of behavior. The machine computation of the mind shows the functional performance of simple to complex machines.

Mental thought and intelligence are considered the mental content of the mind. Mental thought and intelligence as sets of mental events comprise patterns or regularities of mental content that demonstrate the role of the mind in the computation of mental content. The mental content of thought and intelligence consists of mental property or sets of mental events with distinct functional properties. The functional properties of mental content detail the specific patterns or regularities of mental function that comprise core processes of the mind. The functional capacities of mental function for the patterning or regularities of input–output relations demonstrates the robustness of the functional processes of mental thought and intelligence.

The mental processes and function of the cultural mind illustrate the patterns and regularities of culture in the natural world. The cultural thought and intelligence of mental processes facilitate the functional performance of cultural mental content. Cultural mental content arises from the set of mental events with distinct functional properties. The intrinsic and extrinsic properties of cultural mental content detail the autonomous production of functional performance for the purpose of cultural thought and inferential reasoning. The inferential reasoning on the motivational and intentional facets of cultural thought and behavior demonstrates the functional significance of cultural mental content and its real-world advantages. The cultural thought that comprises idealism and imagination illustrates the metarepresentation of intentional action and the functional consequence of cultural mental content on possible world realism.

The cultural mind describes the functional capacities that comprise cultural mental computation. The functional processes of the cultural mind consist of core capacities that show the automated production of cultural mental computation. The functional components of core capacities comprise a set of mental functions that are of importance to the cultural computation of the mind. The core capacities of the cultural mind such as psychological processes or psychological realizers illustrate the specific functional components that contribute to the information processing of the cultural environment. The functional components of cultural processes detail the specific core capacities that are foundational to the effortful and task-based processing of the cultural mind. The core capacities and functional processes of the cultural mind are important to the representation and interpretation of the cultural environment.

The functional processes of the cultural mind consist of effortful and task-based functions that describe the core capacities of cognition and their role in culture and behavior. The effortful processing of cognition comprises sets of cognitive mental function or mental states that entail the automatic and controlled processing of thought and reasoning. The cultural processes of cognition and behavior demonstrate the functional use of cognitive states and their role in cultural thought and behavior. The cultural patterns of rule-based cognition illustrate the differential use of cognitive rules for the purpose of task-based processing. The cultural differences of rule-based cognition demonstrate the cognitive processes that contribute to cultural thought and behavior.

The cultural processing of emotion shows the set of mental states that comprise the generation and regulation of emotion across cultural contexts. The cultural mental states that consist of emotional states detail the representation and expression of emotion as feeling states. Emotion as feeling states encompass states of conscious and unconscious mental activity that accompany the generation, experience and regulation of emotion. The contextualizing of emotion as cultural mental content describes the cultural patterns of feeling states that arise within cultural context. The motivational or social significance of emotion as feeling states facilitates the cultural patterns of feeling states that are consistent with the shared meaning and interpretation of emotion. The cultural understanding of the emotions of others is foundational to cultural thought and behavior.

Cultural processing of motivation and emotion consists of the understanding of the motivational and intentional facets of mental states. The encoding and decoding of cultural mental content facilitates the understanding of the motivations and intentions of others. The mental state understanding of others from social cues demonstrates the cultural processing of the motivation and intention of others. The understanding of mental states contributes to the shared motivation and intention that is fundamental to coordination of social action. The cultural processing of mental state understanding entails the inferential reasoning on the motivational and intentional significance of the actions of others.

Cultural mental computation also comprises the cultural mental content that arises from the automatic facets of cultural processes. The cultural processes of mental content illustrate the cultural mental content that consists of the intrinsic properties of cultural thought and mental function. The cultural patterns of thought detail the automaticity of mental processes that are based on the automatic processing of cultural thought and action. The automaticity of cultural processing is of importance to the representation and interpretation of the cultural environment. Automatic cultural processing contributes to the detection and response of emotional and social cues that are congruent with the cultural context. The cultural processing of sensory cues facilitates the automatic detection and response to the social signaling of the cultural environment.

The cultural mental computation of spontaneous mental activity further illustrates the intrinsic properties of cultural thought and action. The spontaneous mental activity of cultural thought and action details the automatic processing of cultural mental content that facilitates culture and behavior. The cultural mental content of intrinsic properties facilitates the cultural patterning of mental activity that arises based on spontaneous thought that is autonomous. The autonomous activity of mental content demonstrates the maintenance of active representation of cultural processing that is of importance to cultural mental computation. The active representation of cultural knowledge demonstrates the preparedness of the cultural mind for mental activity that is spontaneous and intrinsic to cultural thought and action.

From simple to complex thought, cultural mental computation demonstrates a wide range of mental function that comprises cultural mental states and its functional significance. Cultural patterning of mental states elicits cultural processing that consists of the core capacities of mental activity and their responsivity to the cultural environment. The cultural states of mental activity also demonstrate the role of automatic and controlled processing on the thought and reasoning of culture and behavior. The cultural mental content of thought and intelligence serve as a foundation for the cultural processing that entails the intrinsic and extrinsic properties of mental function.

The cultural mind as a computing machine that is cultural entails the physical realism of the computation of cultural mental content. The computation of minds and machines as cultural implies the functional performance of mental and physical events that show causal significance in the natural world. The cultural computation of minds and machines facilitates the causation of natural realism as mental and physical states that show functional significance. Cultural minds as cultural machines entail the conceptualization of the physical realization of the cultural mind and its causal–functional role. The understanding of the cultural mind as a cultural machine recognizes the importance of the physical implementation of computation and its real-world advantages in the world.

The postulation on the physical realizers of cultural computation suggests a breadth of considerations that are of importance to the understanding of the causal–functional roles of cultural machine computation. The physical realizers of cultural computation connote machines that show a multitude of capabilities that are essential to the functional performance of culture. Multiple physical realizers of cultural computation entail the functional and mechanistic basis of cultural capacities across a wide range of putative mechanisms. From cultural minds to cultural machines, the physical realism of cultural computation abounds of the causal–functional significance of culture and its physical instantiation in the natural world.

The understanding of the cultural patterns of minds and machines serves as a foundation for the understanding of culture in the natural world. The cultural patterns of mind and machines reveal the complexities of the cultural instantiation of mental states and physical events that subserve biological organisms as cultural species. The mechanistic basis of cultural processing facilitates the description of the patterns and regularities that comprise the cultural events in the natural world with causal significance. The cultural life patterns of minds and machines detail the causal interaction of cultural causation of biological organisms and natural realism. The study of culture, mind and machines encompasses a wide realm of complexities that are fundamental to the understanding of the origin and nature of culture.

Culture and Machine Functionalism

The understanding of culture as a system that is comprised of physical and mental states belies the notion of culture as a system with functional significance. Culture is not only a set of physical and mental states in the natural world but also a system of cultural states with functional purpose. Culture as a system of physical and mental states with functional purpose implies the importance of the conceptualization of culture with functional consequence. The study of culture and machine functionalism places emphasis on the scholarly inquiry into the functional significance of culture as a mind and machine. The cultural mind as a cultural machine entails the understanding of the functional performance of cultural systems and its causal significance in the world.

The conceptualization of culture as a system of sets of mental and physical events in the physical world encompasses the fundamentals of culture. Culture as a system of mental and physical events that coincide for a specific functional purpose describes the core capacities of minds and machines with functional performance in the natural world. The functional mapping of mental and physical events describes the functional role of cultural capacities and their functional components. The notion of cultural capacities encompasses the functional mapping of input–output relations that are fundamental to cultural computation. The cultural computation of minds and machines demonstrates the capabilities of cultural capacities for functional performance.

The cultural mind as a computing machine with functional purpose entails the understanding of the functional specialization of the mind and its cultural capacities. The cultural mind consists of core capacities that contribute to the functional performance of culture and behavior. The core capacities of mental function detail the functional components that are fundamental to the cultural processing of the mind. The cultural processing of the mind encompasses mental function with specific purpose. Cultural capacities demonstrate the cultural processing of the mind with functional performance that is culture-specific. The functional mapping of one-to-one relations demonstrates cultural processing that assures specific cultural input–output or cultural input that is specific to a given cultural output. The understanding of cultural processing that entails mental function with a specialized purpose illustrates the computing machine of the cultural mind and its functional specialization.

The functional components of the cultural mind detail the role of cultural capacities with specific function. The cultural processing of the mind is of importance to the representation and interpretation of cultural information. Cultural capacities are central to the recognition and expression of culture. The detection and interpretation of cultural information that is unified and generalizable illustrates the functional capacities of the cultural mind. The generalizability of cultural capacities shows the universal purpose of cultural computation and reciprocally, the cultural-specific function of core capacities. The cultural processing of the mind demonstrates the core capacities of mental function that are foundational as functional components of the cultural mind.

The multidimensionality of culture implies the fundamental dimensions of cultural processing that comprise the representational content of the cultural mind. Culture as fundamental dimensions entails the cultural content of the dimensions of the mind. The cultural content of fundamental dimensions shows the elements of functional components that comprise the cultural mind. Cultural processing as abstract conceptual knowledge details the fundamental dimensions of culture as conceptual structures of abstract knowledge with featural elements. The cultural dimensions of mental content illustrate cultural systems of knowledge as organizational structures that consist of distinct sets of mental and physical events and their featural properties.

Culture as systems of abstract knowledge encompasses the organizational structures of concepts and their featural elements. The conceptual landscape of culture as abstract knowledge systems entails the understanding of fundamental cultural dimensions and their featural elements. The organizational structures of fundamental cultural dimensions comprise conceptual structures of featural elements that describe the semantic distinctions of cultural multidimensionality. The conceptual structures of cultural multidimensionality detail the semantic representations and the specific featural elements that comprise the concept space of fundamental cultural dimensions. The concept-based understanding of cultural dimensions allows for the delineation of the elemental and dimensional components of culture.

The conceptual space of cultural dimensions facilitates the functional mapping of the semantic representation of culture and its causal significance. The concept space of fundamental cultural dimensions allows for the semantic representation of cultural concepts and their featural elements. The functional mapping of semantic representation along cultural dimensions illustrates the relative weighing of featural elements. The generation and maintenance of semantic representations of cultural knowledge facilitates the representation and interpretation of cultural processing. The conceptual structures of semantic representation serve as active representations of cultural knowledge that contribute to the relative weighing of cultural processing and its interpretation into specific cultural output.

Culture as a knowledge system facilitates functional performance with multiple realizability. The multidimensionality of cultural systems contributes to the functional mapping of mental function that entails multiple cultural outputs. The adaptability of the cultural mind for the functional performance of multiple cultural outputs demonstrates the core capacities of cultural thought and intelligence. The functional mapping of one-to-many demonstrates the adaptability of the cultural mind and its functional significance. From simple to complex machines, the capabilities of the cultural mind for a multitude of functional performance illustrates the adaptability of core capacities and the functional role of cognition and higher-level processes on the multiple realizability of cultural thought and intelligence.

The functional basis of the cultural mind further implies cultural capacities with multiple physical realizability. Culture as systems of mental and physical states with multiple realizability demonstrates the adaptability of mind and machines for cultural performance. Cultural minds as cultural machines illustrate the functional mapping of mental and physical states for cultural performance. The functional mapping of one-to-many suggests the notion of culture as the set of mental states with multiple physical states. From biological organisms to complex devices, the multiple physical realizability of cultural computation abounds of the functional and mechanistic basis of cultural processing that demonstrates the feasibility of cultural computation across possible worlds.

The cultural computation of biological organisms as computing machines shows how cultural capacities serve as multilevel adaptation. The cultural capacities of biological organisms demonstrate that cultural capacities illustrate the functional basis of cognition and higher-level processes of biological organisms as cultural species. The cultural computation of biological organisms is foundational to the functional and mechanistic basis of the cognition and higher-level processes of cultural adaptation. The functional and mechanistic processes of cognition and higher-level function detail the adaptive mechanisms that are essential to cultural capacities. The functional machinery of biological organisms shows how adaptive mechanisms comprise a mechanistic basis of the cultural capacities that are of importance to cultural adaptation.

Across biological species, the generalizability of functional machinery for cultural capacities illustrates the conservation of biological mechanisms for the purpose of cultural adaptation. The functional machinery of cultural capacities that are specific to social and emotional processing, for instance, show the functional specialization of biological organisms for the functional processing of social and emotional information that is culture-specific ("cultural facial expression"). The presence of functional mechanisms across distinct species demonstrates a mechanistic basis that shows the conservation of adaptive machinery that is responsible for functional adaptation. The functional adaptation of biological species reveals the biological processes and mechanisms that subserve multilevel adaptation.

The cultural patterns of functional mechanisms of biological organisms detail the multiple realizability of cultural capacities at the computational level. The cultural patterning of complex brain organization demonstrates the physical instantiation of cultural capacities within cortical brain organization across levels of processing. The structural and functional basis of cortical brain organization shows the functional properties of cortical mechanisms that are concordant with cultural capacities. The cultural processes and mechanisms of cortical brain organization demonstrate the functional components of the cultural capacities that arise from the functional activity of neurons and networks of neurons. The cultural patterns of cortical brain organization are a foundation of the cultural computation of biological organisms as cultural species.

The functional use of tools in the environment demonstrates the functional purpose of the cultural capacities of biological organisms. Tool use illustrates the cognition and higher-level function of biological organisms for the interaction with the environment. The cognition and higher-level processes of tool use detail the functional purpose of cultural capacities for niche construction. Tools facilitate changes of the environment that have real-world advantages for biological organisms as a cultural species. The cultural capacities for tool use are important to the enactment of changes to the ecological niche that show the functional purpose of intentional goals and actions. The functionality of tool use shows the multiple realizability of cultural capacities with interaction of the ecological niche.

The scholarly inquiry on cultural life patterns encompasses study of the range of cultural processes and mechanisms across the life system in the natural world. Cultural life patterns demonstrate the natural realism of culture as a life system. The breadth and scope of cultural life patterns across the life system show the range of functional performance of adaptive machinery of biological organisms as cultural species. The cultural capacities of adaptive machinery facilitate interaction of biological organisms with the ecological niche. The cultural construction of the environment demonstrates the cultural capacities of biological organisms for niche construction. The capabilities of biological organisms to change or alter the ecological niche to favor the organism illustrates the cultural capacities that are fundamental to the life patterns of cultural species. The adaptability of biological organisms as cultural species contributes to the understanding of multilevel adaptation.

The feasibility of cultural performance from biological organisms to complex devices that are comparable in functional equivalence suggests the consideration of the causation of cultural patterns and regularities that comprise real-world advantages. The multiple realizability of cultural patterns and regularities as cultural events in the world demonstrates the functional significance of culture as a causal system. The notion of culture as mind and machine is not only a physical realizer with functional purpose but also a computing machine with causal significance. The cultural computing machine connotates on the causal role of the functional performance of mental and physical events that comprise cultural capabilities. The implementation of cultural events across sets of mental and physical events of real and artificial systems in the world implies the functional significance of cultural computation and its causal role.

Culture as a Causal System

Understanding culture as a causal system is of importance to the considerations of how culture enacts causation within the natural world. Culture as laws and principles of the natural world entail the understanding of how culture necessarily occurs in the world. Culture as causal power entails the understanding of how cultural patterns arise from parts and their interaction to govern a complex set of causal effects. Cultural patterns illustrate how the functional components of culture as a causal system demonstrate the causal–functional role of culture. Cultural capacities as functional adaptation detail the emergence of the functional machinery of culture from simple to complex organisms.

The concept of culture as mind and machines abounds of consideration on the causal importance of culture in mental and physical systems. The physical implementation of culture as sets of mental and physical events details how culture acts as a causal mechanism of the occurrences in the natural world. The interaction of mind and the cultural environment demonstrates how culture enacts influence on functional mechanisms. The understanding of culture as causation implies the delineation of the causal significance of types of cultural events in the natural world that matter. The postulation on the cultural events of the natural world as an enactment of the functional purpose and causal significance of cultural performance describes the function of culture as a causal system.

The potency of culture as a causal power arises from the importance of culture as explanation. Culture comprises sets of mental content that are of functional importance to explanatory satisfaction. Cultural systems of knowledge as metaphysics and epistemology provide a foundation of rationality and thought that is purposeful in the tradition of explanation. The metaphysics and epistemology of cultural systems facilitate the understanding of the lawlike principles that govern the world. Cultural systems of knowledge are powerful sources of causal explanation – explanatory causation – that imbue events in the world with shared meaning and significance. The rationality and thought of culture serve as a foundation of the rationale that governs satisfactory explanation of the natural world.

Culture as explanation provides a conceptual basis for prediction and expectation given a set of conditions and outcomes. Explanation as the rationality and thought of prediction and expectation facilitates the evaluation of possible events that occur in the world. The understanding of events as an occurrence within a system of cultural knowledge allows for the generalizability of explanation given a set of possible conditions and outcomes. The cultural understanding of natural occurrence facilitates the causation of events that are consistent within lawlike principles of the cultural system. The demonstration of culture as explanation illustrates the functional role of cultural systems of knowledge in the generation of prediction and expectations that are consistent across possible worlds. The causal power of culture lies in the functional significance of cultural systems of knowledge to distinguish possible conditions and outcomes within the lawlike principles of cultural systems and their truth advantages.

The causal significance of culture as explanatory power encompasses the range of putative explanations that provide a conceptual foundation for the demonstration of the arrangement of events with varying levels of responsibility. The breadth of explanation across cultural systems of knowledge abounds of the rationality and thought that facilitate the rationalizing on the explanation of prediction and outcome. The rational argumentation of logical explanation implies patterns and regularities of occurrences that hold. The development of rational logic as explanation benefits from the conceptual understanding of putative mechanisms that describe how patterns and regularities arise from the interaction of parts and their causal effects or as emergent properties of causal arrangements. The detailed characterization of putative mechanisms across levels of explanation is fundamental to the rationality and thought of cultural systems of knowledge. The understanding of putative mechanisms in the world facilitates the functional and causal significance of the rationality and thought of cultural knowledge.

The notion of cultural computation suggests a causal system of putative mechanisms with causal–functional roles and significance. Culture as minds and machines implies causal systems of functional mechanisms with causal significance. The cultural computing of minds and machines is comprised of the set of mental and physical events that are consistent with the lawlike principles of cultural systems. The arrangement of cultural events that comprise cultural computation illustrates the concordance of mental and physical events that demonstrate the lawlike principles of cultural systems and their emergent properties. Cultural events that constitute cultural mental computation detail how the cultural computing machine arises as an emergent property of the cultural mind. The cultural mind as a computing machine encompasses the cultural events that show functional significance in the world.

The conceptual understanding of cultural mental computation considers the notion of culture as mental causation. The cultural thought and rationality of systems of knowledge comprise ideas and content that enact causal significance in the world. Culture as the power of ideas demonstrates how culture influences social thought and rationality in the world. The power of ideas to change the minds of others illustrates how the spread of cultural ideas is a potent causal influence on the thought and rationality of the world. The cultural spread of ideas demonstrates the causal effect of social influence on cultural systems of knowledge.

The power of culture as mental causation furthers the consideration of the causal–functional role of the mind as a causal influence. Culture as mental causation implies the fundamental notion of mental events as a foundation of causal explanation. The elaboration of causal explanation as mental language within an arrangement of causal explanation details the ascription of a level of responsibility within the causal structure of the world. The mental causation of explanation implies that a specific level of responsibility is necessary and sufficient as a satisfactory explanation. The mental causation of explanation within cultural systems allows for the ascription of responsibility that is necessary and sufficient within causal arrangements that govern and hold in the world.

Culture as mental causation details how the putative explanations of cultural systems facilitate the ascription of responsibility and its conditions within causal arrangements that have functional significance and real-world advantages. Culture as mental causation entails the understanding of the robustness of mechanisms and their causal effects. Cultural mental computation describes how functional mechanisms demonstrate the cultural capacities of biological organisms for the functional performance of culture. The cultural capacities of biological organisms show the causal effect or significance of the mentality and thought of cultural species in the natural world. The cultural thought and mentality of biological organisms demonstrates the causal significance of the cultural mind as a computing machine. The thought and rationality in cultural performance illustrates the causal significance of cultural events as a mechanistic understanding of the world. The cultural computation of functional mechanisms shows how the thought and rationality of the cultural mind serves as a causal power with real-world advantages in the world.

The cultural causation of mental computation that arises as emergent properties describes how the cultural mind generates functional performance. The emergent properties of the cultural mind entail the featural properties of behavioral performance that comprise cultural adaptation. The generativity of the cultural mind for cultural performance demonstrates how the cultural mind is a causal influence on the understanding of the world. The generativity of the cultural mind shows how behavioral performance is a functional basis of cultural causation and its emergent properties. The featural properties of behavioral performance are sufficient to elicit inferences of cultural causation and its functional significance in the natural world. The postulation on the culture of behavior depicts how the cultural causation of the mind enacts cultural changes that have real-world significance.

The cultural patterning of behavior demonstrates how the understanding of culture as emergent properties of interaction contribute to the causal arrangements of the world. Emergent properties describe cultural patterns that have functional significance and that show sufficient explanatory satisfaction. The identity and expression of culture is of importance because it is an emergent property of culture in the natural world. The cultural patterns of behavior are of causal significance as an emergent property of the cultural mind in a causal world. Behavioral performance has cultural significance as an expression of the shared significance and meaning of identity and culture. The emergence of identity and culture is fundamental as an explanation of the shared meaning and significance of cultural life patterns. The understanding of identity and culture broadens the conceptual landscape on the functional significance of culture as an emergent property.

The fundamentals of the laws and principles of culture are foundational to the understanding of the generalization of the laws of culture that govern in the world. Culture as causal system implies that the laws and principles of culture are generalizable across space and time. The lawlike principles of cultural systems depict the relations of parts and their causal effects in the world. The understanding of culture as a causal system relies on the recognition of the causal significance of events in the world that are part of the laws and principles of cultural systems. The detailed characterization of the cultural patterns and regularities in the world that are consistent with the lawlike principles of cultural systems provides a foundation for the understanding of the generalizability of culture as a causal system.

Cultural Mental Property

The cultural computation of the mind is comprised of the functional properties of mental states and their causal–functional role. The cultural computing of minds and machines illustrates the importance of functional properties on the processes and mechanisms of cultural thought. The causal–functional role of the cultural mind is fundamental to the performance of cultural thought and behavior. Culture as mental states and function comprises sets of mental events with functional properties. The cultural thought and mentality of mind and machines shows the functional significance of cultural mental states in the world.

Culture as mind and machines shows how mentality and thought arise from mental states and their functional properties. The functional processes and mechanisms of cultural thought illustrate the cultural computation of the mind as a conduit of cultural performance in the world. The cultural computing of mental function demonstrates the fundamental principles of culture as patterns of thought and action. The cultural patterns and regularities of thought and action describe the functional properties that are of importance to the cultural processes and mechanisms of the mind. Cultural thought and mentality of mind and machines comprise mental states and functional properties that are essential to functional and behavioral performance.

The generalizability of culture in the mind and in the world implies laws and principles of cultural systems that hold. The lawlike principles of cultural systems describe how the computational level of mind and machines serve lawlike principles of culture. The cultural processes and mechanisms of mind and machines show the generalizability of cultural phenomena in patterns and regularities of mental function. The cultural patterns of mental function depict how the lawlike principles govern the cultural processing of thought and action. Cultural patterns of mind and machines detail how culture governs the processes and mechanisms of mental function that hold.

The cultural computing of minds and machines details the role of mental causation on the formulation of cultural thought and action. The cultural patterning of minds and machines shows the interaction of processes and mechanisms on the causal relations of the mind and its causal effects. The mental causation of cultural patterns reveals the higher-level processes of thought and action that arise from the interaction of culture and mental causation. The functional role of minds and machines as physical realizers of cultural computation show how cultural thought and action arise from the higher-level processing of mental causation with the world. The cultural mind as a causal mechanism illustrates the causal effects of higher-level function on cultural thought and action. The causal effects of mental causation reveal the independence of parts and their relational components. The functional role of culture as mental causation serves to effectuate the states of consciousness and higher-level function that show concordance with the self-determination of thought and action.

The mental causation of cultural thought and action details the functional role of higher-level processes as causal mechanism. The mental function of higher-level processes demonstrates the functional role of mental causation on the cultural thought that enacts intentional will and action. Culture as a causal mechanism of deterministic processes facilitates the higher-level function that is fundamental to intentionality and action. Culture as mental causation of deterministic processes shows how consciousness and mind enacts patterns and regularities of culture that govern. The natural realism of intentionality and action illustrates the determinism of cultural patterns and regularities as mental causation.

The functional adaptability of biological organisms for cognition and higher-level function illustrates the importance of multilevel processes on cultural adaptation. The interaction of biological organisms with the ecological niche demonstrates the processes of selection and construction on adaptation. On the one hand, cultural capacities are comprised of emergent properties of functional mechanisms and adaptive machinery that show the functional properties of cultural adaptation; on the other hand, cultural capacities demonstrate the motivated and intentional actions of biological organisms to change or alter the ecological niche. The functional adaptation of cultural capacities illustrates the higher-level processing of mental function for the purpose of cultural adaptation. The adaptive machinery of cultural capacities reveals the causal mechanisms of selection and construction on multilevel adaptation.

The cultural patterns of thought and mentality that are emergent properties illustrate the role of the cognition and higher-level processes of biological organisms on the manifestation of cultural life. The interaction of biological organisms with the cultural environment enacts patterns of cultural life that hold. The cultural processes of thought and mentality demonstrate the functional patterns and emergent properties that are deterministic of the intentional will and action of others. The interaction with the cultural environment manifests the cultural changes that evoke the co-construction of cultural life patterns. The functional role of motivation and intentionality as sources of the mental causation of biological organisms show how cultural patterns of thought and mental function enact cultural causation and its effects.

Cultural mental property serves as a functional basis of the explanatory power of culture in the world. The cultural understanding of the patterning of natural world is essential to the formulation of the mental language of causation and explanation. The cultural property of mental language describes the arrangements of mental events that have causation at levels of responsibility. The cultural processing of the patterns and regularities of the world serves as a functional basis for the ascription of the mental language of the causation and explanation of culture. The mental language of cultural causation ascertains on the levels of responsibility of mental causation in the world that govern. The cultural understanding of mental language and its causal effects details the scaffolding of causal explanations that imbue mental events with causal significance.

The mental language of cultural causation ascertains the cultural mental property that has functional or causal significance. Culture as a causal power describes arrangements that are generalizable across space and time. The mental property of culture details the mental language and its functional properties that enact the causal power of culture. The power of cultural causation implies the apparent or actualities of natural realism in the arrangements of events of the world. The cultural understanding of the possibilities of apparent or actual realism subserves functional properties that are of importance to the explanation and prediction of the cultural mind in the world. The mental language of self-determinism illustrates the functional and emergent properties of the mind that detail the functional role of the cultural mind and its causal effects. The cultural mental property of determination demonstrates how the mental causation of other minds serves as a functional basis for cultural causation. The mental causation of other minds facilitates understanding of how the cultural mind enacts mental causation with others. The self-determination of motivation and intentionality is a necessary and sufficient causal power of the mental language and action of culture. The autonomism and determinism of mental causation is fundamental to the cultural causation of other minds.

The cultural mind is an important foundation of cultural causation in the natural world. Across levels of responsibility, cultural mental property serves as a functional basis of the cultural mind and its causal effects in interaction with the world. From cultural patterns to causal mechanisms, the processes of the cultural mind show the multitude of sources of cultural mental causation and its functional role. The understanding of the multilevel mechanisms of cultural causation demonstrates how the cultural mind serves as a fundamental basis of the enactment of the patterns and regularities of culture in the world. The cultural computation of minds and machines illuminates how culture as a causal power is fundamental to the autonomism and determinism of cultural life in the natural world.

Cultural Physical Property

The multiple realizability of culture as mind and machines encompasses a wide range of considerations regarding cultural physical property and its causal–functional significance. Culture as sets of physical states and physical state transitions that are essential to the automated production of functional performance describes culture as a computing machine. The conceptual notion of culture as a computing machine serves as a foundation of the understanding of the causal and functional roles of cultural computation across multiple physical realizers. The physical implementation of the computational mechanisms of culture across real and artificial systems implies that the physical property of culture supervenes on natural and supernatural realism. The multiple physical realizers of cultural property demonstrate the feasibility of cultural computation across a wide range of physical systems. The plausibility of causal processes and mechanisms of culture within distinct physical systems illustrates the adaptability and complexity of cultural systems and its functional purpose. From biophysical mechanisms to complex life patterns, culture as processes and mechanisms demonstrates the wide range of capabilities of complex systems. The cultural computation of simple to complex mechanisms demonstrates the functional role of the adaptability across levels of system complexity. The understanding of the physical realization of culture as a manifestation of causal mechanisms across levels of responsibility facilitates the patterns and regularities of culture that hold.

The notion of cultural computation as a computing machine recognizes the conceptualization of culture in the mechanistic understanding of the world. The notion of cultural computation as sets of physical states with a functional purpose details the parts and properties of computing machines that show a causal role in the computation of culture. The computing machine demonstrates how the mechanical states of physical systems comprise the representational content of the states of cultural computation. The mechanical states of physical systems as representations of the states of cultural computation demonstrate how machine computation is sufficient as a physical realizer of the functional performance of culture. The feasibility of cultural machine computation as a physical realization of the capability and functionality of cultural systems is a fundamental basis of cultural computation.

The early notions of analog computation illustrate the notion of the depiction of the computation of functional relations based on the mechanical states of simple physical devices. The physical states of computing machines are sufficient as analog representation of the machine computation and its relational states of physical systems. The description of physical states as componential to the computation of functional relations depicts the role of analog representation in the formulation of states of machine computation. Machine computation as a means of the representation of the mental states of computation as the physical states of computing machines in the world shows the recognition of the plausibility of the functional performance of computing machines as physical realizers of the intentionality of the mind in the world.

The concept of the cultural mind in the world brings forth a sense of metaphysical realism that broadens the understanding of states of mind and their purpose. The actualization of mental states as a metaphysical realism shows a causal power of culture in the world. The notion that states of mind are part and particular of culture in the mind and in the world connotes on the cultural states of mind that matter. The recognition of machine computation as a representation of mental computation broadens the causal power of the culture of the mind into the world. The functional use of machine computation in the representation and production of culture shows the adaptability and complexity of cultural systems.

The demonstration of electronic computation or the modern computer details the functional equivalence of machine computation within electromechanical states of the digital machine. The capabilities of digital computers for electromechanical computation shows the feasibility of the functionality and adaptability of the digital representation of mental computation within multiple physical realizers. The distinct physical states of computing machines and the range of capabilities and functional performance demonstrate the plausibility of a multitude of physical systems that show the functional capabilities of cultural systems. The illustration of physical states of digital computers as physical realizers of mental computation is of importance to the understanding of the parts and particulars of cultural thought and intelligence.

The modern computer illustrates the importance of functional capabilities on the physical realization of machine computation. The early notions of machine computation postulate on the mechanical states of computing machines that showed the functional purpose of computation. The demonstration of the feasibility of computation within a physical machine was sufficient to establish a standard of machine computation that was comparable to mental computation. The formulation of theory building on mind and machines placed emphasis on the functional basis of computation and the capabilities of the design of its physical implementation.

The contemporary notion of modern computation demonstrates how machine computing is sufficient as an explanatory source of thought and intelligence based on the compatibility of the functionality and capabilities of computing machines with real-world advantages. The modern computer contains machine intelligence because of the compatibility of the functional capabilities of mental and machine computation. The machine computation of physical devices serves as an expansion of the causal significance of mental computation. The functional role of machine computation as a tool of mentality and thought illustrates the capabilities of the mind to broaden its physical realism in the natural world. From analog machines to modern computers, the causal realism of mental computation as multiple physical systems depicts a range of physical states that have material or functional significance in the world. The functional equivalence of mental and machine computation illustrates how multiple physical realizers show the functionality and capability of intelligent machines. The modern computer illustrates how the functional performance of machine computation is sufficient as a causal source of cultural thought and intelligence. The automation of machine computation with functional performance for automated cultural production demonstrates the functionality and capability of physical devices for cultural computation. The capabilities of physical devices for automated cultural production ("automated language generation") and other computing programs that show the functional performance of cultural machine computation illustrate the causal role of physical devices and their functional purpose. The notion of modern computers with cultural intelligence ("CULTURE AI") implies that physical devices have functional capabilities for cultural performance that is comparable to that of mental computation.

The feasibility of functional capabilities of physical devices for cultural performance suggests the plausibility of the generation and simulation of cultural thought and intelligence with machine computation. The notion of modern computers that show cultural intelligence ("machines that think culture – CULTURE THINK") or computer programs that generate and simulate cultural performance ("simulated cultural performance") suggests the practical applications of computational models that explain the functional or causal significance of cultural computation. The functionality of modern computers is of importance to the generation and simulation of cultural intelligence that shows the adaptability and complexity of cultural change in real and artificial systems. The capabilities of modern computing to show functional compatibility with the cultural change of real and artificial systems is beneficial to the manifestation of physical devices that demonstrate automated cultural production consistent with natural realism. The functionality and capability of modern computing is of importance to the understanding of cultural computation and its range of causal significance.

The cultural physical property of simple to complex biological organisms demonstrates the cultural computing of the biological machine. From simple mechanisms to complex organisms, the cultural computation of biological organisms shows how the computation of complex species facilitates the functional performance of multilevel adaptation. The functional adaptation of biological organisms shows that the gradation of adaptation has causal significance on the emergence of adaptive mechanisms with functional purpose. The adaptive mechanisms of biological organisms for cultural capacities serve as functional processes and mechanisms. The functional machinery of cultural capacities illustrates the functional and emergent properties of adaptive mechanisms across simple to complex organisms. The functional adaptation of the cultural mind and its causal mechanisms entail the adaptive machinery of biological organisms as cultural species.

The physical states of biological computation consist of the functional mechanisms that are important to multilevel adaptation. The functional mechanisms of the nervous system describe adaptive machinery that demonstrate the functional capacities of cultural adaptation. The functional processes and mechanisms of the brain entail the core capacities that show the functional specialization and functional integration of cortical brain organization for behavioral and cultural adaptation. The functional mapping of mental states and physical states of biological organisms shows how the computation of minds and biological machines demonstrate a functional correspondence that has causal significance to functional performance. The functional correspondence of the cultural computation of biological organisms across systems details the complementarity of functional roles of mental processes and brain function.

The physical property of the functional mechanisms of biological organisms shows the fundamental principles of culture across levels of processing. The functional properties of cortical mechanisms demonstrate the fundamental principles of dynamical systems that subserve the computation of cultural systems. The autonomous and self-organizing principles of dynamical systems describe relations of physical states that detail the functional and mechanistic basis of patterns and regularities that hold. The principles of autonomy and self-organization describe the feasibility of the generation and production of cultural patterns and regularities that are essential as fundamental principles of complex systems. The characterization of the functional properties of cortical mechanisms of biological organisms as consistent with dynamical systems illustrates a physical realizer that is capable of the autonomous production of cultural patterns and regularities.

The interaction of biological organisms with the cultural environment further illustrates the role of adaptive mechanisms as cultural causation. The intentionality and action of the cultural mind serves as functional components of the mental causation of culture in the world. The states of consciousness and mentality of the cultural mind imbue intentionality and action with a sense of moral purpose that enacts a rationale of cultural thought for moral action. The moral sense of intentionality and action predicates on the generation and maintenance of cultural representation in the mind and in the world. Cultural thought as moral action relies on the understanding of the functional purpose of cultural patterns in thought and behavior. Cultural thought and moral action serve as an assurance of the fundamental principles of culture in the world. The realization of cultural thought as moral action arises from the causal power of culture to enact patterns and regularities that have functional and causal significance that govern and hold. The interaction of the cultural mind in the world is a part of the functional role of thought and mentality of the mind in interaction with the world.

The emergent properties of the adaptive mechanisms of biological organisms details how the generativity of cultural thought and mentality evokes functional and behavioral patterns of culture. The cultural computation of the mind as sets of mental states enacts the functional and behavioral performance of culture in the world. Culture as linguistic and behavioral repertoires illustrates the functional performance of cultural patterns. Cultural expression in language and behavior depicts the cultural representation of the mind in the world. The behavioral and linguistic expressivity of the cultural mind demonstrates how cultural patterns of functional and behavioral performance are sufficient as a material realism of culture.

The cultural patterns of the functional mechanisms of language and behavior comprise functional components of core capacities that are concordant of the functional properties of adaptive machinery. The cultural and brain basis of language and behavior demonstrates how adaptive machinery facilitates the cognition and higher-level processes of brain function. The emergent properties of cortical brain regions that are responsible for language and behavior show the functional role of specialized adaptive machinery. The characterization of the functional and emergent properties of cortical brain organization is essential to the understanding of the role of adaptive machinery in multilevel adaptation.

The understanding of culture as physical property abounds of considerations regarding the multiple realizability of culture in the world. Culture as physical property demonstrates how physical states comprise states of culture. The notion of culture as physical property recognizes that culture is part and particular of physical systems and that the physicalism of culture is essential to the understanding of the physical realism of cultural states. Physicalism entails that culture is comprised of mechanisms and their parts. The conceptualization of culture as physical property further suggests how the physical states of causal mechanisms are part and particular of states of culture in the mind and in the world.

The understanding of culture as physical property facilitates the elaboration on the functional mapping of the mental and physical properties of culture in the mind and in the world. Culture as mental and physical states describe the sets of mental and physical events that comprise cultural thought and action. Culture arises from the representation and production of the functional mapping of input–output relations. The cultural property of mental and physical states implies how the computation of mind and machines facilitates the implementation of states of culture. The description of the physical property of the world as cultural property allows for the mechanistic understanding of the world that is cultural.

From computing machines to complex biological organisms, the causation of the physical states of cultural computation in the world encompasses a broad range of mechanisms and physical systems. The conceptualization of fundamental mechanisms of physical systems details the understanding of the causal role of mechanisms. Culture as a mechanism of physical systems implies that culture serves a causal role in physical systems. Cultural causation entails how cultural states exert a causal influence on physical states. The causation of the cultural mind illustrates how mental and physical states of culture enact causal effects of the mind and of the world.

Culture and computation

The scholarly and scientific interest on culture and computation is beneficial to the humanistic and scientific understanding on the culture and computation of minds and machines. The postulation on the lawlike principles of culture and computation contribute to the understanding of culture as patterns and regularities in the causal structure of the world. The fundamentals of the computational principles provide a conceptual basis for the understanding of the cultural patterns and regularities of the mind and of the world. The demonstration of the multiple realizability of culture and computation illustrates the inherent naturalism of cultural computation. The cultural patterns of mental and machine computation illustrate the generalizability of computational principles of cultural systems. The understanding of culture and computation as complex systems facilitates understanding of the cultural patterns of real and artificial systems.

The cultural patterns and regularities of the mind demonstrate the functional basis of the performance of culture and behavior. Cultural patterns of the mind show the functional performance of cultural capacities that are based on functional processes. The functional processing of mental function demonstrates the functional components that contribute to the culture and behavior. Cultural patterns of the mind show the functional significance of cultural thought and behavior. The cultural capacities of the mind for functional performance is essential to the understanding of the importance of cultural computation on cultural thought and behavior.

The study of cultural patterns of cognition and brain function demonstrates the importance of culture as a mechanistic basis of cultural capacities. Cultural patterns of cognition and higher-level function show the functional significance of processes and mechanisms on behavioral performance. The cultural patterns of cognition and brain function illustrate how higher-level processes contribute to the thought and mentality of culture and behavior. The cultural patterns of functional brain activity show the causal mechanisms that subserve the functional processes of complex behavior. The cultural patterns of cognition and higher-level function serve as a functional and mechanistic basis of the cultural capacities that are essential to multilevel adaptation.

The cultural capacities of simple to complex biological organisms illustrate the functional significance of culture as patterns of life systems. Cultural life patterns of biological organisms show how functional adaptation contributes to the complexity of life systems. The adaptation of cultural capacities of biological organisms contributes to the multilevel adaptation of cultural species. The functional basis of cultural capacities shows how the patterning of cultural performance of the mind and of the brain contributes to the preparedness and responsiveness of biological organisms for cultural adaptation. The functional performance of cultural capacities is essential to the adaptive machinery of biological organisms.

The functional adaptation of cultural capacities contributes to the functional performance of biological organism to alter or change the cultural environment. The functional performance of cultural adaptation illustrates how cultural capacities serve as a functional basis of the construction of the cultural environment. The functional basis of cultural capacities provides a causal mechanism for the alteration or change of the cultural environment. The interaction of biological organisms with the cultural environment facilitate the adaptability of functional machinery that is essential to multilevel adaptation. The functional basis of cultural capacities is essential to the adaptive machinery of biological organisms.

The cultural life patterns of biological organisms illustrate the cultural capacities for the complexities of cultural and mental life. The cultural computation of the mind demonstrates the functional basis of the patterning of adaptive machinery for cultural performance and mental function. Cultural mental computation is fundamental to the cultural patterns that facilitate the understanding of other minds. The cultural computation of mental function contributes to the functional basis of the cultural processing of social information. The understanding of the cultural patterns of mental life is fundamental to the characterization of the higher-level processes of complex organisms as cultural species. The cultural life patterns of biological organisms are essential to the understanding of the functional adaptation of higher-level processes. Cultural patterns of life systems contribute to the understanding of the functionality and adaptability of complex organisms.

The cultural patterns and regularities of life systems contribute to the characterization of the culture and computation of real and artificial systems. The cultural patterns of thought and behavior demonstrate the computation of minds and machines that has multiple physical realizability in the natural world. The demonstration of the multiple physical realizability of culture and computation of the mind and its biological basis shows the functional performance of biological organisms as part and particular of life systems. The feasibility of culture and computation as biological organisms shows how cultural patterns and regularities of life systems are a manifestation of the complexities of culture as a causal system.

The causation of culture and computation is foundational to the understanding of the cultural patterns of life systems and their causal effects. The patterns of culture as mental function and adaptive machinery illustrates the parts and particulars of cultural patterns that hold. Cultural patterns of the mind and the brain demonstrate culture and computation that arises as emergent property. The interaction of the mind with the cultural environment shows the functional components of mental computation that demonstrate the cultural patterns of mental life. The causal mechanisms of culture and computation illustrate the levels of cultural processing that show causal effects in the natural world.

The understanding of the mechanistic basis of culture and computation facilitates the design of practical applications that show cultural patterns as artificial systems. The design of computational tools and technologies that demonstrate the feasibility of culture and computation illustrates the functional significance of cultural patterns as the computation of complex devices. The culture and computation of complex devices show how computing machines and artificial devices demonstrate a functional performance of cultural computation. Computing machines and devices that show the functional performance of culture and computation relies on the use of culture as a functional basis of the algorithmic function of computational performance. The design of computing machines that show the functional performance of culture illustrates the multiple realizability of culture and computation in the world. The understanding of the mechanistic basis of culture and computation is foundational to the design of practical applications of culture as artificial systems.

The ubiquity of culture and computation of real and artificial systems shows the functional equivalence of biological and artificial computing and its causal–functional significance. The functional equivalence of the cultural computation of biological and artificial computing implies that the mechanistic basis of culture and computation are comparable across dimensions of functional performance. The efficiency and accuracy of the functional performance of culture and computation provide a functional basis for the understanding of the computational performance of biological and artificial computing. The demonstration of the culture and computation of real and artificial systems broadens the understanding of the mechanistic basis of cultural computation.

The generalizability of the lawlike principles of culture as a causal mechanism in the natural world provides a foundation for the understanding of the causal–functional significance of cultural computation. The detailed understanding of the laws and principles of culture as part and particular of complex systems facilitates the characterization of the mechanistic basis of culture and computation. From causal mechanisms to emergent property, culture as causal mechanism shows the multiple realizability of cultural states across physical realizers. The generalizability of cultural patterns and regularities in the world demonstrates the inherent naturalism and scientific realism of culture and computation.

Discussion

The scholarly study of cultural computation broadens the understanding on the mechanistic basis of the computational level of culture. Research on cultural computation investigates how culture of minds and machines demonstrates the functional performance of culture and behavior. The cultural computation of the mind illustrates the functional components of the performance of cultural tasks. Cultural mental computation is of importance to the understanding of the functional and mechanistic basis of cultural computation. The characterization of the functional mechanisms of cultural processing contributes to the understanding of the functional processes of culture and behavior.

Research on cultural machine computation is essential to the demonstration of the computing machine as a physical realizer of culture and its functional significance. The cultural mind as a computing machine details the functional performance of the computing machine as a physical realizer that shows the capabilities of cultural performance. Cultural machine computation elaborates on the consideration of how machines demonstrate cultural computation as sets of physical states and its physical state transitions. Cultural machine computation contributes to the understanding of the mechanistic basis of cultural computation and its functional significance.

The physical instantiation of cultural computation of biological organisms shows the functional adaptation of cultural capacities as a causal mechanism of cultural adaptation. The functional adaptation of cultural capacities shows how simple to complex biological organisms illustrate the functional mechanisms of cultural capacities. The adaptation of functional mechanisms of cultural capacities is of importance to the understanding of the adaptability and functionality of cultural computation. The manifestation of the functional mechanisms of cultural capacities reveal how the functional adaptation of complex organisms is sufficient as a conduit of the cultural processing of multilevel adaptation. The characterization of the multilevel mechanisms of cultural capacities shows how cultural computation arises from the functional and mechanistic basis of cultural computation.

The cultural neurocomputation of the brain shows the functional and mechanistic basis of the adaptive machinery of biological organisms for cultural computation. The cultural brain serves as a mechanistic basis of the biophysical mechanisms of cultural adaptation. The functional machinery of the cultural brain illustrates how structural and functional properties of the nervous system encompass the causal mechanisms that are of importance to the adaptation of cultural capacities. The structural and functional properties of the cultural brain demonstrate the functional basis of cortical brain organization for cultural adaptation. The biophysical mechanisms of the cultural brain demonstrate the functional mechanisms of adaptive machinery that contribute to the cultural computation of biological organisms.

The study of cultural computation of minds and machines facilitates the understanding of the functional and mechanistic basis of cultural computation. The characterization of the cultural computation of biological organisms contributes to the understanding of the functional mechanisms of cultural capacities. The cultural capacities of biological organisms show the functionality and adaptability of biological machinery for multilevel adaptation. The demonstration of the cultural capacities of biological organisms as higher-level processes contributes to the understanding of the functional and mechanistic basis of culture and behavior. The higher-level processes of simple to complex organisms illuminates the functional adaptation of biological organisms as a cultural species.

Conclusion

The study of cultural computation provides a conceptual foundation for the understanding of the computational performance of minds to machines. From simple to complex machines, cultural computation demonstrates the feasibility and generalizability of the computation of culture across multiple physical realizers. The cultural processing of machine computation shows the importance of computational principles in the understanding of the fundamentals of cultural computation. The functional equivalence of cultural performance across physical devices and machines demonstrates the generalizability of functional properties that comprise cultural computation. Research on cultural computation contributes to the conceptual language on the mental and machine computation of culture.

The scholarly and scientific inquiry on cultural computation broadens the understanding of the cultural computation of minds and machines. Cultural computation contributes to the considerations on the multiple realizability of culture in the mind and in the world. The study of cultural computation facilitates the scholarly and scientific enrichment on the cultural computation of minds and machines. Understanding how minds and machines perform cultural computation contributes to a mechanistic understanding of the computational level of culture. The scholarly and scientific endeavor in the characterization of the multilevel mechanisms of cultural computation of biological organisms contributes to the understanding of the inherent naturalism of culture in the world.

The scholarly and scientific enrichment on cultural computation is beneficial to the building of societal and public understanding of culture and technology. Research on cultural computation is beneficial to the understanding of the cultural mind as a computing machine. The scholarly and scientific study of cultural computation provides a foundation for the understanding of how culture serves as a causal mechanism in the world. The study of cultural computation is beneficial to broaden scientific and societal awareness of the benefits of culture and computation.

3
CULTURAL-MENTAL COMPUTATION

Introduction

Scholarly interest on the origin and nature of culture and the mind has long sought to determine the ethereal realism of consciousness and mental life. Ancient philosophical notions have long postulated on the ideal and material facets of cultural and mental life. The ideal notions of cultural and mental life place emphasis on the spiritual idealism of the mind. The mind as a material and immaterial being comprises the material forms of life and afterlife and encompasses the breadth of spiritual realism that constitutes consciousness and the mind. The social and cultural understanding of the idealism and materialism of cultural thought illustrates the range of phenomenological and material presence that determines the nature of culture and mental life.

The conceptual breadth of scholarly interest on the philosophy of the mind shows the depth of considerations on the nature and origin of culture and the mind. From robots and caricatures to bats and beasts, the breadth of phenomenological parts and particulars that comprise the mind illustrate the range of illuminous notions that constitute cultural and mental life in the world. An understanding of the cultural and mental experience of the world illustrates the importance of the phenomenological states of the cultural mind. The material being of the mind in the world connotes a spiritual and material realm of being and an existence that is a compendium of the phenomenological inquiry on the shared purpose and meaning of cultural and mental life in the natural and supernatural world.

DOI: 10.4324/9781003384236-4

The understanding of the cultural foundations of mental life broadens scholarly inquiry into the fundamentals of the mind to ascertain the shared purpose and meaning of mental being and existence. The cultural capacities of mental life highlight the sense of shared purpose and meaning of being and existence that are fundamental to the material and immaterial being. Cultural capacities detail the mental states that are foundational to the shared understanding of the material and immaterial being. The cultural interests of mental life in the mind and in the world illuminate on the streams of consciousness and thought that entails the understanding of culture in the mind and in the world.

The goal of this chapter is to provide a review on the foundations of cultural mental computation. The chapter discusses the fundamentals of cultural mental computation and its role as a functional basis of cultural computation. Thematic topics of cultural mental computation include culture and functionalism, cultural affective computation, cultural cognitive computation and cultural social computation. The study of the functional basis of the cultural computation of the mind contributes to the broader understanding of culture in the mind and in the world. Implications of research on the cultural mental computation for practical applications in health, medicine and related fields are discussed.

Cultural mental computation

The study of cultural mental computation is the characterization of the cultural computation of the mind. Research on cultural mental computation encompasses a broad range of thematic topics that entail the understanding of the functional components of the cultural mind at the level of computation. The functionalism of the cultural mind connotes on the functional processes of cultural capacities and its causal–functional role. Culture as mental computation illustrates the cultural states that enact mental states and their functional significance. The identification of the cultural computation of distinct functional components of the mind demonstrates the breadth of cultural idealism and materialism in the mind and in the world.

The functional machinery of the cultural mind implies that the realization of culture in the mind and the world has a specific kind of functionality. Culture in the mind and in the world that pertains to mental function entails that mental events of functional significance to culture serve as psychological realizers of culture and mental life. The cultural mind as a behemoth of psychological realizers depicts mental processes and function as materialization of the functional components of cultural computation. The psychological realizers of the cultural mind are the functional components of cultural computation in the mind and in the world. The understanding of the psychological realism of functional processes that comprise cultural computation illuminate on the role of cultural mental computation in the world.

The psychological realism of the cultural mind depicts how psychological processes subserve as mental function that is essential to culture and behavior. The manifestation of the cultural mind as psychological realizers illustrates how mental processes are a core component of cultural capacities. The functionality of the mental processes as core components of cultural capacities details the specific psychological processes that are of importance to the functional performance of the cultural mind (Figure 3.1). Cultural capacities as a range of specific psychological processes encompasses the functional operationalization of the cultural mind as a psychological realizer. The psychological realism of the cultural mind allows for an understanding of the patterns and regularities of mental function that are essential to cultural computation in the mind and in the world.

The cultural patterns and regularities of mental life abound of the causal arrangements of parts and particulars that hold. The cultural patterning of the mental function depicts the causal-functional role of specific psychological processes that are essential to mentality and thought. The causal arrangement of the psychological processes that govern cultural thought and behavior illustrate the causal-functional role of mental causation on the cultural patterning of mental life. The understanding of the causality of culture and mental life is fundamental to the study of cultural mental computation and its effects on the world.

Cultural mental computation consists of the functional components of the mind that are of importance to cultural capacities. The functional components of the mind that are dedicated to cultural capacities consist of specific functional tasks that are culture-specific. The core capacities of the mind are essential to the representation and transformation of the sensory environment into multilevel adaptation. The functional components of the cultural mind show dedicated and specialized functional processes that are

(a)

(b)

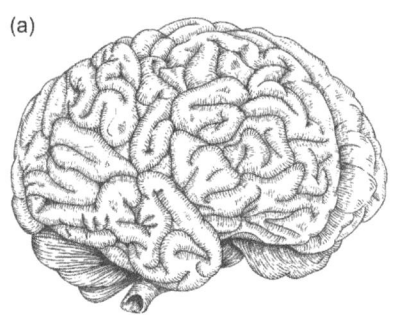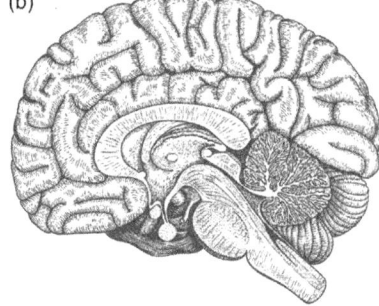

FIGURE 3.1   Brain Anatomy

consistent of functional tasks that are automatic and innate. The dedicated and specialized functional machinery of the cultural mind illustrates the functional components of cultural capacities.

Cultural capacities refer to the functional processes that demonstrate the cultural processing of the mind. The cultural processes of mental function detail the specific functional tasks that are culture-specific and that are of importance to the automatic and controlled processing of cultural capacities. The automaticity of mental processes that comprise cultural capacities illustrate the prepotent response of core mental function to the cultural environment. The cultural capacities that show the automatic processing of mental function illustrate the highly responsive and adaptive functional machinery of the cultural mind that is illustrative of multilevel adaptation.

The functional components of the cultural mind describe the functional processes that are culture-specific. The functional components of the cultural capacities are responsible for the functional mappings of core components that perform cultural input–output. The cultural processing of the functional tasks that are culture-specific describes the functional mappings of input–output relations that are componential of culture. The functional mappings of cultural input–ouptut describes the functional processes of cultural capacities that are essential to the functional performance of the cultural mind.

The functional components of cultural capacities show how the functional processes of mental function are culture-specific. The functional components of culture-specific tasks demonstrate the importance of culture-based task processing on the functional processes of the mind. Cultural processing describes how functional components show automatic response to the cultural environment. The changes in the functional processes that are responsive to the cultural environment demonstrate the culture-based processing of task activity that is specific to culture. The cultural processing of task activity illuminates on the functional basis of mental function that is culture specific to the environment.

The automaticity of cultural processing entails the automatic response of the functional processes to cultural capacities. The detection of sense data of the cultural environment elicits the automatic response of functional processes that are essential to adaptation. The cultural processing of the environment contributes to the automatic response of functional capacities that show the dedicated and specialized processing of core capacities. The automatic responsivity of functional processes to the cultural environment illustrates the automatic processing of cultural capacities that are essential to multilevel adaptation.

Cultural capacities that are fundamental to the controlled processing detail the role of effort on mental function. The cultural capacities that are based on effortful processing describe the role of task-based activity on mental function. Cultural processing that relies on effortful activity describes the importance of cognition on mental function. The functional processes of culture and cognition depict a wide range of functional tasks that subserve mental life. The functional specificity of culture and cognition details the role of the functional use and availability in the functionality of mental life.

The levels of cultural processing in the functional components of cognition illuminate on the depth of processing that arises from task-based function. The levels of functional capacities describe the depth of processing that is specific to culture and task-based activity. Cultural processing shows the levels of mental activity that are distinct across levels of functionality. The levels of functionality demonstrate the types of functional processes that are essential to task-based activity that is culture-specific. The cultural specificity of task-based function illustrates how functional capacities show the depth of cultural processing of a functional process as levels of functionality.

The cultural processing of task-based effortful activity details the functional significance of cognition on the control or regulation of mental function. The cultural processing in the control or regulation of mental function shows the functional role of culture and cognition. Cultural capacities demonstrate the specific functional tasks or rule-based cognition that is emblematic of culture and cognition. Culture as rule-based cognition describes the functional use of rules in the deliberation or conscious processing of cognition and mental function. Cultural capacities show the functional use of rule-based cognition in semantic processing and language comprehension as functional components of abstract cognition and reasoning. The functional reliance on rule-based cognition illustrates a higher-level process of consciousness and mental function that comprise cultural capacities.

The abstract representation of semantic processes is fundamental to the functional processing of cultural capacities. The abstract representation of semantic understanding contributes to the inferential reasoning on the shared purpose or meaning of mental events that comprise mental life. The semantic comprehension of abstract reasoning is essential to the representation and interpretation of social and emotional events and the understanding of their motivational or social significance. The semantic processing of language contributes to the distributed processing of semantic meaning that is of importance to the selection of interpretation of mental events. The abstract processing of semantic understanding illustrates a higher-level process of cognition and reasoning that is fundamental to the functional processing of cultural capacities.

The cultural capacities of task-based activity that is effortful illustrates the role of cognitive control or inhibitory control on the functional processing of core capacities. The reliance on cognitive control that is culture-specific is of importance to the understanding of how culture contributes to the control or regulation of mental function and behavior. Cultural control or regulation of cognition is of importance to the adherence to the societal norms and social conventions of society and civilization. The cultural control of cognition demonstrates how functional capacities of the mind show social conformity in the cultural patterning of thought and behavior. The functional capacities of the mind that demonstrate social conformity to societal norms illustrate how cognition and regulatory function are of importance to the social and emotional regulation of cultural life. The functional capacities of the mind to subserve cultural patterns of thought and behavior illustrate how cultural control of cognition is essential to an understanding of the social and cultural facets of mental life.

The cultural capacities of the mind illustrate the functionality and adaptability of mental function for multilevel adaptation. The cultural processing of functional components of the mind detail the functional specificity of mental activity that is essential to cultural capacities. The breadth of functional processes that comprise cultural processing illustrate the complexity of mental function. The role of consciousness and higher-level function of cultural capacities detail the functional significance of mental function as cultural adaptation. The adaptive function of higher-level processes that subserve cultural capacities show the functional importance of mental life to culture in the mind and the world.

The cultural mind as a computing machine

The cultural mind as a computing machine connotates on the notion of the cultural mind as a machine capable of the functional performance of cultural computation. The concept of the cultural mind as a computing mind introduces conceptual language that broadens the landscape of the mind as a computing machine. The concept of the cultural mind as a computing machine elaborates on the notion of culture in the mind and the world pertaining to a functional performance with specific purpose. The mentality and thought of the cultural mind demonstrate how the mind as a cultural computing machine shows the functional capabilities of the performance of cultural computation.

The mind as a cultural computing machine implies that the mind is a psychological realizer of cultural computation. The psychological realization of cultural computation as a psychological stance entails that mental computation is of importance to the functional performance of culture. Psychological realizers, the set of mental states and its changes based on functional performance, illustrate how the mind serves as a psychological realizer of culture. The laws and principles of mental computation are necessary and sufficient as generalizations that describe how the mind as a psychological realizer enacts laws of culture. The generalizations of mental computation describe the mental states as part of the interaction of the mind in the causal structure of the world. The cultural computation of the mind details how cultural states as psychological realizers entail parts and particulars of the mind in causal interaction with the world. Mental states as states of the mind are the fundamental elements of cultural mental computation. The functional performance of psychological realizers illustrates how the cultural mind arises from the cultural computation of the mind.

Psychological realizers as a fundamental element of cultural mental computation entail specific functional components that hold as a part of the interaction of culture in the mind and in the world. The psychological states or mental states that comprise cultural patterns of thought detail the mental states that contribute to the functional performance of culture. The detailed characterization of the set of mental states that comprise cultural processes illustrates the role of functional processes of culture and the mind. The psychological realism of cultural mental computation contributes to the functional specificity of processes that are essential to culture. The manifestation of mental states as cultural patterns of thought demonstrate the role of psychological realizers as functional components of culture as a computing machine. The functional role of cultural processes encompasses the mental states that are elemental as cultural patterns of thought. The functional performance of the cultural mind as mental states belies the role of specific processes as a functional basis of culture and the mind.

The postulation of the cultural mental computation as a set of psychological realizers implies that mental states serve a causal-functional role. Cultural mental states describe states of the mind that comprise cultural patterns of thought. The manifestation of cultural states as mental states entails that the functional performance of cultural computation is pertinant to mental function. The set of cultural states as mental states implies computation of cultural performance as cultural mental function. Cultural mental function entails the sets of mental states such that the functional processes of cultural computation are instantiated within psychological realizers. The psychological states of cultural computation as comprised of cultural mental states details the functional role of the cultural mind as a computing machine.

The cultural mind as a computing mind entails the notion of the set of mental states that comprise cultural computation. The cultural computation of the mind suggests that the mind is a computing machine of cultural states in the mind and in the world. The mind as a machine that is capable of cultural performance belies the functional role of the mind in the production of cultural states. The states of cultural mental computation detail the sets of cultural states that are instantiated as mental states or psychological realizers. The states of cultural computation as mental states or psychological states uniquely demonstrate how cultural states manifest as a physical realizer with a functional role. The manifestation of cultural states as part of the computation of the mind as a machine illustrates how states of cultural computation arise from mental states.

The cultural computation of the mind further demonstrates how culture as mental computation shows states of culture as states with a multitude of functional roles. Culture as mental computation is comprised of the mental states and its transitions in the functional performance of culture. Culture as mental states detail the functional processes that comprise the functional mapping of cultural input ("sense data") to output ("motor output") relations. The internal state transitions manifest from algorithms or rule-based states that facilitate the transition of initial states to final states of culture. The mental computation of culture consists of internal states as mental states with a functional role and the transitions to internal states that arise as a function of cultural computation. Culture as mental computation encompasses the set of initial and final states and its transitions that are fundamental to the functional performance of culture.

Culture as mental states illustrate the changes of the internal states that arise in the functional processing of culture. The functional mapping of cultural input–output relations suggests a multitude of functional roles that are fundamental to the performance of culture. The demonstration of cultural input–output relations show multiple routes to the functional mapping of culture. The description of cultural input–output relations is the set of state transitions that demonstrate a functional mapping of initial to final states of culture with its transitions. The set of cultural states as mental states and its transitions comprise cultural algorithms or rule-based states that show the multiple sets of cultural input–output relations that are capable of the production of culture. The multiple routes of the functional performance of culture shows the adaptability of the cultural mind.

The functional performance of culture and the mind implies not only that mental states of culture contribute to the generalizations of culture in the world that hold, but also show actual occurrence in the world. Mental computation as states of actual occurrence in the world describes states of the world that show plausibility as modes of psychological realism. Cultural mental computation as cultural states of the mind in the world demonstrates how mental states comprise cultural states that have occurrence in the actual world. Culture as states of mind is not only a generality, but an actuality – a psychological realism. The plausibility of culture as mental states illustrates the actualities of the world that contribute to a naturalism of the world.

Culture as mental states describes the cultural states that are based on the mind, which may occur in independence of the actualities of the world.

Psychological realism as a bounded in a wide realm of mental states that are distinct from those of naturalism ascertains on those mental states that comprise the mind as independent of the world. Cultural mental states consist of those mental states that are of importance to the cultural representation of the mind that are independent of that of the world. Cultural states of the mind also arise as consistent with the aggregation of the parts and particulars of mental states; cultural mental states comprise a higher-level function of the mind. Culture as mental states of the mind further depict states of the mind that are emergent and comprise a psychological realism, rather than an inherent naturalism.

The manifestation of culture as a higher-level function of the mind describes the states of mental function or consciousness that are important to the cultural mind. The manifestation of cultural states as states of mental experience illustrates sets of cultural mental states that reflect the actualities of psychological realism. Consciousness as a higher-level process describes states of mental experience that comprise mental states. Culture as mental states that are actual occurrences show the unique, distinct and private realms of culture as mental experience. The actualities of mental experience describe distinct mental events that uniquely describe the mental states of the cultural mind. Culture as mental experience that is private depicts mental states bounded within the cultural states of the mind. Mental states as states of mental experience depict higher-level functions of the mind that are of importance to the understanding of culture of the mind in the world.

The culture of the mind in the world implies not only a psychological realism, but also an inherent naturalism. The concordance of cultural states of the mind and the world illustrates the shared sense of naturalism that arises from the understanding of the inherent naturalism of culture and the world. The states of culture in the mind and in the world depict the cultural representation of idealism and materialism. The cultural representation of ideas and material culture shows the importance of psychological and material realism as conduits of cultural experience. Cultural representation illustrates how culture arises as mental and physical states that are emblematic of cultural experience. The extent to which cultural states of the mind are dependent with the world illustrates the inherent naturalism of culture.

The mental states that entail the cultural patterning of thought illustrate the functional role of mental states as a causal power of culture. Mental states as a psychological realizer of culture show the causal arrangement of mental states as an interaction of the cultural mind and the world. The interaction of the cultural mind and the world facilitates the mental causation of culture and its effects. Culture as mental causation illustrates the causation of culture as part of the system of mind and its interaction with the world. Culture as a psychological realizer details how mental states exert a causal role on mental states and their functional role. The causality of culture as mental states in the structure of the world illustrates the causal interaction of the cultural mind in the world.

Cultural mental causation

Culture as a causal system implies that culture describes states in the world with causal effects. Culture as mental causation encompasses the mental states of culture in the world that hold. The cultural causation of mental states entails the states of mental function that have causal effect. Cultural states of the mind as mental states with a causal effect that is cultural illustrates culture as mental causation in the world. Cultural mental states as cultural causation demonstrates how the mind in cultural performance has causal effect. The cultural expression of the mind as a functional performance is an example of the cultural causation of the mind. The states of the mind that enact cultural causation detail how the mind is a causal system that is cultural.

Culture as mental causation illuminates on the causal impetus of mental events with a functional or causal significance. Culture as a causal power implies that mental events have causal significance that is cultural. The states of culture in the mind and in the world show the causal effect of patterns of thought that hold. Cultural causation of the mind shows how patterns and regularities of cultural thought demonstrate mental states with functional significance. The patterns and regularities of cultural thought show the causal significance of the shared meaning or significance of mental states that are cultural. Culture as mental states that have functional or causal significance encompass cultural patterns of thought that show how states of culture hold in the mind and in the world.

The cultural causation of the mind as patterns and regularities shows the lawlike principles of culture that govern. The cultural patterns and regularities of thought detail the mental states that have a causal significance that holds across space and time. The causal effects of the cultural patterns of thought entail generalizations of mental states and their causal-functional significance. The generalization of mental states as cultural thought describes patterns of the natural world that enact arrangements of events. The cultural patterning of the natural world as regularities of mental states illuminates the causation of mental states as thought and action. Culture as a system of mental causation facilitates the role of mental states as a causal impetus of states of culture in the mind and in the world.

The cultural causation of mental states is of importance to the complexity of causal effects that entail systems of cultural thought. The mental causation of cultural patterning illustrates the causality of mental states that encompass cultural systems of thought. Cultural patterns of thought describe the parts and particulars of states of culture that comprise cultural thought. The parts and particulars of cultural states as mental states illustrate the fundamental basis of the mental causation of cultural patterns of the mind. Mental states as parts and particulars of cultural states depict the manifestation of cultural causation as systems of cultural thought.

Cultural states of the mind encompass causal-functional properties that detail how higher-level features are emergent from lower-level features. The cultural causation of mental states underlies the functional properties of emergent features that comprise mental states of culture. The functional properties of the mental states of culture describe emergent features that enact mental states as cultural states. The functional properties of higher-level features show how the aggregation of lower-level features encompass emergent properties. Higher-level features demonstrate the emergence of lower-level features that entail the aggregation of functional properties. The mental causation of culture illustrates the functional and emergent properties that contribute to the causal-functional properties of cultural states of the mind.

The independence of the cultural and mental states describes the mental causation that comprises psychological states as a social realism. Cultural causation of the mind is a realism of the world because of the psychological realism of mental states. The causal effects of the mental states as a causal mechanism of social realism depicts the independence of the culture of the mind from that of the world. The causal realism of cultural mental states is necessary and sufficient as a causal impetus of social realism. The cultural mental causation of psychological states brings forth a social realism that depicts the states of culture as psychological states of the mind in the world.

The cultural mental causation of psychological states as social realism details the causal independence of states of culture. The causal independence of psychological states illustrates the functional significance of mental states as cultural states that reveal a psychological realism that is inherently social. The distinction of cultural mental states that show causal independence implies the functional properties of the mind that are discrete to the mental states of psychological realism. The mental states of culture show that states of the mind have causal and functional significance because mental states are discrete and distinguishable as a causal mechanism that is independent of the world.

Psychological states as a functional basis of causal independence shows how psychological realism is a causal mechanism of states of the mind that have functional and emergent properties. Culture as mental states with causal and functional significance show emergent properties that are distinct as psychological states of the cultural mind. The functional and emergent properties of culture illuminate on the causal-functional significance of the mental states of the mind that is independent of the world. Cultural mental states as functional properties that hold demonstrate how mental states as causal mechanisms illustrate emergent properties of the mind. Cultural causation as an emergence of the mind entails the functional and emergent properties of the cultural mind that hold.

The interaction of the cultural and mental states shows the core capacities that arise based on the functional properties. Cultural and mental states facilitate the information processing of core capacities that are of importance to the causal interaction of functional processes of the mind. The functional properties of the functional processes are essential to the cultural and mental states that are dependent and that facilitate cultural capacities. The interaction of the cultural and mental states show mental causation of cultural capacities that is dependent on the interaction of the world. Cultural mental states as causation illustrate the states of the mind that are dependent on those in the world. The interaction of the cultural and mental states show the multilevel processing of cultural capacities that illustrates the functional basis of cultural mental causation.

Culture as mental causation describes mental states that show interaction with the world. Cultural states of the mind are representative of those in the world. The state of culture as a mental state has causal effect in the world. At the same time, mental states as cultural states illustrate the cultural representation of the mind in the world. Culture as mental states illuminate the causal influence of the representational content of the cultural mind as interactional with the world. Culture as causal mental states depict the mental representation of culture and its causal effects. Cultural states of the mind that interact with the world depict the states of mental causation that show the concordance as states of culture in the mind and in the world.

The interaction of culture as a causal system depicts the states of the cultural mind and the world that enact psychological realism. The cultural states of the mind reveal as states of the world as a state of cognition and higher-level function. The cultural causation of the mind as a psychological realism depicts the states of thought and reasoning that show functional significance. The cognition and higher-level function of culture as mental states illustrate the causal influence of cognition and higher-level processes as a causal mechanism of states of the mind and of the world. The mental states of culture as higher-level processes illuminate on the functionalism of culture and the mind.

Culture as a causal system of mental inference illustrate the functional significance of the mental causation of culture. Culture as mental causation details the states of mental inference that have a causal-functional role in the processes of cultural capacities. Cultural mental causation demonstrates the causality and functionality of mental states that are fundamental to cultural processing. The cultural causation of mental inference depicts the wide range of causal effects based on inferential thought and reasoning. The cultural causation of mental inference encompasses a broad understanding of the importance of thought and reasoning as a causal and functional basis for intentional action.

Cultural causation as mental inference is fundamental to the thought and reasoning on intentional action. The causal inferences of cultural thought illustrate the functional significance of inferential thought and reasoning as a functional basis of intentionality and action. The causality of cultural thought and reasoning shows the causal significance of intentionality as a precursor to action. Mental states as intentional states detail how the discernment on states of the cultural mind serves as a functional basis for the inferential thought and reasoning of intentional action. Cultural states as mental states show that the causal effects of cultural thought encompass functional processes of mental inference that contribute to intentional action.

Cultural causation of the mind implies the functional significance of culture as a causal system. The states of cultural causation entail the mentality and thought that comprise functional performance. The cultural thought of the mind has functional and causal significance that illustrates the importance of culture as mental states. The cultural causation of mental states detail the functional performance of intentional action that leads to ascription of levels of responsibility. The understanding of the mental states of culture as a causal impetus for intentionality and action that enacts a sense of responsibility that is foundational to the functional significance of culture as a causal system. Cultural states as mental states demonstrates the causal role of mentality and thought on the reasoning on levels of responsibility. Culture as a causal system shows the functional significance of the cultural mind.

The determination of the causal role of responsibility on reasoning arises from cultural patterns of thought. Culture as causal reasoning shows the functional role of mental causation in the patterning of cultural thought. Cultural patterns of thought detail how distinct states of cultural mental states demonstrates an attribution of responsibility that distinguishes between objects and the cultural environment. The differential attribution of responsibility based on thought and reasoning illustrates how culture serves as a functional basis of the causal reasoning on the cultural environment. Cultural states as mental states show how causal reasoning is of importance to the functional significance of culture as a causal system.

Functionalism of culture and the mind

Functionalism as a stance on culture and the mind is foundational to an understanding of the functional role of mental states as psychological states. The study of the functional role of psychological states entails a characterization of mental states as physical states with a functional significance. The functionalist stance explores the wide realm of psychological states that show a functional role that is specific to mental states as physical states. The functional mapping of mental states as physical states implies a functional basis of culture and the mind in the world. The functionalism of culture and the mind depicts the multitude of mental and physical states that comprise psychological states as a causal mechanism of behavior and its functional significance.

The functional role of mental states as psychological states illustrates the psychological states that serve a functional basis. The notion of mental states as physical states reflects on the relations of psychological and physical realizers. Mental states as physical states implies that psychological states have a functional significance in the world. The functional role of psychological states as a physical state of the world entails the functional basis of the cultural mind in the world. The mental states of culture as psychological states show a functional significance of cultural states as a psychological realism. Cultural mental states serve as physical states that have functional significance.

Cultural mental states depict the functional basis of psychological states that have a functional role. Cultural states of the mind as psychological states show the functional mapping of psychological and physical states of the mind. The cultural mental state as a psychological state is comprised of a functional mapping of psychological and physical states that describes the functional role of cultural states of the mind. The functional mapping of psychological and physical states as cultural states illustrates the functional role of the mind as a physical instantiation of culture. The cultural mental states as psychological states describe a functional basis of the cultural states of the mind.

Cultural states of the mind illustrate the functional mapping of psychological and physical states as a multitude of relations. The relations of the cultural mind consist of a multitude of functional relations of psychological and physical states with a causal significance. The functional role of one-to-many mapping details the psychological states that map onto multiple physical states of the cultural mind. The functional one-to-many mapping of the functional basis of culture and the mind consists of the relations of psychological and physical states that comprise culture, mind and behavior. The functional mapping of psychological and physical states shows how psychological states arise as a multitude of physical states with a causal-functional significance.

Culture as a multitude of relational states brings forth postulation of the phenomenological basis of functional states. Culture serves as a functional basis of phenomenological states and its causal effects. Mental states comprise phenomenological states that emerge from the functional relations of psychological and physical states. Cultural mental states enact functional relations that comprise phenomenological states of experience. The cultural mind demonstrates the functional role of psychological states as phenomenological realism. The functional role of cultural mental state as phenomenological states illustrates the functional basis of the cultural mind in the world.

The functional basis of culture and the mind describes the causal-functional significance of the relations of psychological and physical states as cultural states. Cultural states as relational to psychological and physical states show how mental states serve as a causal mechanism of culture. Cultural mental states have causal effects on the states of culture in the mind and in the world. Cultural states of the mind show the causation of cultural states as mental and physical states of behavior. The cultural causation of mental and physical states manifests of the causal-functional significance of behavior. The cultural states of the mind as causation of physical states illuminates the causal relations of the cultural mind on behavior.

The relational states of culture show causal influence on the cultural mind and behavior. Cultural states as relational to mental and physical states demonstrate how the cultural mind serves as a causal mechanism. The cultural mind is a causal impetus of the states of culture in mind and behavior. Cultural mental states comprise the causal mechanism that is of culture of the mind and of behavior. Cultural states as states of the cultural mind with causal-functional significance shows the functional role and causal influence of cultural states that are relational. The cultural states that are relational illustrate the functional role of culture and the mind on behavior.

The functional relations of the psychological and physical states of culture and the mind show causal-functional significance. The functional relations of the cultural mind depict the relations of psychological and physical states that serve a functional role on behavior and its causal effects. The functional relations of cultural mental states describe the set of psychological and physical states that show the functional significance of behavior. Cultural mental states that have functional significance demonstrate the causal-functional role of cultural states as relational sets of psychological and physical states of culture and the mind.

The relational sets of culture as psychological realizers demonstrate cultural mental states and their causal effects. The cultural states of the mind as psychological states depict the mental states of culture that have causal role on behavior. Behavioral states are a manifestation of cultural states of the mind. Cultural states as states of mind illustrate the causal role of mental states on behavior. Culture as mental states details the manifestation of culture as behavioral states. The cultural mind serves as a causal role of the functional basis of mental and behavioral states.

Functionalism of culture and the mind illustrate how culture serves as a functional basis of behavior. The functionalist stance entails the understanding of culture as psychological and physical states with functional significance. The functional mapping of psychological and physical realizers shows the manifestation of culture as states of the mind and the world. Culture as a functional basis of mental states of behavior demonstrates the functional significance of the cultural mind. The cultural mind is a causal mechanism of the functional basis of behavior. The study of culture and the mind illuminates on the functionalist stance and its implications for culture and behavior.

Culture as a psychological realizer

The functionalist stance on culture and the mind postulates on culture as a psychological realizer. The manifestation of culture as a psychological state entails the understanding of the causal role of the cultural mind. Culture as a psychological realizer details the manifestation of cultural states as psychological states of the mind. The psychological states of the cultural mind comprise states of culture that serve a causal-functional role. The understanding of culture as a psychological realizer demonstrates the functional performance of the cultural mind as psychological states of behavior.

Culture as a psychological realizer entails the notion of culture as generalizations of lawlike principles. Culture as psychological states shows how the computation of culture enacts generalizations of laws and principles that hold. Cultural states as psychological states serve as a functional basis of the laws and principles of culture. The psychological realism of culture illustrates how the lawlike principles of culture serve as actualities in the world. Cultural states as psychological states demonstrate the functional basis of the cultural patterns and regularities of the mind. Culture as a psychological realizer illustrates how mental states contribute to the cultural patterns and regularities of the world.

The psychological realizer of culture depicts the modes of psychological realism that are of importance to the actualities of the patterns and regularities of culture. Culture as patterns and regularities of the mind describe the occurrences of cultural thought that are of importance to the understanding of the causal effects of culture in the world. Cultural patterns as mental states illustrate the causal-functional role of culture in the mind and in the world. The cultural patterns and regularities of mental states show the functional significance of the cultural mind. Culture as a psychological realizer demonstrates the patterns of cultural thought that hold.

The psychological realism of culture encompasses a wide realm of mental states. Culture as mental states consists of the phenomenological and physical states of the mind that show functional performance. The functional basis of the phenomenological and physical states of the cultural mind demonstrates the multitude of functional relations that comprise culture as a psychological realizer. The phenomenological states of culture show how culture is a functional performance of the mind and its higher-level function; the physical states of culture and its transitions detail how culture is a functional performance of the mind as a computing machine. The functional basis of the cultural mind encompasses the phenomenological and physical states of functional performance.

The psychological states of the cultural mind contribute to the performance of the functional processes of cultural states of mind. Cultural mental states demonstrate the distinction of cultural and mental states as functional processes of behavior. The cultural mind as a functional basis of behavior illustrates core psychological processes that underlie behavior. Culture as psychological processes demonstrates the functional basis of the cultural mind. The core capacities of psychological processes of the mind illustrate the functional basis of culture and behavior.

The cultural mind shows how the manifestation of cultural states is distinct as a psychological realizer of the cultural mind. The cultural mind demonstrates a manifestation of culture comprised of states of mind and its functional and emergent properties that are specific to psychological states of behavior. The functional and emergent properties of the cultural mind show that psychological states are a causal impetus of behavior. The psychological states of the cultural mind detail the wide range of cultural and mental states that comprise distinct psychological processes of the cultural mind. The manifestation of culture as psychological states demonstrates the functional role of cultural and mental states of behavior.

The functional processes of the cultural mind show how the core psychological processes serve as a functional basis of culture and behavior. From emotion to social behavior, core psychological processes demonstrate the functional basis of the mind and behavior. The functional basis of the core psychological processes are of importance to an understanding of the functional performance of the mind at the level of computation. The functional basis of the core psychological processes depict the information processing that shows the functional machinery of cultural capacities. The core psychological processes are foundational to an understanding of the information processing mechanisms of culture and behavior.

The functional machinery of cultural capacities are comprised of the core processes of psychological states. The core processes of psychological states serve as information processing mechanisms that contribute to the functional performance of the cultural mind. The information processing mechanisms of the cultural mind illustrate the functional machinery of cultural capacities. Culture serves as a causal influence on the core psychological processes of behavior. Cultural states of the mind enact causal effects on psychological processes and their functional consequence. Culture serves as a causal influence on the core processes of psychological states of behavior.

Culture demonstrates the specialized processing of specific psychological realizers and the functional processes that facilitate the performance of cultural capacities. Culture as sets of psychological states describes the core processes that contribute to the functional machinery of culture and behavior. Cultural capacities show the core processing of psychological states that facilitate the functional performance of input–output relations. The fundamental basis of psychological processes details the functional performance of cultural input–output relations as the information processing of culture and behavior. The core processing of cultural capacities demonstrate the specialized processing of psychological realizers and their causal effects on the functional and behavioral states.

The cultural and functional basis of behavior contributes to the core psychological processes that are of importance to behavior. The functional performance of the core processes shows the cultural and functional basis of behavior. The function of the core psychological processes demonstrates the psychological states that are of importance as a functional basis of culture and behavior. Culture as psychological states serve as functional processes that are foundational to behavior. The core psychological processes of the cultural mind contribute to the functional processing of culture and behavior.

Culture as a psychological realizer demonstrates the functional role of culture as psychological states. Culture as states of the mind and behavior details the causal-functional role of cultural states and their functional significance. The cultural mind is comprised of psychological processes that serve as a functional role and causal influence of behavior. Culture as a functional basis of behavior demonstrate the functional machinery of core processes that are essential to behavior. The understanding of culture as a psychological realizer contributes to the characterization of the multiple realizations of culture in the mind and in the world.

Cultural affective computation

Research on cultural affective computation consists of the study of the cultural computation that is of importance to the functional processes of emotion. Cultural computation demonstrates the functional components of the core capacities of psychological processes such as emotion. The study of cultural affective computation details the understanding of the cultural computation of emotion in the mind and its functional significance on the core capacities of emotion. The functional components of core capacities describe the psychological processes that contribute to emotional and cultural capacities. The understanding of cultural affective computation contributes to the detailed characterization of the functional components of emotion as a mental process and a psychological realizer of culture in the mind and in the world.

The study of cultural affective computation consists of a range of thematic topics that are of importance to an understanding of the cultural computation of emotion. Cultural processing of emotion relies on the functional components of mental processes that are of importance to core capacities. Emotion as a psychological realizer describes functional components of core capacities that are essential to the automatic and controlled processes of emotion. The cultural processing of emotion details how the functional processes of the core capacities arise based on the interaction of parts and particulars. The control or regulation of emotional processes demonstrates the interaction of cognition and emotion in cultural context. The detailed characterization of the psychological processes of emotion illustrates the functional basis of cultural computation of emotion.

Emotion as cultural affective computation details the role of the functional processes of emotion as components of cultural capacities. The functional components of emotion describe the core capacities that constitute the functional processes of emotion and affect. The functional processes of emotion and affect entail the feeling states that arise based on the automatic and controlled processing of core capacities. The cognitive representation of emotion details the representation of sense data of the environment ("emotion percept") and its transformation into motor output ("emotional expression"). Emotion demonstrates the automatic and controlled processing of the functional processes of the mind and its response to the cultural environment. The processes of emotion as feeling states in response to the cultural environment illustrates the functional role of emotion as a component of cultural capacities.

The automatic response of emotion to the cultural environment demonstrates a prepotent and reflexive response that has functional significance. The automaticity of the emotional processes of the cultural mind shows the preparedness of the mind in interaction with the cultural environment. The automatic and prepotent response of functional processes depicts how the functional machinery of the mind is capable of adaptive performance. The functional performance of adaptiveness is a fundamental element of the core capacities that are illustrative of the functional processes of the mind and its capabilities. Emotion shows the functional processes of the mind that are adaptive and responsive to the cultural environment.

The controlled processing of emotion depicts the regulation of emotional experience and its functional significance. The controlled processing of emotion illustrates how the bidirectional information processing of emotion affects the functional processes of higher-level function and behavior. The bidirectional processing of emotional information facilitates the automatic and controlled processing of emotion processes. The control of emotional processing contributes to the reinterpretation of the motivational and social significance of emotional experience. The reinterpretation of the motivational and social significance of emotional experience is of importance to emotion regulation. The interaction of cognition and emotion is foundational to the control and regulation of emotion processing.

The functional basis of emotion processes is of importance to the control and regulation of cultural capacities. Cultural factors serve as a top-down influence on the bidirectional processing of emotion. Culture enacts controlled processing that facilitates the inhibition and regulation of emotion. The control and regulation of cultural capacities facilitate the functional processes of emotion in the cultural context. Functional processes of emotion demonstrate how inhibition and control of emotional experience facilitates the processes of culture. The bidirectional processing of emotion illustrates the interaction of the functional processes of emotion and cognition in the cultural context.

Functional processes show the role of inhibition and control of emotion on the cultural processes. Emotion regulation is of importance to a wide range of cultural processes. From emotion perception to emotional expression, cultural processes facilitate the control and regulation of emotion. Culture affects the perception of emotion of facial and bodily expression. Cultural differences in emotion perception show the functional role of top-down influences on the perceptual and cognitive processing of emotion. Culture affects the perceptual and cognitive processing of emotional expression. The functional modulation of the processes of emotion based on culture demonstrates the role of inhibition and control on emotion and cultural capacities.

Cultural differences affect the control and regulation of emotion expression. Culture describes societal norms that are of importance to the regulation and control of emotional expression. The shared understanding of the motivational and social significance of emotion affects the expression of emotion. Culture is of importance to the shared meaning of emotional expression and its functional significance. The shared meaning of emotional expression imparts the functional role of culture on the motivational and social salience of emotion. Culture is fundamental to the understanding of the motivational and social significance of emotion.

The study of cultural affective computation depicts the identification of the functional processes of emotion that are of importance to cultural capacities. At the computational level, the functional components of the core capacities depict the task-based activity that is culture-specific. Emotion processes are fundamental to the perception and expression of culture. The automatic and controlled processing of emotion illustrate how psychological realizers are fundamental to the functional processing of emotion. The processes of emotion detail the interaction of emotion and cognition as a functional basis of the perception and expression of culture. The perception and expression of culture illustrates the importance of the functional processes of emotion to cultural capacities.

Cultural cognitive computation

The study of the cultural cognitive computation is of importance to the systematic investigation of the functional processes of cognition and higher-level function as cultural capacities. Research on cultural cognitive computation details the functional components of cognition that are of importance as mental processes and higher-level function of cultural capacities. Cognition as a mental process and higher-level function demonstrates how the core components of psychological processes such as cognition and language are essential to cultural capacities. The study of cultural cognitive computation details the functional components of cognition and higher-level function that are fundamental to the cognition and computation of culture.

The systematic study of cognition as cultural computation illustrates how the core capacities of cognition contribute to the mental activity that is task-based processing and culture-specific. The functional use of cognition in the functional performance of cultural tasks demonstrates the role of cognition and mental function in cultural computation. The reliance on effortful or task-based mental activity facilitates levels of mental function that show mental processes with functional significance. Culture as mental states illustrate levels of mental function that are effortful and based on task-relevant activity that is culture-specific. The differential levels of mental function that is task-based describes the cultural states of mental function. The functional use of cognition in cultural tasks illustrates the core processes of cultural cognitive computation.

Cultural tasks that are reliant on cognition and higher-level processes show the role of functional performance on task-based activity. The differential levels of mental activity of distinct types of mental tasks demonstrate the distinct modes of mental function that are characteristic of cognition and mental function. Cultural tasks show the cultural processing of task-based and spontaneous mental activity that comprise cognition and higher-level function. The task-based and spontaneous mental activity of cognition and higher-level activity detail the modes of mental function that are of importance to the functional performance of cultural tasks. Cultural tasks contribute to the functional use of cognition and higher-level processes with functional significance.

Research on cultural cognitive computation demonstrates the functional processes of cognition and culture at the computational level. The functional components of cognition detail core capacities that are fundamental to cultural computation. The functional processes of cognition describe multiple routes to states of cognition that describes algorithms or rule-based processes that contribute to functional performance. The functional use of rule-based cognition as cognitive processes illustrate the transitions of mental states that contribute to cultural computation.

The functional processes of cultural computation show how cognition and higher-level processes contribute to the shared understanding of the motivational or social significance of cognition and behavior. The functional basis of cognition and higher-level processes are of importance to the semantic representation of language ("conceptual knowledge") and its role in the shared meaning of the functional significance of cognitive and cultural processing. The semantic representation of conceptual knowledge facilitates the shared understanding of the motivational or social significance of language and behavior. The semantics of behavioral and linguistic expression depict the conceptual basis of the representation and interpretation of the functional significance of cognition and behavior. The semantic processing of the conceptual knowledge provides a functional basis for the cultural computation of cognition and higher-level processes.

The cognition and higher-level processes of semantic understanding facilitates the interpretation and reinterpretation of the shared meaning of language and behavior. Semantic comprehension is of importance to an understanding of the pragmatics of linguistic expression. The cognitive processing of conceptual knowledge provides layers of interpretation and meaning ("abstract representation") to the higher-level processing of culture and language. The computational level of information processing is essential for the interpretation and reinterpretation of the culture and language of cognition and higher-level processes. The semantic processing of culture and language show the cognition that is of importance to the shared understanding of linguistic and behavioral expression.

The functional processes of cognition are of importance to the control and regulation of cultural capacities. Cognitive processing demonstrates the functional role of inhibition and control on cognition as a higher-level function. The bidirectional processing of information facilitates the inhibition and control of levels of processing that is of importance to the control and regulation of higher-level processes. The inhibition and control of cognition comprise functional processes that demonstrate the functional role of feedback inhibition on levels of processing. The inhibition and control of functional processes is essential to the control and regulation of cultural capacities.

The control and regulation of cultural capacities is fundamental to the cognition and higher-level function. The functional processes of cognition and higher-level function show the levels of processing that are essential to the inhibition and control of cultural capacities. Higher-level function facilitates the top-down processing that exerts control or regulation on the bottom-up processing of cognition. The top-down and bottom-up processing of information contributes to the changes of levels of processing that demonstrate a functional basis of cognitive control and regulation. The functional basis of cultural capacities facilitate a depth of processing of information that entails the control and regulation of culture.

The cognitive control and regulation of the core capacities detail the role of functional processes on the changes of cognition that are essential to cultural capacities. The functional processes of cognitive control and regulation serve as a functional basis for the interpretation and reinterpretation of cultural and mental states. The relaying of information processing that is unidirectional contributes to the changes in levels of processing that is of importance to the interpretation of mental states. The inhibitory processing of mental states that is bidirectional facilitates the reinterpretation of mental states. The cognitive control and regulation of core capacities shows the functional processes that facilitate the interpretation and reinterpretation of cultural and mental states.

The control and regulation of cognition is fundamental to the understanding of the functional significance of cultural capacities. The inhibition and control of cognition describes a functional basis for the higher-level processes of the cultural mind. The functional significance of cultural and mental states imbues core capacities with the functional states of the cultural mind that comprise the shared meaning on understanding of others. The functional states of the cultural mind demonstrate the sets of cultural and mental states that comprise the shared meaning and understanding of others. The functional processes of the control and regulation of cognition facilitates information processing that is essential to the functional significance of cultural capacities. The control and regulation of cultural capacities illustrate the cognition and higher-level processes of the cultural mind.

The cognition and higher-level function of consciousness illustrate functional processes that are essential for an understanding of the phenomenological states of the cultural mind. The functional processes of cultural and mental experience detail the functional states that are of importance as the functional machinery of multilevel adaptation. Cultural and mental experience comprise states of knowledge that facilitates functional adaptation. Cultural and mental experience as phenomenological states describe knowledge states of the mind that is interactional with the world. Phenomenological states as knowledge states illuminate on the functional processes of cultural and mental experience that serve as a functional basis of the cultural mind.

Consciousness details the functional processes that are of importance to cognition and higher-level function. The functional processes of the streams of consciousness describe states of mental experience that comprise functional and cultural adaptation. The phenomenological states of culture and mental experience facilitate the functional states of the mind that are of importance to cultural states. The cultural and mental states of consciousness comprise states of phenomenological being that are illustrative of functional adaptation. The functional processes of consciousness detail the cognition and higher-level function of culture and mental life.

The functional components of cognition and mental life describe the core capacities that are foundational to the effortful and task-based processing of higher-level processes. The computational level of cognition and higher-level function depicts the task-based processing of culture that is specific to functional tasks. The core processes of cognition and mental life illustrate the causal-functional role of higher-level function as a functional and cultural adaptation. The study of cultural cognitive computation is fundamental to the delineation of the cognition and higher-level processes of culture and mental life.

The study of cultural cognitive computation is fundamental to an understanding of the computational principles of culture and cognition. The characterization of the multilevel processing of cognition and higher-level function demonstrates the role of functional processes on cultural computation. The understanding of the functional basis of cultural computation entails the specific functional processes that comprise cognition and higher-level processes. The functional components of cognition illustrate how higher-level processes facilitate cultural capacities. The scholarly interest on cultural cognitive computation provides a foundation for the scientific inquiry on cultural mental computation.

Cultural social computation

Research on cultural social computation investigates the cultural computation of the functional processes that are of importance to culture and social behavior. The study of cultural social computation details the functional components of cultural capacities that are fundamental to social processes and cultural computation. The cultural computation of functional processes contribute to the cultural capacities that are essential to the understanding of the cultural mind in social context. Functional processes of cultural computation contribute to the social processing of cultural capacities. The fundamental elements of cultural capacities are of importance to an understanding of the social and cultural world.

The scholarly interest on cultural social computation demonstrates the functional processes that are of importance to the functional components of cultural and social processing. The cultural computation of the mind shows how cultural and social processing arises from the functional performance of core processes of behavior ("social behavior"). The functional performance of task-based processes contribute to the social processing of cultural social computation. The functional components of cultural and social processing illustrate the core processes of social behavior. The functional basis of social processes demonstrate the functional machinery of cultural computation.

The study of cultural social computation generates knowledge of the cultural computation of social processing. The cultural capacities of social processes describe the processing of cognitive and social information in context. The social processing of cultural capacities illustrates the functional role of motivational and social significance on the understanding of the social and cultural environment. The functional components of cultural capacities detail the cognitive and social processes that are essential to information-processing mechanisms. The functional processes of cultural social computation are foundational to the adaptive machinery of cultural capacities.

The fundamental components of cultural capacities consist of the functional processes that comprise cognitive and social processing. Cultural capacities facilitate the cognitive and social processing that are specific to the causal-functional role of functional mechanisms. The functional basis of cognitive and social processing ("social cognition") are of importance to the cultural capacities and its functional significance. The processing of cognitive and social information demonstrates the functional role of cultural states as mental and behavioral states. The characterization of the functional mechanisms of cultural and social capacities is fundamental to understanding of the cultural social computation.

The functional machinery of cultural and mental states illustrate the social processing that is of importance to the functional components of the mind. The functional basis of social processes facilitates the detection and understanding of other minds. The processing of social information ("social recognition") demonstrates the core processes that are essential to social capacities. Cultural states of the mind illustrate the causal influence of culture on the social processing of others. The cultural processing of social information facilitates the detection and understanding of other minds that is culture-specific. The cultural and social processing of functional mechanisms depict the detection and understanding of other minds that is based on core capacities.

Cultural processing of social information shows the distinct types of social processing that are essential to the core social capacities. The social capacity of the detection and recognition of other minds is a fundamental social process. The mental representation of other minds as a set of mental states that hold shows the multilevel information processing that comprises social capacities. The cultural and social processing of the mind illustrates the distinct types of functional response that are of importance to an understanding of other minds. The mental representation of other minds as a functional basis encompasses a wide realm of functional components that comprise social information processing.

Cultural and mental states show the functional basis of the multilevel information processing of social capacities. The cultural and social processing of the mind details the functional basis of causal mechanisms that contribute to the depth of processing of social capacities. Cultural and social processing facilitates the encoding and decoding of social information of the cultural environment. The depth of processing of cultural and social information ("social percepts") depicts the functional specificity of social information processing. The representation and transformation of social information details the functional processes that are essential to social and cultural understanding. The functional processes of cultural and social processing facilitate the multilevel information processing of social capacities.

The multilevel processes of social information processing show how the cultural processing of social information serves as a functional basis of the multilevel processing of core social capacities. The cultural processing of social percepts illustrates the depth of processing of social information mechanisms. Culture serves as a causal influence on the multilevel processing of social information. The cultural representation of social information details the functional characteristics of social percepts that are fundamental to social processing. The functional components of social capacities demonstrate how multilevel processing is essential to the cultural representation of social information.

The functional basis of multilevel processing details the core components of social and cultural capacities. The distinct types of processing of featural characteristics shows how functional components facilitate the production of cultural input–output from social information. The concordance of cultural and social processing depicts the functional specificity of cultural and social capacities. The functional performance of core components show the functional correspondence of cultural and social processing. The cultural specificity of social processing demonstrates the functional performance of cultural input–output that is based on social information. The core components of social and cultural capacities are essential to the functional performance of social processing.

The multilevel processing of social information demonstrates the cultural causation of social processes. The functional performance of the core components of social capacities depict the causal-functional role of social processing. The functional basis of social processing shows how social information serves as a causal mechanism of the functional components of mental state understanding. The detection of other minds relies on the encoding and decoding of the social information of the cultural environment. The functional processing of social information details the functional specificity of cultural processing that is of importance to the understanding of the mental states of others. The mutlilevel processing of social capacities depict the cultural causation of social processes.

The cultural processing of social information affects the depth of processing of social capacities. The core components of social processing show how culture affects the depth of processing of social information. The functional processes of social information depict the level of social processing that is of importance to social and cultural understanding. The cultural processing of social information facilitates the level of processing that is necessary and sufficient for the understanding of the cultural and mental state. The depth of processing of social information faciltiates the understanding of the cultural and mental states of others.

The encoding and decoding of social information processing mechanisms is fundamental to an understanding of the mental states of the cultural mind. The functional mechanisms of cultural capacities encompass the cognitive and social processing of mental state understanding. The cultural computation of cognitive and social processing describes the functional basis of mental state inference. The cultural capacities of social information processing facilitate the understanding of other minds. The social processing of cultural capacities contribute to the functional basis of the mental state understanding.

The understanding of mental states is fundamental to the functional processes of the cultural mind. The functional processes of cultural mental states entail the cognitive and social processing on other minds. Cultural mental states consist of the mental states that are of importance to the inferential reasoning of other minds that is cultural ("social concepts"). The inferential reasoning on other minds is fundamental to the understanding of cultural and mental states. Cultural and mental state understanding reflect the interaction of the functional processes of mental state understanding and the cultural environment. The functional processes of mental state inference shows functional dependence on cultural processing. The cultural capacities of social thought processing facilitate the mental state inference that is essential to the understanding of other minds.

Cultural and mental state understanding describes functional processes that show the interaction of social and cultural processing. The cultural and mental states that depict other minds show the processing of social and cultural processing in context ("social context"). The contextualizing of social and cultural processing illustrates the inferential reasoning on other minds that is contextual. The context-based processing on social and cultural information demonstrates functional processes that show the malleability of the cultural mind. The understanding of mental states as a functional process illustrates the interaction of social and cultural processing in context.

The functional components of cultural and mental state understanding show how cultural and mental states demonstrate the manifestation of culture in the mind and in the world. The detection and understanding of other minds arises from the encoding and decoding of cognitive and social information of the cultural environment. The inferential reasoning on other minds enacts mental states that are dependent on cultural states. The understanding of mental states shows the functional role of cultural states as a conduit of mental state inference. Cultural states contribute to the decoding and encoding of mental states that shows functional dependence on culture. The interaction of cultural and mental states illustrates how culture manifests as parts and particulars of states of the mind and of the world.

The functional dependence of cultural and mental states demonstrates cultural and social processing that has causal-functional significance. The interaction of cultural and mental states illustrates functional processes that show the causal influence of mental state in the world. The cultural and mental states of inferential reasoning comprise a casual-functional role that is part of the causal structure of the mind in the world. The cultural states of mental state inference show the functional significance of the mind and its causal effects on the world. The functional performance of cultural and mental states describe the causal influence of the cultural mind that is dependent on the world. Cultural and mental states encompass cultural and social processing that shows functional dependence and its causal-functional significance.

The functional processes of cultural and mental state inference encompass the cognitive and social processing of understanding other minds. The social processing of mental state understanding illustrates the functional independence of mental state inference. The understanding of other minds describe the inferential reasoning on mental states that is inherently social ("fundamental social processes"). The inferential reasoning on mental states depicts the understanding on other minds that is based on social processing that shows functional independence. Mental state understanding that is independent demonstrates inferential reasoning on other minds that is general and describes mental states of others that are consistent across space and time.

Culture and mental state understanding also illustrate the functional role of mental state inference in the understanding of other minds that reflect social processing. Mental state understanding describes the inferential reasoning that arise based on the social processing of the cultural mind. Cultural and mental states arise because of the social and cognitive processing of the cultural mind. The social processing of the cultural mind is necessary and sufficient to generate inferential reasoning on other minds that is independent of the world. The cultural and social processing on other minds entail the inferential reasoning on others that shows the functional independence of the cultural mind from that of the world.

The functional independence of cultural and social processing demonstrates functional properties of the cultural mind that are emergent. Cultural and social processing on mental states shows functional properties of cultural states that are specific to mental states that are independent of the causal structure of the world. Cultural states as mental inferences depict inferential reasoning that is specific to the cultural mind and its causal effects. Cultural and social processing on mental states details the functional properties of the cultural mind that have causal significance that is inherently social. The social processing on cultural and mental states encompasses functional properties of the mind that are emergent and independent of others.

Cultural and social processing that is emergent describes a functional performance of cultural and mental states that holds. The cultural and social processing on other minds details the social processes that serve as a functional basis of cultural capacities. The functional performance of cultural and social processing illustrates how social processes have functional and emergent properties. Functional properties of cultural processing depict the understanding of cultural and mental states that is a functional performance of the emergent properties of the cultural mind.

Cultural capacities demonstrate the cultural and mental state that serve as a functional basis of behavior. The cultural and mental states of social processing describe the functional machinery that is of importance to the core processes of behavior. The cultural processing of functional processes detail the social information processing that is essential to functional and behavioral states. The functional correspondence of cultural and social processing with causal mechanisms illustrates the functional basis of behavior. The concordance of cultural and social processing with behavioral states illustrates the functional role of cultural and mental states as a fundamental basis of behavior.

The cultural and social processing of the mind is a conduit of the functional basis of the mental causation of behavior. The formulation of cultural and social mental states is essential to the functional processes of social behavior. Cultural and social mental states ("intentionality"") serve as a causal impetus of simple to complex behavior ("action"). Cultural and social processing of the mind illustrates how social mental states as higher-level processes enact causal effects on behavior. Cultural and social patterns of thought demonstrate the functional processing of complex behavior (""intentional action"). Cultural and social thought serve as a causal mechanism of patterns of behavior. The functional correspondence of cultural and social patterns of thought and behavior show how functional processes underlie the social capacities of the cultural mind. The functional performance of the cultural and social processing of mental states illuminates on the higher-level processing of mental causation and its function.

The cultural causation of social mental states describes functional processes that have causal-functional significance. The mental causation of cultural and social processing demonstrates the functional role of core social processes and their causal effects. Cultural and social capacities show the cultural causation of social processing and its causal-functional significance. The cultural causation of social processes depicts the functional role of social mental states as a causal influence of cultural and social patterns of thought and action. The cultural mental causation of social processing arises based on the causal patterning of cultural and social thought.

Research on cultural social computation provides a characterization of the functional basis of the cultural computation of social processing. Cultural and social processing comprise functional components that serve as a causal mechanism of cultural computation. The social processing of cultural capacities illustrates the importance of social processes as a functional basis of the cultural mind at the level of computation. The study of the cultural social computation demonstrates the functional components of the cultural and social processing of the mind. The scholarly study on cultural social computation generates knowledge on the cultural computation of social processing and its functional significance.

Discussion

The study of the cultural mental computation broadens the understanding of the cultural computation of the mind. The cultural computation of the mind shows how the cultural mind performs as a computing machine and its functional significance. The scholarly inquiry on the culture of the mind and world contributes to an understanding of the cultural causation of the mind and its causal-functional roles. Research on cultural mental computation is of importance to the detailed characterization of cultural computation as functional processes of the cultural mind. The functional processes of the cultural mind show the cultural processing of the mind across core capacities. The study of cultural mental computation facilitates the understanding of cultural computation in the mind and in the world.

Research on cultural mental computation contributes to the understanding of the foundations of the cultural computation of the mind. The empirical study of cultural mental computation illustrates cultural states as mental states in the mind and in the world. Cultural states as mental states demonstrate the causal-functional role of the cultural mind in causal interaction. The cultural causation of the mind shows the fundamental principles of culture as mental states. The study of the causal-functional role of the cultural mind contributes to the understanding of the functional significance of cultural states as mental states.

Cultural mental computation is of importance to the broader understanding of culture and mental life. The cultural computation of the mind illuminates on the philosophical and scientific inquiry on culture, minds and machines. Culture as a manifestation of psychological realizers demonstrates the functional basis of core capacities as essential to cultural adaptation. The cultural computation of psychological states illustrates the functional processes of core capacities that contribute to the adaptive machinery of culture and behavior. The study of cultural mental computation contributes to an understanding of the functional machinery of culture, mind and behavior.

Conclusion

Research on the cultural mental computation describes the cultural computation of the mind and its functional significance. The study of the cultural mental computation is of important to an understanding of the cultural mind as a computing machine. The detailed characterization of the cultural mind as a computing machine describes how the functional performance of the mind contributes to the production of culture. The scholarly interest on cultural mental computation provides a foundation for an understanding of culture as mental causation and more broadly, culture as a causal system. The demonstration of culture as mental states or psychological realizers shows the functional significance of culture in the mind and in the world as an inherent naturalism.

The scholarly and scientific study of cultural mental computation is foundational to an understanding of the origin and nature of culture and mental life. The generation of scientific knowledge on cultural mental computation contributes to the fundamentals on cultural computation of the mind and the world. The understanding of the functional performance of the cultural mind contributes to the study on culture and mental computation. The functional significance of cultural states as mental states shows how psychological states are part and particular of the manifestation of the cultural mind. The philosophical and scientific postulation on the cultural states of mentality and thought contributes to the conceptual understanding of the cultural foundations of mental life.

Implications

The study of the cultural mental computation has practical implications for health, medicine and related fields of study. The practical applications of the cultural mental computation demonstrate the scientific and techno-logical innovation on the cultural computation of the mind. The scientific and technological innovation on cultural mental computation illuminates the practical and societal benefits of a broad understanding of the computational principles of the cultural mind. Research on the cultural mental computation facilitates the scientific and educational enrichment on how cultural mind performs as a computing machine. The broad understanding of cultural mental computation contributes to the building of scientific and educational enrichment on the cultural computation of the mind.

Practical applications of cultural mental computation consist of the development of scientific and technological innovation of computational tools and technologies that are beneficial to the study of culture and the mind. The design of computational tools and technologies serves as an advancement of the technological innovation that is beneficial to the scientific and societal understanding of the cultural mind. Computational tools and technologies provide advancement of scientific and technological innovation that is beneficial to the scientific discovery in health, medicine and related fields. The design of practical applications of cultural mental computation is of importance to the scientific discovery of cultural computation.

The broad implications of the development of computational approaches to the study of the cultural mind illustrate the wide interest on culture in mind and machine. Since antiquity, scholarly and scientific interest on the ethereal and spiritual realm has brought forth pontification on the phenomenological and physical basis of the cultural mind in the world. Scholarly and philosophical inquiry on culture in the mind and the world has broadened understanding on the shared meaning and purpose of being and existence from the ancient to the modern world. The contemporary interest on the study of computation brings forth novel approaches to the scholarly and empirical study of culture and the mind. The study of cultural mental computation is foundational to the generation of knowledge on computational approaches to the scholarly inquiry on the nature and origin of culture and the mind.

Societal and educational enrichment on cultural and mental life illuminate on the broader societal understanding of cultural mental computation. The intellectual and societal interest on the cultural computation of minds and machines builds societal interest of the implications of scientific and technological advancement. The contemporary notion of cultural computation captures the imagination of culture as minds and machines. The scholarly and scientific inquiry on cultural computation serves as a contemporary notion that is beneficial to the exploration of interests on cultural and mental life.

4

CULTURAL MACHINE COMPUTATION

Introduction

From the 17th to the 20th centuries, the early notions of the computing machine illuminate the scholarly and scientific interest for the development of the computing machine that demonstrates the functional capabilities of computation. The scholarly and scientific interest on the computing machine reflects on the contemporary notion that computation demonstrates the high-level performance and real-world advantages of minds and machines in the world (Copeland, 2004; Turing, 1947). The computation of mentality and intelligence illustrates the sophistication of thought and mental function that serves as a functional basis of the computation of the mind and machine in the world. The notion that computing machines have the capabilities of functional performance shows the equivalence of mental and machine computation that presents considerable real-world advantages. The functional operationalization of the mechanistic basis of the computation of minds and machines illustrates the functionality and adaptability of the computing machine as a physical realizer of computation.

The demonstration of the computing machine as a conduit of the functional performance of mental computation is fundamental to the understanding of the notion of the mind as a computing machine. The historical development of the computing machine led to the consideration of the multiple capabilities of computation as mental and machine function. The functional capabilities of machines for computational performance relies on the development of machine capabilities that show functional equivalence with that of the core capacities of mentality and intelligence. The parallel

DOI: 10.4324/9781003384236-5

considerations of the computation of minds and machines illustrates the functional significance of the causation of computation and its multiple physical realizability in the world.

The cultural computation of the mind demonstrates the feasibility of the core processing of the mind as a functional basis of cultural capacities. The cultural processing of the mind demonstrates a higher-level function of mentality and thought that is essential for cultural capacities. The functional performance of the mind as a functional basis of cultural capacities illustrates the cultural processing of the mind that is fundamental to the cultural mind and its functional significance. Cultural capacities contribute to the functional processing that is essential for the functional performance of culture. The understanding of the cultural processing of the mind illuminates on the core processing of cultural capacities that comprise cultural computation.

The feasibility of the cultural computation of the mind implies the plausibility of cultural machine computation. The notion of the computing machine as a functional basis of cultural computation suggests that the functional capabilities of the computation of minds and machines have multiple realizability. Cultural processing as a higher-level function implies a level of complexity of machine computation. The cultural processing of biological organisms demonstrates how the adaptive machinery of cultural species serves as a mechanistic basis of culture as a biological computing machine. The considerations of cultural machine computation contribute to the broader understanding of the functional role of adaptive machinery as a functional and mechanistic basis of cultural capacities.

The goal of the chapter is to provide a comprehensive review of cultural machine computation. Research on cultural machine computation explores the notion of culture as a computing machine. The study of cultural machine computation elaborates on the functional basis of cultural computation and its causal–functional significance. The understanding of the functional and mechanistic basis of culture as a computing machine provides insight into the adaptive machinery of biological organisms. The cultural computation of minds and machines is foundational to the study of the cultural computation of biological organisms as a cultural species.

Cultural Machine Computation

Cultural machine computation considers the cultural computation of the computing machine and its functional significance. The cultural computation of machines connotes on the feasibility of the computing machine as a functional and mechanistic basis of the functional performance of culture. From biological organisms to synthetic devices, cultural machine computation entails the understanding of the functional and mechanistic basis of cultural processing. The notion of culture as a computing machine encompasses considerations of the functional performance of the computing machine and its capabilities for cultural production. The understanding of cultural machine computation provides a foundation for the further consideration of adaptive machinery as a functional and mechanistic basis for the cultural computation of biological organisms as a computing machine.

The cultural computation of the computing machine implies that minds and machines show the functional capabilities of cultural performance. The functional capabilities of the computing machine as a conduit of the functional performance of culture illuminates on the physical states of the computing machine as a basis for cultural states. The physical states of the computing machine serve as a mechanistic basis of computation. The consideration of the physical states of the computing machine as cultural states illustrates how the mechanistic states of the computing machine enact the functional performance of culture. The physical states of computation demonstrate the feasibility of culture as a computing machine.

The adaptability of the computing machine as a mechanistic basis of cultural performance is essential to the functional basis of physical and cultural states. The cultural computation of the computing machine implies the physical states and its transitions as a basis of cultural states. The functionality of the computing machine as sets of physical states and its functional performance encompasses a broad range of physical states and physical state transitions that demonstrate functional capabilities. The functional capabilities of the computing machine to show the functional significance of physical states as cultural states entails the functional basis of sets of physical and cultural states. The cultural computation of minds and machines show the adaptability of function that is fundamental to cultural performance.

The physical instantiation of minds and machines as computing machines illustrates how machine computation is a functional basis of cultural performance. The physical realizability of culture as machine computation shows how sets of physical states are necessary and sufficient as a functional basis of cultural performance. The physical states of computing machines illustrate a mechanistic basis for the transformation of information processing into cultural production. The transformation of information processing as physical states to the functional performance of culture elaborates on the causal role of algorithmic function on the production of physical states as cultural states. The computation of minds and machines shows the functional significance of algorithmic function as a basis of the transformation of initial states into final states of cultural output.

The representation and transformation of information processing as machine states demonstrate a functional basis for the cultural computation of minds and machines. The physical states of machine computation show

how information processing arises as a set of transitions of initial states to final states. The production of information processing demonstrates how physical state transitions facilitate the transformation of initial to final states of machine computation. The cultural processing of machine computation implies the sets of physical states that undergo transformation from initial to final states of machine computation and its functional performance. The cultural computation of minds and machines suggests that the sets of physical state and physical state transitions are sufficient as a mechanistic basis of cultural performance and its functional significance.

The cultural computation of minds and machines entails the understanding of culture as a computing machine. The notion of culture as a computing machine implies a physical realizer of culture that demonstrates the functional performance of cultural capabilities. The feasibility of a physical realizer of culture implies that cultural computation is a set of functional capabilities that manifests as sets of physical states. The cultural computation of minds and machines implies the sets of physical states as cultural states and its transformation from initial to final states. The functional performance of computing machines as physical realizers of culture shows the efficiency and accuracy of functional performance based on the physical realizability of the computing machine.

The cultural computation of minds and machines demonstrates how high-level processes contribute to the functional basis of cultural capacities. The cultural processing of the mind shows how cognition and high-level processes are essential to the functional performance of cultural capacities. The core capacities of cultural processing illustrate how cognition and higher-level processes facilitate functional processes that show the complexities of core capacities. From memory and cognition to language and culture, the core capacities of cognition and higher-level processes serve as a functional basis of the cultural computation of the mind. The cognition and higher-level processes of memory and language facilitate the functional performance of mental thought and function that is essential to culture and behavior.

The functional performance of computing machines shows levels of capabilities based on the functional components of computation. The computation of computing machines benefits from functional capabilities that show a multitude of functional performance. The levels of capabilities of computing machines illustrate the importance of functional performance that is based on the design of functional capabilities. The integration of functional capabilities contributes to the high-level performance that is of importance to the functionality and adaptability of the computation of minds and machines. The integra components of levels of capabilities illustrate how the performance of functional components facilitates machine computation. The functional capabilities of high-level performance are essential to the complexities of the computation of minds and machines.

The levels of capabilities of computing machines illustrates the functional role of internal components as a functional and mechanistic basis of high-level performance. Internal components of machine computation facilitate the levels of capabilities that are essential to computation. The functional components of machine computation such as memory and central processing facilitate the capabilities of machines that learn from experience and demonstrate problem-solving. Functional components of computing machines illustrate the capabilities of computation that show functional performance of the efficiency of high-level computing. The internal parts of machine computation contribute to the functional basis of high-level performance that is essential to computing machines.

Other internal components of machine computation that are essential to high-level performance include the capabilities of programming. Machines that demonstrate programming of functional capabilities show how internal components contribute to the functional performance of machine computation. The programming of basic to algorithmic function of machine computation illustrates the capabilities of the computing machine for machine intelligence. The high-level performance of computing machine demonstrates the algorithmic functioning of computers that shows the capabilities of programming their own functions. The programming of functionality demonstrates the adaptability of computing machines for the performance of machine intelligence and algorithmic function.

The functional capabilities of computing machines for distributed processing further demonstrate the role of internal components as a mechanistic basis for the performance of high-level capabilities. The internal components of distributed processing facilitate the distribution of central processing that facilitates parallel functioning that is of importance to the efficiency of information processing. The parallel functioning of distributed processing illustrates the capabilities of machine computation for multiple information processing streams. The distributed functioning of memory and central processing units that comprise computation facilitates the high-level efficiency of functional performance. The distributed processing of memory and information is essential to the high-level performance of computing machines.

The internal architecture of machine computation serves as a conceptual basis from which to understand the functional machinery of the computation of minds and machines. The functional capabilities of computing machines show the multiple physical realizability of computing machines that perform simple to complex functions from mentality and thought to programming and intelligence. The internal components of machine computation illustrate the functional machinery that is essential to the performance of mentality and intelligence. The feasibility of the computation of machines connotes on the notion of computing machines as a demonstration of the plausibility of high-level performance from the functional machinery that is integral to the causal mechanisms in the structure of the world.

The cultural computation of the computing machine implies that the functional performance of machine computation is sufficient as a physical realizer of culture in the world. Cultural computation entails the functional use of internal components of machines for the purpose of cultural performance. The functional capabilities of computing machines for cultural computation demonstrate the functional role of internal components as a mechanistic basis of mental function and intelligence. The functional architecture of machine computation demonstrates the functional significance of internal components as a mechanistic basis of cultural performance.

Culture as a computing machine entails the functional components that are essential to the computation of culture. The internal architecture of machines such as memory and central processing serve as a mechanistic basis for the functional performance of cultural computation. The functional role of memory and central processing of computing machines demonstrates the functional capabilities of the computing machine for the functional performance of culture. Other functional components such as the capabilities to learn from experience and to perform problem-solving show how the computing machine shows the functional performance of high-level function. The functional components of computing machines demonstrate how the internal architecture of machine computation is sufficient for the high-level function of cultural computation.

The demonstration of the capabilities of the computing machine for cultural performance details how the internal architecture of machines is of importance to the high-level function of culture. The internal parts of machine computation demonstrate the capabilities of machine computation for cultural performance. The computing machine as comprised of memory and central processing with functional capabilities for learning and problem-solving is sufficient as a functional basis of cultural performance. The computing machine shows that the level of capabilities of mentality and thought are a functional basis of the high-level performance of culture. The demonstration of the cultural computation of the computing machine shows how the computer is a physical realizer of culture.

The cultural computation of the mind as a computing machine connotes on the functional capacities that are fundamental to the performance of culture. The cultural mind as a physical realizer of culture implies that the functional processes of mentality and thought are sufficient as a functional basis of cultural performance. The functional processes of the cultural mind comprise mental function and higher-level processes that are essential to cultural capacities. The mental function of thought and intelligence demonstrates the functional role of higher-level processes as a functional basis of the cultural mind. The physical realizability of cultural computation of the mind illustrates the functional role of mental function as a functional basis of cultural performance.

The cultural computation of the mind entails the functional processes and mechanisms of mentality and thought. The functional basis of the cultural mind encompasses a wide range of core processes that are essential to cultural capacities. Functional processes of the cultural mind demonstrate how cultural processing illustrates levels of function that detail the causal–functional role of mentality and thought on cultural capacities. The cultural computation of the mind encompasses processes and mechanisms that contribute to cognition and higher-level processes. The functional processes of cognition and higher-level function detail the functionality and adaptability of the cultural mind.

The cultural mind as a computing machine shows how functional processes are sufficient as a physical realizer of culture. The functional processes of cultural capacities detail how the functional components of cultural tasks demonstrate the core capacities for task-based activity. The functional components of task-based activity that is cultural show how the computational level of functional processes illustrates the information processing that is essential to cultural capacities. The functional basis of cultural capacities illustrates the functional components of the cultural mind that contribute to the core capacities of higher-level function.

Cultural processing shows how mental processes facilitate the functional basis of cultural capacities. The functional processes of cultural capacities are of importance to the processing of cultural information. The functional components of task-based activity that are cultural contribute to the cultural processing of information that is essential to the problem-solving of cultural tasks. The processing of cultural information as cultural input furthers the computation of core capacities for the production of cultural output. The functional performance of cultural input–output relations illustrates the cultural processing of core capacities.

The functional components of cultural capacities rely on the functional processing of cultural tasks. The functional processing of task-based activity that is cultural shows the processing of cultural tasks that is based on algorithmic function. The functional performance of cultural tasks that is algorithmic demonstrates the functional use of rule-based tasks as a functional basis of cultural information processing. The functional role of rule-based algorithms for the problem-solving of cultural tasks shows the cultural processing of algorithmic function that is of importance to the problem-solving of rule-based cultural task activity. The functional processing of cultural tasks that is rule-based shows the processing of cultural information that is specific to algorithmic function and its higher-level processing.

The functional processing of cultural task activity shows the types of cultural processing that are of importance to the functional performance of cultural capacities. The algorithmic processing of cultural task activity demonstrates the types of cultural processing that facilitate the representation and transformation of cultural information. The cultural processing of algorithmic function details the functional performance of task activity that is specific to culture. The cultural specificity of algorithmic functioning shows the types of cultural processing that facilitate the transformation of cultural representation into cultural task performance. The functional processing of cultural task activity depicts the processing of cultural information that is specific to cultural capacities.

The computational level of cultural capacities demonstrates the functional significance of cultural processing. Cultural processing is essential to the functional significance of cultural computation. Cultural processing facilitates the representation and transformation of cultural information into cultural input–output. The representational content of cultural information processing shows how the functional states of mental processes contribute to the transformation of cultural information into cultural output. Cultural processing includes the representation and interpretation of the functional states of core capacities. The representation and transformation of cultural information are essential to the functional significance of cultural processing.

The cultural representation of information processing shows the information processing of cultural capacities. The representation and interpretation of cultural information demonstrate how functional components transform cultural input into cultural output. The interpretation and reinterpretation of cultural information illustrate the multiple constraint satisfaction of information processing mechanisms. The representation and interpretation of cultural processing show how the interpretation of cultural states of information satisfies the multiple constraints of layers of information processing. The representation and transformation of cultural information demonstrate the functional role of cultural processing on cultural capacities.

The cultural processing of the mind shows how the mind is a physical realizer of culture. The cultural information processing of functional processes details how the mind is a computing machine. The cultural computation of the mind demonstrates the functional capacities of processes and mechanisms that contribute to the performance of culture. The cultural processing of the mind is of importance to the representation and transformation of cultural information into cultural input–output. The functional processes of the cultural mind are essential to the processing of cultural information. The functional basis of cultural capacities encompasses the processes and mechanisms that are essential to the cultural information processing of the mind and its functional significance.

The cultural computation of biological organisms further demonstrates the functional role of adaptive machinery as a biological computing machine. The adaptive machinery of biological organisms shows the functional adaptation of cultural capacities. The functional adaptation of cultural capacities illustrates the functional significance of adaptive machinery for the higher-level processes of cultural computation. The functional role of adaptive machinery as processes and mechanisms that show cultural capacities illuminates on the functional purpose of multilevel adaptation. The functional processes and mechanisms of cultural capacities demonstrate the feasibility of adaptive machinery as a physical instantiation of the cultural computation of biological organisms.

The manifestation of the cultural computation of adaptive machinery of biological organisms illustrates the processes and mechanisms of cultural adaptation. The functional processes and mechanisms of cultural capacities show the causal–functional role of cultural processing on multilevel adaptation. The functional adaptation of cultural processing serves as a causal mechanism of the adaptive machinery of cultural capacities. The adaptive machinery of cultural capacities facilitates the causality of processes and mechanisms that comprise cultural capacities. Cultural processing as a causal mechanism demonstrates the causal–functional significance of cultural capacities as adaptation.

The demonstration of the cultural computation of complex organisms as biological mechanisms shows the functional significance of multilevel adaptation. Cultural computation at the computational level of biological mechanisms implies the causal–functional role of processes and mechanisms as a mechanistic basis of cultural capacities. The functional processes of biological mechanisms show the states of biophysical mechanisms that depict the functional relations of cultural adaptation. The processes and mechanisms of cultural capacities detail the functional significance of cultural computation as a higher-level process of cultural adaptation. The cultural computation of biological mechanisms demonstrates the functional role of cultural capacities on multilevel adaptation.

The cultural capacities of biological organisms demonstrate the cultural computation of the biological computing machine. The computational level of biological mechanisms reveals the functional components that comprise the core capacities of the biological computing machine. The cultural computation of the biological computing machine shows the levels of functional performance that illustrate the functionality and adaptability of simple to complex organisms as a physical realizer of culture. The functional components of cultural processes and mechanisms detail the internal architecture of culture as a biological computing machine. The understanding of the adaptive machinery of simple to complex organisms as cultural species is essential to the understanding of the cultural computation of the biological computing machine.

Culture as states of biological mechanisms depicts the functional role of the cultural processing of biological mechanisms. Cultural states as the information processing of biological mechanisms detail the functional relation of causal mechanisms that comprise the cultural processing of biological organisms. The functional relation of cultural states as biological states shows the functional correspondence of the information processing of causal mechanisms. The biological mechanisms of cultural capacities detail the adaptive machinery that is essential to cultural processing. Biological mechanisms as cultural information processing show how biological states encompass the functional states of cultural processes. The understanding of cultural states as part and particular of biological mechanisms reveals the functionality of the cultural processing of biological mechanisms.

Cultural states as brain states illustrate the causal–functional role of the brain as a mechanistic basis of cultural capacities. Cultural states as brain states demonstrate how the functional mechanisms of the brain correspond with cultural states. The cultural states of information processing show correspondence with the functional processing of brain states. Cultural states of brain mechanisms detail the functional processes that contribute to the cultural processing of the brain. The cortical mechanisms of the brain serve as a functional basis of the processing of cultural information. The cultural processing of brain mechanisms depicts the causal–functional role of the brain as a mechanistic basis of cultural capacities.

The functional correspondence of cultural states as brain states is of importance to the understanding of culture as a biological computing machine. The functional relation of cultural states and brain states implies the machine states that facilitate the functional performance of culture. Brain states serve as a mechanistic basis of the functional performance of cultural computation. The functional significance of cultural states as brain states shows the machine states of cultural computation and its causal effects. The machine states of cultural computation belie the causal–functional significance of the functional performance of culture. The demonstration of cultural states as brain states illustrates how culture manifests as the functional performance of the biological computing machine.

The manifestation of culture as brain states illustrates the emergent properties of the brain basis of culture. Culture as brain states shows how functional mechanisms reveal emergent properties based on the cultural computation of the brain. Culture as brain states and its functional and emergent properties demonstrates the functional significance of the cultural computation of the brain. The functional and emergent properties of the cultural brain show the cultural states of the brain that arise based on the cultural computation of the brain. The manifestation of cultural states as brain states reveals the sets of cultural states that serve as emergent properties of the brain.

The brain basis of culture illuminates on the adaptive machinery of biological organisms as a cultural species. The brain as adaptive machinery shows the functional adaptation of biological organisms as a cultural species. The brain basis of culture demonstrates the structural and functional properties of cortical brain organization that are of importance to cultural capacities. The brain basis of culture demonstrates how the structural and functional properties of the brain are essential to cultural processing. The cultural processing of cortical brain mechanisms shows the functional processes that contribute to the cultural capacities of biological organisms as cultural species.

The adaptive machinery of biological organisms as cultural species shows how the cognition and higher-level processes of the brain serve as a functional adaptation of cultural capacities. The cognition and higher-level processes of the brain demonstrate the functional and mechanistic basis of cultural capacities. The functional relation of cultural states as brain states depicts the functional role of cultural processing as causal mechanisms of the brain. Cultural states as brain states illustrate the functional significance of cultural processing as an integral part of the causal effects of brain mechanisms. The cognition and higher-level processes of the brain demonstrate how the functional and mechanistic basis of cultural capacities contribute to multilevel adaptation.

The cognition and higher-level processes of the brain further demonstrate the functional adaptation of complex organisms as cultural species. The brain basis of cognition and higher-level function details the functional mechanisms that contribute to cultural processing. The functional mechanisms of cognition and higher-level processes show how the processes and mechanisms of the brain facilitate the processing of cultural information. The higher-level processes of cognition show the brain states that are concordant with cultural information processing. The functional adaptation of cognition and higher-level processes facilitate the causal mechanisms of the brain as a mechanistic basis of the functional adaptation of complex organisms.

The functional states of cultural processes demonstrate the cultural processing of functional mechanisms. The functional states of cultural processes and its causal effects show how cultural processing contributes to the emergent properties of functional mechanisms. The functional basis of cultural processes shows how cultural states are functional states of causal mechanisms. The cultural processing of functional states details the cultural states that arise based on the emergent properties of functional mechanisms. Cultural processing as functional states facilitates the understanding of the functional basis of cultural processing.

Culture and Machine State Functionalism

The consideration of culture as a system with functional significance entails the understanding of culture as minds and machines. The study of culture and machine state functionalism implies the scholarly and scientific inquiry into the cultural states of minds and machines and their functional purpose. The functional basis of cultural capacities entails the demonstration of the cultural states of the mind and machine. Culture as systems of mental and physical states encompasses a wide realm of causation as a functional basis for the causal–functional role of other mental and physical states. The mind as a computing machine shows a functional performance of culture that facilitates the understanding of culture as minds and machines. The postulation on the functionalism of the culture of minds and machines is essential to the philosophical understanding of the functional purpose of culture as systems with causal–functional significance.

The notion of culture and machine functionalism entails the understanding of culture as sets of mental states with causal–functional significance. Culture as mental states serve as a causal impetus for other mental and physical states with functional purpose. The functional basis of culture as mental states shows causation to other mental and physical states that are of importance to the functional performance of culture. Cultural states are not only mental states as physical states, but also mental and physical states with functional purpose. The functionalism of cultural states broadens the functional relation of mental and physical states to consider the causal effects of cultural mental and physical states as part of the functional performance that encompasses the sets of mental and physical states of culture.

Culture as sets of machine states entails physical states of the computing machine and its transitions. The cultural computation of machine states demonstrates the functional performance of physical states and its state transitions that are based on functional relations that are predetermined. The deterministic performance of machine states illustrates the functional basis of cultural machine computation. Cultural computation arises based on the functional performance of sets of physical states that show the functional relations of cultural input–output that is deterministic. The sets of physical states and machine state transitions show the levels of capabilities of the computing machine for the functional performance of cultural computation.

Cultural machine computation also encompasses the machine states that show the conditional performance of the computing machine. The determination of the sets of physical states and machine state transitions as conditional or probabilistic to the satisfaction of specific criteria illustrates the conditional performance of machine states that is based on machine computation. Cultural machine computation that is conditional implies that the production of machine states is based on the presence or absence of specific conditions or set of criteria. The cultural computation of conditional performance shows the feasibility of machine computation that performs at a predetermined rate rather than as a deterministic production of machine states.

Cultural computation as machine states demonstrates the functional significance of cultural states as physical states of the computing machine. The determination of the physical states of machine computation as a functional performance of cultural input–output relations is necessary and sufficient as a precursor to cultural machine computation. The functional purpose of cultural states as physical states shows how physical states of machine computation facilitate the performance of cultural input–output relations. The manifestation of machine computation as physical states illustrates the functional performance of the cultural machine computation. The states of machine computation show how cultural states as physical states illustrate the functional performance of computing machines.

Culture as mental states implies the cultural states of the mind with functional purpose. The notion of cultural states as mental states encompasses the manifestation of cultural mental states with a function. The cultural states of the mind detail the specific functional processes that contribute to the functional basis of cultural capacities. Culture as mental states shows the functional significance of cultural mental content. Cultural states as mental states illustrate how the functional specificity of mental states serve a functional role. The functional role of cultural mental states as a causal impetus of other mental states that are essential to cultural capacities shows the sets of functional processes that comprise culture.

The functional role of cultural states as mental states implies that cultural mental states show the automation of cultural performance. Cultural mental states that are based on the computation of machine states show the automatic processing of culture. Cultural states as the automatic processing of mental states reveal the sets of mental states that are based on the functional performance of cultural states that is automatic. The automatic processing of mental states that are cultural shows the functional processes of cultural capacities. The automatic processing of cultural information as mental states facilitates the functional basis of cultural processing that is based on the automation of cultural performance.

Cultural states demonstrate the functional processes that show the importance of the control or regulation of the processing of cultural thought. The control or regulation of the processing of cultural information details the functional role of cultural states as mental states that perform a specific mental function. The functional role of cultural states as the control of information processing illustrates the cultural mental states with functional purpose. The controlled processing of mental states that serve a causal–functional role of other mental states facilitates the functional basis of culture and machine computation. Cultural processing that is controlled demonstrates the causal influence of culture on the functional processes of cultural capacities.

Cultural mental states illustrate the casual–functional significance of cultural mental content. Cultural states as mental states show the mental causation of culture. Cultural mental states serve as a functional basis of other mental states that are essential to the inferential reasoning on the mental causation or culture. Culture as mental states implies that cultural states have a functional purpose that contributes to the ascription of intention in mentality and thought. The mental causation of cultural states illustrates how intentionality as mental thought and function imbues cultural states with mental causation. The manifestation of cultural states as mental states with causal effects illustrates how cultural mental states have functional purpose.

The implicature of cultural states as mental states with functional purpose demonstrates how intentional thought as mental causation contributes to the manifestation of mental states with functional purpose. Culture as mental states that show causal effects of intentional thought implies the manifestation of cultural states that are a causal mechanism in interaction with the world. Cultural states as intentional states demonstrate the mental causation of intentional thought with functional purpose. Cultural states as mental states of intention show the causal effects of intentional thought on action that is interactional and responsive to the world. The manifestation of culture as mental states depicts the causal effects of mentality and thought that have functional purpose.

Culture as phenomenological states shows the functional significance of cultural states. The phenomenological states of culture illustrate the qualia and experience that comprise the functional processes of cultural capacities. Cultural states as phenomenological states depict how the states of qualia and experience serve as a functional basis of cultural capacities. The phenomenological states of culture as the qualia and experience of mental states demonstrate the functional relation of mental states and phenomenological experience. The functional correspondence of mental states and phenomenological states demonstrate the functional dimensionality of the mental states of culture. Cultural mental states encompass the wide realm of phenomenological states that are essential to the high-level processing of culture.

The phenomenological states of culture show the functional relation of qualia and experience as cultural states. Culture as states of qualia and experience shows the functional relation of cultural mental states and cultural states of qualia and experience. The determinism of the functional relation of cultural states and states of qualia and experience reveals the functional significance of culture as mental states. Culture as mental states serves as a functional basis for the states of qualia and experience that imbue shared meaning and purpose to being and existence. The determination of cultural states as mental states relies on the functional relations of cultural states with states of qualia and experience that show the generalizability of the causal effects of cultural states in the world.

Culture as phenomenological states show the functional role of qualia and experience. Cultural states as phenomenological states depict the states of qualia and experience that comprise the high-level processing of biological organisms as cultural species. Culture as phenomenological states detail the functional significance of cultural states as an adaptive byproduct of multilevel adaptation. The manifestation of culture as phenomenological states shows how qualia and experience illustrate the high-level processing of cultural adaptation. The states of qualia and experience facilitate cultural states that comprise the mental experience of biological organisms as complex species. The phenomenological states of cultural states depicts the functional purpose of qualia states as mental experience.

The functional purpose of culture as a functional basis of mental experience shows the causal–functional significance of high-level processing. Culture as mental experience demonstrates the functional adaptation of biological organisms as a cultural species. Culture as mental experience serves as a functional basis of the emergent properties of the cultural mind. The functional adaptation of mental experience serves as a source of knowledge that is of importance to the emergence of the functional basis of the cultural mind. The functional purpose of culture as mental experience facilitates the high-level processing of cultural capacities.

The conceptualization of culture as mental experience serves as a functional basis of the mental causation of culture. The states of qualia and experience serve as a causal influence on the mental computation of culture. Cultural states as states of experience imbue the computation of minds and machines with thought and mentality that are based on the emergent properties of mental experience. The functional performance of minds and machines based not only on thought and mentality that are predetermined, but also responsive to the emergent properties of qualia and experience that shows how cultural states facilitate the causal effects of the states of qualia and experience. Culture as states of qualia and experience serve as a functional basis of the causal effects of mental states that show the phenomenological states of cultural capacities.

Culture as mental and physical states with functional significance shows the functional relation of cultural states of the mind and of the world. The manifestation of culture as mental and physical states shows the functional correspondence of the mentality and thought of cultural capacities as the computation of machine states. The cultural processing of the mind and brain as mental and physical states shows the functional relation of mental states and brain states of culture. The functional mapping of mental and physical states shows the one-to-many correspondence of culture as mental and brain states. The functional relation of culture as mental and brain states demonstrates the causal role of mental and brain states as a functional mapping of cultural states. The functional basis of cultural states as a multitude of mental and brain states demonstrates the multiple realization of the functional basis of cultural computation.

The functional relation of mental and brain states of culture demonstrates a functional purpose of mental states as a causation of other mental states and brain states. The causation of cultural mental states shows how the cultural mind shows the functional significance of mental and brain states that contribute to the functional performance of cultural capacities. The cultural causation of mental states illustrates the causal role of mental and brain states as a functional basis of cultural processing. Cultural processing as mental and brain states depicts the functional purpose of mental states as a conduit of the processing of cultural information. The functional relation of mental and brain states as a functional basis of cultural processing shows the functional purpose of mental states as a precursor to the mental and brain states of culture.

Culture as behavioral states with functional purpose encompass the sets of cultural states that show functional relation with behavioral states. The notion of culture and machine computation facilitates the demonstration of behavioral states as the final states of cultural computation. Cultural processing as the sets of initial and final states of machine computation detail how mental states of cultural processes facilitate the functional performance of states of behavior. Cultural processing as the functional basis of behavioral states depicts the mental and machine states that serve as a causal mechanism of culture and behavior. The functional performance of cultural processing illustrates the functional purpose of cultural and mental states as a causal mechanism of behavioral states.

Culture as Functional States

The manifestation of culture as functional states demonstrates the functional significance of cultural states as mental states. The transformation of representational content from cultural input to cultural output relies on the cultural processing of information from initial to final states of computation. The functional processing of cultural capacities shows how the cultural processing of information contributes to the changes of cultural states into mental states with functional purpose. The manifestation of cultural mental states as internal states of computation demonstrates the functional processing of cultural capacities.

The casual role of cultural states as a functional basis of mental states shows the functional significance of the cultural processing of information. The representation of sensory stimulus and its transformation into motor output shows the cultural processing of the mind. The changes of cultural states as mental states with functional purpose show the states of mental function that comprise cultural computation. The functional basis of cultural mental states illustrates how mental states enact cultural states with functional purpose. The causal–functional significance of cultural states as a functional basis of mental states illustrates the cultural processing of minds and machines.

Culture as mental states shows the functional role of mental states as a causal impetus of other mental states. Cultural processing of minds and machines comprises the flow of information that encompasses a wide realm of states of cultural computation. The functional role of culture as mental causation implies that cultural states show functional relation with other cultural mental states. The mental causation of cultural states and mental states depicts the functional significance of cultural mental states and their causal effects. Cultural processing as sets of cultural states and their causal effects encompasses the states of mental causation that are of importance to the functional significance of cultural computation.

The functional relation of cultural mental states with other states of the mind encompasses the manifestation of phenomenological states of qualia and experience. Culture as mental states comprises the phenomenological states of qualia and experience that serve as a functional basis of cognition and higher-level function of biological organisms. The phenomenological states of qualia and experience show the mental states that show the functional processing of cultural information. Cultural states as phenomenological states depict how qualia and experience manifest as the functional processing of the cultural environment. The manifestation of cultural states as phenomenological states shows the functional capacities of cognition and higher-level function that are of importance to functional adaptation.

The phenomenological states of qualia and experience comprise a functional basis of cultural states. The states of qualia and experience of cultural capacities show the higher-level function of cognition that is essential to functional adaptation. Cultural states of qualia and experience formulate a functional basis of the phenomenological qualities of cognition and higher-level function. The cultural capacities of qualia and experience contribute to the learning of the cultural environment. The functional processes of cognition and higher-level function contribute to the mental states that facilitate the functional adaptation of biological organisms.

The phenomenological qualities of cognition and higher-level function comprise a functional adaptation of biological organisms as cultural species. The functional adaptation of consciousness and higher-level processing shows how phenomenological states as qualia and experience formulate adaptive machinery of biological organisms. The functional capacities of consciousness and higher-level processing depict the qualities of mental states that correspond with the cultural causation of the mind and brain. The adaptive function of the cultural mind and its functional role in consciousness and higher-level processing shows the functional significance of functional processes on multilevel adaptation.

The states of consciousness and higher-level processing of biological organisms as a cultural species show the adaptive preparedness of biological organisms for multilevel adaptation. The functional basis of consciousness contributes to qualities of mental states that comprise cultural capacities. The functional processes of consciousness facilitate the qualia and experience of mental states with causal–functional significance. The functional capacities of biological organisms for the cultural causation of the mind and brain are essential to the adaptive function of higher-level processing. The functional basis of consciousness and higher-level processing demonstrate the functional significance of consciousness and qualities of mental states as adaptation.

Cultural states also serve as a functional basis of biological mechanisms. Culture as mental states show the causal role of functional states as a precursor to the computation of biological mechanisms. Cultural mental states comprise the cultural processing that facilitates the functional basis of biological mechanisms. The changes of mental states that comprise cultural processing demonstrate the functional basis of cultural mental states as a causal influence on biological mechanisms. The functional role of cultural mental states and its casual–functional significance is of importance to the understanding of the functional basis of biological mechanisms.

Cultural mental states show how biological states change as a function of mental states. The changes of mental states serve as a causal mechanism of changes to biological states. Culture as mental states shows the causation of mental states and biological states. The functional role of mental states as a causal mechanism of biological states demonstrates how mental states enact changes to biological states. Cultural states as mental states show the causal effects of culture as a causal mechanism of biological states.

The changes of biological states also demonstrate how mental states serve as a causal influence on the changes of cultural states. The manifestation of biological states demonstrates the causal–functional role of functional mechanisms as a mechanistic basis of cultural processing. The changes of biological states illustrate how the information processing of functional mechanisms facilitates the performance of cultural capacities. The levels of processing of biological mechanisms contribute to the functional processing of cultural capacities. The changes of biological states show the functional basis of mental states as a causal influence on cultural processing.

Culture as a Biological Computing Machine

The study of culture as a biological computing machine facilitates the understanding of the cultural patterns of the functional basis and biological mechanisms of multilevel adaptation. The conceptualization of culture as a biological computing machine highlights the causal role of biological mechanisms as a functional basis of cultural processing. The cultural patterns of biological mechanisms demonstrate a culture as a physical realizer. The biological basis of culture depicts the computational level of functional processes and causal mechanisms that contribute to cultural processing. The cultural processing of biological mechanisms facilitates the functional processes that are of importance to the adaptation of biological organisms as a cultural species.

Culture as biological states entails the understanding of the functional mapping of cultural and biological states. The functional mapping of cultural and biological states implies a functional relation of culture as biological states. The notion of biological states as a physical realizer of culture encompasses the states of biological mechanisms that contribute to the cultural patterns of biological adaptation. The characterization of the multilevel mechanisms of cultural processing shows the functional basis of cultural and biological mechanisms. The demonstration of the cultural patterns of biological mechanisms illustrates the functional role of basic mechanisms as a conduit of the cultural processing of complex organisms.

The cultural patterns of biological mechanisms further demonstrate the emergent properties of biological mechanisms that comprise cultural processing. The cultural patterns of biological mechanisms show the emergence of the functional processes and mechanisms of cultural processing. The emergent properties of biological mechanisms and cultural processing facilitate the manifestation of the cultural capacities of complex organisms. Emergent properties reflect the cultural processing of biological organisms that show causal–functional significance. The emergence of cultural processing as a higher-level function shows the cultural patterns of biological organisms that hold.

The cultural computation of the biological computing machine considers the functionality and causality of the cultural processing of biological mechanisms. The functionality of cultural processing consists of the processes and mechanisms that comprise the machine states of biological mechanisms. The cultural computation of biological organisms details the causal role of functional mechanisms as a conduit of cultural processing. The functional capacity of biological mechanisms for cultural processing illustrates a mechanistic basis of the functional purpose of cultural adaptation. The causal mechanisms of cultural computation further show the functional significance of biological mechanisms as a physical realizer of culture.

Cultural computation is a higher-level process of the cultural capacities of biological organisms. The adaptive machinery of biological organisms demonstrates the functional adaptation of biological organisms in response to the alteration or change of the cultural environment. The functional adaptation of biological organisms shows the cultural processing that comprises the higher-level processing of cultural capacities. The adaptive machinery of biological organisms demonstrates the functional basis of causal mechanisms that comprise cultural information-processing mechanisms. The characterization of the cultural computation of biological mechanisms shows the higher-level processing of cultural capacities as a functional adaptation.

The adaptive machinery of the brain shows the functional architecture of biological organisms as a cultural species. The structural and functional organization of the brain details the fundamental principles of cortical brain organization that contributes to the understanding of the multilevel processing of cortical brain organization. The adaptive machinery of biological organisms illustrates the functional processes and basic mechanisms of cortical brain organization that contribute to cultural adaptation. The cortical brain serves as a biological basis for the cultural capacities of complex organisms. The functional mechanisms of the cortical brain demonstrate the internal architecture of biological organisms that is essential to the neurocomputation of cultural capacities.

The cultural patterns of brain function detail the functional processing of cortical brain mechanisms that show functional relation with cultural capacities. The functional processing cortical brain mechanisms illustrates the cultural patterning of brain function that shows functional correspondence with behavior. The cultural patterns of brain–behavior relationships illustrate how the functional processing of cultural capacities arise from cortical brain function. The functional processes of cultural capacities detail the specific functional relations that are of importance to the characterization of the cultural patterning of brain–behavior relationships. The functional relation of cultural and cortical processing shows multitude of cultural patterns that comprise the functional relations of culture, brain and behavior.

From neurons to networks of neurons, the multilevel processing of cortical brain organization shows the functional processes and mechanisms that contribute to cultural adaptation. The representation of cultural information as patterns of cortical brain function shows the functional significance of the functional processing of multilevel mechanisms. The characterization of the multilevel mechanisms of cultural processing details the causal–functional role of cortical brain function on the functional processing of culture, brain and behavior. The demonstration of the causal–functional role of cortical brain function on cultural processing illustrates the functional significance of processes and mechanisms on multilevel adaptation.

The characterization of the multilevel mechanisms of cortical brain function and cultural adaptation detail the functional processes and mechanisms of cultural processing. Neurons comprise the biophysical mechanisms that are essential to the electrophysiological signaling of information processing mechanisms. The cultural patterns of neuronal mechanisms demonstrate the encoding and decoding of the cultural representation of the environment. The encoding and decoding of cultural information as patterns of neuronal mechanisms demonstrates the cultural processing of neuronal mechanisms. The patterns of neuronal mechanisms show the representation and transformation of cultural information processing of the environment.

The representations of neuronal mechanisms demonstrate the cortical representation of cultural information processing. The patterns of neuronal response of multilevel mechanisms characterize the cortical representation of the cultural environment. The cultural processing of cortical mechanisms the encoding and decoding of cultural information across cortical brain areas. The cortical representation of cultural information processing contributes to the patterning of neuronal response of multilevel mechanisms. The cortical representation of cultural information processing mechanisms facilitates the coding of cultural information. The cultural processing of neuronal mechanisms shows the patterning of neuronal response that is of importance to multilevel adaptation. The cultural patterns of neuronal response depict the mechanistic basis of cortical representations of the cultural environment. The cultural processing of cortical mechanisms facilitates the cultural information processing of the environment.

The cortical representation of cultural processing that comprises neuronal response demonstrates the neurocomputation of cultural information. The patterns of neural coding as internal representations of the cultural environment demonstrate the neurocomputation of cultural information. The computation of neuronal response shows the functional relation of machine states as neuronal activation states. The patterns of neuronal activation demonstrate the cultural computation of neurons as states of neuronal activation. The representation and transformation of cultural processing as states of neuronal activation shows the patterns of neuronal response that comprise cultural neurocomputation. The cortical representation of cultural processing as neuronal response shows the machine states that comprise the computation of cultural information processing.

The patterns of neuronal mechanisms show the functional role of neuronal activation as causal mechanism of states of neuronal coactivation. The patterns of activation of neurons demonstrate the causation of states of neuronal activation to states of neuronal coactivation. The coactivation of the patterns of neuronal response demonstrates the facilitation of levels of processing that are based on states of neuronal coactivation. The states of neuronal activation to states of neuronal coactivation describe the causal mechanism of patterns of neuronal response and its functional significance. The patterns of neuronal mechanisms as states of activation to states of coactivation that serve as a functional basis of levels of processing illustrate the states of machine computation that comprise higher-level processes.

Cultural patterns of neuronal mechanisms facilitate the states of activation that comprise learning and experience. The cultural patterns of neuronal activation show the causation of states of neuronal activation to states of neuronal coactivation. The cultural patterns of neuronal response demonstrate the types of processing that facilitate states of neuronal activation. The cultural processing of neuronal activation states shows the types of processing that contribute to the facilitation of the states of coactivation of neuronal mechanisms. The states of neuronal coactivation of cortical mechanisms describe a functional basis for the processing of cultural information.

The states of neuronal coactivation as cultural patterns of information processing illustrate a functional basis for the facilitation of cultural processing that is of importance to cultural learning and experience. The cultural patterns of neuronal coactivation describe the states of coactivation of neuronal mechanisms that are concordant with the information processing of the cultural environment. The cultural processing of learning and experience contributes to the facilitation of information processing that is essential to the functional performance of cultural capacities. The cultural patterns of information processing demonstrate the facilitation of functional performance that arises based on the levels of cultural processing. The states of coactivation of neuronal mechanisms demonstrate the facilitation of levels of cultural processing that are of importance to the functional performance of cultural capacities.

The levels of cultural processing comprise the functional performance of cultural capacities that show a causal influence on the states of coactivation of neuronal mechanisms. The functional performance of cultural capacities describes states of mental activity on cultural tasks that are concordant with the states of coactivation of neuronal mechanisms. The cultural processing of task activity details the mental states that are of importance to cultural learning and experience. The levels of processing of cultural capacities show the mental states and its functional role on the patterns of coactivation of brain function and higher-level processes. The functional performance of cultural processing depicts the states of mental activity that correspond with brain function and higher-level processing. The cultural processing of task activity shows the mental states that are essential to cultural learning and experience.

The networks of neurons of cortical brain regions show the functional patterns of cortical brain activity that comprises cultural processing. Neural networks characterize the structural and functional connectivity of cortical pathways that are of importance to cultural capacities. Cultural patterns of neural networks describe the cortical brain activity that comprises cultural processing. The functional patterns of cortical brain activity describe the population-level activity of cortical brain regions that show functional significance to the cultural processing of core capacities. The functional processing of cortical brain activity is of functional significance to the interconnectivity of cortical pathways and their functional relation to cultural capacities.

Cultural patterns of functional brain activity show the processing of cultural information. The patterning of functional brain activity illustrates the cultural information processing of neural networks. The cultural processing of patterns of functional brain activity demonstrates the relaying of information that comprises the functional pathways of behavior. The functional pathways of culture and behavior detail the role of functional brain activity as a mechanistic basis of the levels of processing of cortical brain regions and their interconnection. The levels of processing of cortical brain regions are of importance to the cultural processing that characterize patterns of functional brain activity. Cultural patterns of functional brain activity are of importance to the understanding of the functional and mechanistic basis of cultural processing.

The cultural patterns of cortical brain function show the states of activation that comprise the processing of cultural information. The activation states of cortical brain regions reflect the levels of processing that are of importance to the functional processing of culture and behavior. The states of functional activation show the depth of processing that is of importance to the pattern completion of cultural processing. The functional patterns of cortical brain regions demonstrate the activation states of cultural processing. The functional patterns of cortical brain regions show the states of brain activation that comprise cultural capacities. The cultural patterns of cortical brain function comprise sets of activation states that facilitate the information processing of cultural capacities.

The brain activation states of cortical brain function show the states of activation that facilitate the initial and final states of the functional performance of cultural processing. The representation and transformation of initial and final states of functional performance as activation states shows the functional relation of machine states as functional brain activation states. During cultural processing, the patterns of functional brain activation states show the representation and transformation of cultural information to completion of the final states of the functional performance of cultural computation. The pattern completion of functional brain activation shows the final states of the computation of activation states that facilitate the completion of cultural processing. The transformation of representational content from initial to final states of computational performance demonstrates the causal role of functional brain activation as machine states of cultural computation.

The functional role of functional brain activation as machine states of cultural computation demonstrates the functional performance of cultural processing. From initial to final states of functional performance of cortical brain mechanisms, the computation of functional brain activation states shows the patterns of activation that comprise the representation and transformation of cultural information into cultural input–output. The computation of functional brain activation demonstrates the states of activation that facilitate the transformation of the representational content of cultural processing. The transitions from initial to final states of machine computation illustrate the changes of activation patterns that facilitate the final states of cultural processing. The activation patterns of functional brain activation show the states of activation that entail the functional relations of cultural input–output.

The cultural computation of neural networks demonstrates the causal role of brain activation as a precursor to other states of brain activation. The causal–functional role of states of brain activation to other brain activation states shows the functional relation of the initial state of functional brain activation to final states of functional brain activation. The changes of patterns of functional brain activation to other states of functional brain activation demonstrates the causal role of cortical brain activation as activation patterns. The changes of activation patterns of neural networks show functional correspondence with the changes of the patterns of cultural information processing. The changes of activation patterns of functional brain activation show that states of brain activation serve a causal role to other states of brain activation that comprise final states of cultural computation.

The cultural computation of brain activation states shows the functional relation of states of activation with states of cultural processing. The functional relation of states of brain activation and states of cultural processing demonstrates the causal role of functional brain activation on cultural information processing. The functional correspondence of functional brain activation states and mental states of cultural processing details the functional significance of activation patterns on the functional performance of cultural processing. The levels of activation of functional brain states correspond with the levels of processing of cultural capacities. The functional relation of levels of activation of functional brain activation states and levels of processing of mental states illustrates the correspondence of the functional performance of cultural processing.

The functional correspondence of brain activation states to behavioral states details the functional relation of the brain–behavior relationships. The information processing of states of brain activation facilitates the production of states of behavior. The causal role of brain activation states as a conduit of behavioral states shows the functional significance of functional brain activation as a precursor to behavioral states. The characterization of the functional activation patterns that comprise behavioral patterns shows the functional correspondence of brain–behavior relationships. The functional relation of brain states as behavioral states demonstrates the transformation of internal representation into patterns of motor output. The understanding of the functional relation of functional brain activation states and behavioral states is essential to the characterization of the machine states of behavior.

The cultural processing of functional brain activation shows functional relation to culture and behavior. The representation and transformation of cultural information processing into behavioral patterns of cultural expression shows the functional role of activation states as a causal impetus of behavioral states. The processing of cultural information entails cultural patterns of functional brain activation that contribute to the production of cultural patterns of behavior. The cultural processing of functional brain activation illustrates the representation and transformation of cultural processing as cultural input. The demonstration of the functional relation of cultural patterns of functional brain activation and behavior shows the functional significance of cultural processing on brain–behavior relationships.

The cultural computation of biological organisms demonstrates the multitude of physical realizers that comprise cultural processing. The understanding of the mechanistic basis of cultural computation facilitates the characterization of the multilevel mechanisms that comprise cultural capacities. The cultural life patterns of biological organisms show the lawlike principles of culture that hold. From biological mechanisms to cultural life patterns, the cultural patterns and regularities of functional mechanisms entail the functionality and causality of adaptive machinery for multilevel adaptation. The cultural computation of biological organisms shows the adaptiveness and responsiveness of complex organisms as cultural species.

The Cultural Computer as Machine Computation

The multiple realizability of cultural computation implies the characterization of machine states as a mechanistic basis of cultural processing. The understanding of the cultural computation of computing machines further demonstrates the plausibility of the functional performance of culture. From simple devices to modern computers, the manifestation of cultural computation as synthetic devices demonstrates how computing machines serve as a conduit of the functional performance of cultural computation. The demonstration of the level of capabilities of computing machines for cultural performance shows the role of scientific and technological innovation on the research and development of cultural machine computation. The postulation on the design of computational tools and technologies that show levels of capabilities for functional performance of cultural thought and intelligence is of importance to the demonstration of computing machines that perform cultural computation.

The history of computing illuminates on the plausibility of the design of computing machines that demonstrate the functional performance of mentality and intelligence. The development of the computing machine details the functional capabilities that facilitate the performance of simple to complex operations. The computation of machines that show the functional capabilities of mental function is based on the internal architecture of computing machines. The internal components of computing machines illustrate the functional capabilities that are essential to the high-level performance of functional operation. The wide realm of functional operations of computing machines reveals the high-level performance of computation that constitutes mentality and intelligence.

The functional significance of the computing machine as a device that is capable of functional performance demonstrates the real-world advantage of computation as high-level performance. The efficiency of computation that has levels of capabilities shows the importance of functional operation on the functional performance of computing machines. The demonstration of computing machines as devices that demonstrate high-level performance facilitates the plausibility of the design of simple to complex devices that show functional equivalence on the levels of capabilities of computation. The functional equivalence of computing machines with mentality and intelligence details the multiple realizability of computation.

Early computing machines built on mechanical components show the feasibility of mechanical computation as a functional basis of computational performance. Mechanical engines as a model of the early computing machine demonstrates the feasibility of analog representation as a functional basis of the computation of simple to complex functions. The notion of analog representation as a composite of mechanical components that function as computation based on the structural components of the computing machine illustrates the rationale of early computing machines that show the functional performance of computation. The physical states of the structural components of the computing machine as analog representation formulate the functional basis of the performance of computation. The design of the analog computer that performs computation based on analog representation demonstrates how the mechanical states of computing machines and their internal components serve as a functional basis of the functional performance of computation. The demonstration of analog computation as mechanical devices is sufficient as a manifestation of the functional performance of computation.

Mechanical computation as a basis of the computing machine shows the functional use of analog representation that performs cultural computation. The cultural computation of mechanical devices contributes to the understanding of the role of structural components as a basis of the mechanical computation of simple to complex function. The design of mechanical devices that perform algorithmic function that is cultural shows the feasibility of analog computation as a functional basis of the performance of cultural computation. The role of rule-based function as algorithms of analog computation demonstrates the mechanical states of cultural computation. The functional use of mechanical components as analog representation of algorithmic function explores the notion of structural components as a physical realizer of cultural computation.

The early digital computer built on electromechanical components shows the functional components of computing machines that are based on digital computation. The digital computer that performs on internal components of electromechanical units that show high-level computing details how electromechanical states comprise the fundamentals of digital computation. The digital representation as a functional basis of simple to complex operation shows the performance of electromechanical states as high-level computation. The electromechanical states of elemental units such as vacuum tubes demonstrate the relay of data processing that contributes to the high-level computing of digital computation. The level of capabilities of digital computers for high-level computing demonstrates the functional performance of digital computation that is based on the integration of electromechanical components.

The development of the modern computer as a computing machine that shows the levels of capabilities of the functional performance of high-level computation shows the importance of internal architecture on electronic computation. The modern computer with internal components that facilitate memory, central processing and algorithmic function facilitates the high-level performance of electronic digital computation and demonstrates the levels of capabilities of modern computation. The development of internal components of electronic computation that consist of a multitude of functional components ("software", "hardware") connotates on the functional significance of internal architecture as a functional basis of high-level performance computation. The design of internal components that facilitate electronic computation shows the intricacies of the design of internal circuits and hardware that are of importance to the data processing of electronic digital computation.

The development of digital electronic computers as modern devices with capabilities of high-level processing shows the functional performance of cultural computation. The design of modern computing devices comprised of electronic digital representation shows the level of capabilities of internal components that comprise electronic computation. The electronic computation of functional operations and algorithmic function that are cultural shows the functional role of electronic states as a physical realizer of cultural computation. The electronic digital representation of cultural data processing shows the functional use of electronic states for the high-level performance of cultural computation. The design of functional components of computers that perform cultural computation entails the development of internal architecture that shows high-level capabilities for the performance of cultural computation.

The internal components of modern digital computers show the levels of capabilities that comprise high-level computational performance. Modern digital computers that consist of internal components such memory, central processing and autonomous programming show the levels of capabilities of functional performance that are of importance to high-level processing. The innovation of modern computers that demonstrate the high-level processing of digital computation facilitate the functional performance of computation that shows functional equivalence with the high-level performance of mentality and intelligence. The levels of capabilities of modern computers detail the high-level data processing that is essential to the computation of simple to complex operations.

Electronic digital computers that show the capabilities of computation demonstrate the role of digital representation as a functional basis of cultural performance. The design of software that shows the functional capabilities of cultural programming demonstrates the feasibility of internal components that show high-level performance that is cultural. The development of software programs that facilitate the programming of cultural tasks illustrates the functional capabilities of digital computation for the purpose of cultural computation. The research and development of software programs that are based on algorithmic functions ("cultural heuristic", "cultural search engine") that are cultural demonstrate the functional capabilities of cultural programming. The innovation of software programs that further demonstrate the capabilities of computation to perform cultural programming that is autonomous illustrates the role of digital computation as machine states of cultural performance. The functional capabilities of digital computation for functional performance that is cultural illustrates the high-level performance of digital computers that is of importance to cultural computation.

The cultural computer is an example of the design of a simple to complex device with a high-level performance of cultural computation. The design of a modern computer that shows functional capabilities for cultural performance illuminates on the notion of the high-level performance of cultural computation. The functional capabilities of modern computers illustrate the importance of high-level performance as an essential component of the performance of functional operations that is cultural. The design of software and hardware with internal components of digital representation and data processing based on algorithmic function that is cultural demonstrates digital computation that serves as functional basis of cultural computation.

The internal components of memory and high-level processing of modern computers are of importance to the functional performance of cultural computation. The functional capabilities of memory of digital computation demonstrates the functional operation of machine states that perform with the capabilities of high-level processing. The performance of functional operations as machine states shows internal states of data processing that are essential to the completion of the final states of machine computation. The internal components of memory and high-level processing facilitate the storage and functional operation of internal states of machine computation with final states.

The internal architecture of computing machines facilitates the performance of cultural computation. The memory and high-level processing of computing machines is essential to the functional performance of cultural computation. The internal components of memory with the storage of internal states of cultural processing facilitate the completion of functional operations. The high-level processing of cultural computation with memory facilitates the transformation of cultural input–output relations that shows the efficiency of computing machines. The capabilities of internal components of computing machines are of functional importance to the high-level processing of cultural computation.

The development of the cultural computer with digital environments that facilitate the interaction of mind and machines demonstrates the functional capabilities of the modern computer to inform the performance of cultural tasks. The design of digital environments that show capabilities for the high-level processing of cultural tasks contributes to the use and availability of computational tools and technologies that facilitate the performance of cultural capacities. The development of cultural programs that facilitate the performance of cultural tasks contributes to the development of computational tools and technologies that show functional capabilities that are culturally and ethnically appropriate. The development of computational tools and technologies that facilitate the interaction of minds and machines shows the functional capabilities of digital computation to contribute to the performance of cultural tasks.

The functional operationalization of cultural computation as functional capabilities for high-level processing facilitates the scientific and technological innovation of the concept of the cultural computer. The cultural processing of machine computation shows the types of representations that are essential to the transformation of cultural input into cultural output. The design of computing machines with digital technologies that are of importance to high-level processing contributes to the functional performance of cultural computation. The types of representations of machine computation show levels of performance based on the complexity of internal components. The design of computing machines with digital technologies that are integral to the internal components show the functional capabilities of cultural computation.

Practical Applications

The understanding of the multiple realizability of cultural computation is beneficial to the research and development of computational tools and technologies that facilitate the humanistic and scientific exploration of cultural computation. The development of computational tools and technologies that contribute to the scientific discovery of the cultural computation of minds and machines is beneficial to the understanding of the functional and mechanistic basis of cultural capacities. The research and development of scientific and technological innovation on cultural machine computation explores how minds and machines perform cultural computation. The functional significance of minds and machines as physical realizers of cultural computation shows the feasibility of functional performance that is based on the physicalism of cultural processing.

From analog to digital computation, the design of computing machines that perform functional operations that are cultural contributes to the understanding of the functional capabilities that are essential to cultural performance. The breadth of computing machines that show the functional capacities of machines states illustrates the functionality and adaptability of culture in minds and machines. The feasibility of analog to digital computation as a functional basis of cultural processing shows the functional equivalence of machine states as cultural states. The high-level performance of computing machines that perform the functional operations of cultural tasks demonstrate the high-level processing of minds and machines that comprise cultural capacities.

From real to artificial systems, cultural computation demonstrates the feasibility of machine computation across levels of systems of the natural world. The understanding of the types of representations is essential to the understanding of the mechanistic basis of cultural capacities. The demonstration of the machine computation of biological organisms shows the importance of cultural processing as the multilevel processing of biological mechanisms. The distributed processing of cultural information across functional mechanisms of biological organisms shows the functional significance of cultural processing on the biological computing machine. The considerations of synthetic devices that show comparable functional performance with that of biological computation entails the understanding of the functional equivalence of the capabilities of machine computation of real and artificial systems.

Discussion

The scholarly and scientific interest on cultural machine computation explores the multiple realizability of the cultural computation of minds and machines. The understanding of cultural computation and its causal–functional significance contributes to the understanding of culture as a physical realizer. From biological organisms to complex devices, the cultural computation of minds and machines illustrates the functional performance of computing machines as a mechanistic basis of cultural capacities. The cultural computation of minds and machines shows how the functional mechanisms of adaptive machinery reveal the higher-level processes of cultural capacities. The manifestation of culture as a physical realizer illustrates the functionality and adaptability of culture as a computing machine.

The understanding of the cultural mind as a computing machine is fundamental to the demonstration of the functional processes of cultural capacities. The cultural mind shows the functional capacities of cultural performance. The functional processes of cultural capacities comprise the functional components that facilitate the task-based activity of culture. Cultural capacities demonstrate how the mind as a computing machine shows the levels of capabilities for problem-solving that are cultural. The functional performance of cultural capacities shows how the mind as a computing machine serves as a functional basis of cultural processing. The conceptualization of the cultural mind as a computing machine shows how the functional performance of the mind that is cultural illuminates on the higher-level processes of cultural computation.

The notion of culture as a biological computing machine harkens on the plausibility of biological mechanisms as a causation of cultural computation. Culture as a biological machine comprises the cultural states and biological states that show functional significance. The cultural computation of the biological organisms demonstrates the adaptive machinery that is essential to the computational level of biological mechanisms. The cultural computation of biological organisms shows how biological mechanisms and their causal effects show the functional capacities of multilevel adaptation. Cultural computation is a higher-level process that demonstrates the causal–functional significance of multilevel adaptation and its real-world advantages. The functional significance of cultural and biological states as a functional relation demonstrates the importance of culture as a biological computing machine.

The cultural computation of real and artificial systems depicts the feasibility of the functional performance of culture across multiple physical realizers. The machine computation of the cultural mind shows the scientific realism of culture as a computing machine. The cultural computation of the mind demonstrates the cultural events in the world that comprise cultural machine computation. The machine computation of biological organisms further considers the notion of culture as a biological computing machine. The functional significance of biological mechanisms as a mechanistic basis of cultural computation details how biological organisms show the functional capacities of complex organisms as cultural species. The cultural computation of complex machines demonstrates the feasibility of the computation of complex machines that is cultural. From the cultural mind to the cultural brain, the functional basis of cultural computation shows how the high-level process of mentality and thought is cultural. The functional performance of complex machines that is cultural demonstrates how the high-level performance of computing machines is sufficient as a functional and mechanistic basis of cultural computation.

The multiple realizability of cultural computation shows the functional significance of culture and computation in the world. The cultural computation in the mind and in the world depicts the causal–functional roles of cultural states as part and particular of the causal structure of the world. The understanding of the causal mechanisms that comprise cultural computation illustrates how the physical realization of cultural capacities manifests the set of cultural and physical states that are of importance to machine computation. The characterization of the multilevel mechanisms that comprise cultural capacities further contributes to the understanding of the functional significance of the cultural computation of real and artificial systems.

The cultural computation of biological organisms demonstrates the adaptive machinery of complex organisms as cultural species. The functional adaptation of cultural capacities shows how functional machinery comprises the processes and mechanisms that facilitate higher-level function with causal–functional significance. The functional adaptation of cultural capacities as a higher-level function shows the functional components of cultural performance and its functional purpose. The adaptation of biological organisms for cultural capacities facilitates the computational level of adaptive machinery and its preparedness and responsiveness. The cultural computation of biological organisms shows the adaptive processes and mechanisms that comprise the functional performance of culture. The understanding of the cultural computation of biological organisms contributes to the scholarly and scientific understanding of the functional and mechanistic basis of culture in the mind and in the world.

Conclusion

Cultural machine computation investigates the cultural computation of the computing machine. The cultural computation of the computing machine illustrates how the cultural processes and mechanisms of adaptive machinery contribute to the computation of functional mechanisms. The cultural performance of machine computation shows the functional capabilities of high-level performance and real-world advantages of minds and machines. From the cultural mind to the cultural computer, the functional performance of culture shows the functional role of mechanisms as a causal impetus of cultural computation. The understanding of culture as a computing machine contributes to the broader understanding of the multiple realizability of culture in the world.

The cultural computation of minds and machines demonstrates the functional significance of the computation of cultural capacities. Cultural computation as part and particular of minds and machines facilitates the understanding of the computational principles of culture that describe the computation of minds and machines. The cultural patterns and regularities in the mind and in the world illustrate the lawlike principles that contribute to the manifestation of culture across multiple physical realizers. The understanding of cultural processing and its causal effects contributes to the characterization of the multilevel mechanisms of cultural capacities.

The scholarly and scientific study of cultural machine computation contributes to the cultural performance of machine computation. The cultural processing of machine computation illustrates the functional capabilities of computing machines for the high-level performance of culture. The functional components of computing machines demonstrate how fundamental elements of mentality and thought serve as a functional basis of machine computation. The functional capabilities for computation show how functional components of mental function are essential for the high-level performance of cultural computation. The understanding of the capabilities of simple to complex machines for cultural performance is foundational to the scientific and technological innovation of cultural computation from simple and complex organisms to high-level performance devices and computers.

Implications

The conceptualization of cultural machine computation as the broader notion of computing machines that show functional performance that is cultural illuminates on the functional significance of cultural computation. The scientific realism of cultural computation and its multiple realizability shows the plausibility of culture as a physical realizer in the world. The cultural computation of minds and machines demonstrates the manifestation of cultural states as physical states and its functional significance. The pontification on the high-level performance and real-world advantages of cultural computation – machines that think cultural – promotes the understanding of the causal–functional significance of the shared meaning and purpose of culture in the mind and in the world. The scientific and societal appreciation for the functional significance of cultural computation is beneficial to the broader understanding of the causation of culture of minds and machines.

Practical applications of the study of culture, minds and machines broaden understanding of the implications of scientific and technological innovation on culture and computation. Research and development on culture, minds and machines facilitate the scientific and technological innovation of computational tools and technologies that are beneficial to the understanding of the cultural computation of minds and machines. The generation of scientific and scholarly knowledge on cultural computation builds conceptual approaches to the systematic investigation of the cultural capacities of biological organisms as cultural species. The scientific and technological innovation on culture and computation is beneficial to the broadening of humanistic and scientific understanding on the practical benefits of culture and computation for science and society.

References

Copeland, B J. (Ed.). (2004). *The essential turing*. Oxford University Press.
Turing, A. M. (1947). Lecture on the automatic computing engine. In B. J. Copeland (Ed.), *The essential turing*. Oxford University Press.

5

CULTURAL NEUROCOMPUTATION

Introduction

Empiricism as philosophical and social thought illuminates on the role of logical inference and scientific discovery as conduits of scientific theory. Empiricism as philosophical and scientific inquiry broadens insight into the role of logical inference and empirical observation in the building of scientific theory and the theory verification. The building of scientific theory provides a conceptual realm from which to postulate on scientific concepts and hypotheses, which allow for prediction and explanation of the natural world. The central consideration of theory building as a conceptual foundation of hypothesis generation facilitates the conceptual language and scope that is fundamental to scientific theory.

The interest on scientific discovery allows for the role of scientific observation to inform the logical inferences on scientific theories and hypotheses that contribute to the prediction and explanation of the natural world. The compendium of plausibility in the understanding of causation and explanation belies the interests in the understanding of the role of scientific observation on the truth confirmation of the states of the mind and the world. The truth confirmation of empirical observation and the falsification of hypothesis testing provide sufficient means for the understanding of fundamental mechanisms and their causal significance. The advancement of scientific discovery to inform the formulation of social and philosophical thought on empiricism and naturalism illustrates the importance of empirical observation on the understanding of causation and explanation of the natural world.

DOI: 10.4324/9781003384236-6

The scientific developments of the 18th and 19th centuries demonstrate the impact of empiricism and naturalism on the humanistic and scientific understanding of being and existence. The scientific interest in the development of biology as a scientific and scholarly feat of the 18th and 19th centuries arises from the interest in the role of scientific discovery on the philosophical and social thought of empiricism and naturalism. Biology as scholarly and scientific inquiry illustrates interest in the study of the origin of life. The scientific development in biology broadens scholarly interest on the material being and the fundamental elements that comprise the origins of life and its mechanistic basis. The development of neuroscience in the 19th century further shows interest in the role of scientific discovery on identification of basic mechanisms of the nervous system as a mechanistic basis of the function and adaptation of simple to complex biological organisms.

The scientific feasibility of empirical observation as a means to further understanding of the ideal and material facets of being illuminates on the role of scientific discovery on the metaphysical understanding of the ideal and material being. The scholarly interest on the material notions of being and existence illustrate the importance of culture on humanistic and scientific understanding. Empirical observation as a basis of scientific discovery brings forth a wide range of approaches to the scientific inquiry on the fundamental principles that are foundational to the understanding of culture. Scientific empiricism allows for scientific and philosophical inquiry on lawlike principles of culture in the natural world.

The scientific discovery of the adaptative machinery of biological organisms as cultural species illustrates the functional basis of cultural capacities. The functional machinery of biological organisms demonstrates how biological mechanisms serve a causal role in the functional performance of cultural capacities. The demonstration of the feasibility of biological mechanisms as cultural capacities illustrates the physical instantiation of culture as a biological computing machine. The cultural computation of the biological organism shows the functional role of adaptive machinery on culture and behavior. The biological plausibility of culture and adaptation as mechanisms of biological organisms shows a functional significance of the adaptive machinery of culture in the natural world.

The goal of the chapter is to investigate the processes and mechanisms of cultural neurocomputation. The characterization of the functional processes and mechanisms of cultural neurocomputation contribute to the understanding of the functional and mechanistic basis of culture and adaptation. Research on cultural neurocomputation describes the cultural information-processing mechanisms that are fundamental to the adaptive machinery of simple to complex organisms. The study of the multilevel mechanisms of the cultural brain contribute to the understanding of the functional machinery of cultural adaptation. The understanding of the cultural neurocomputation of the cultural brain at the level of computation demonstrates how the functional machinery of biological organisms enact the behavior of cultural species.

Culture as a Physical Realizer

Culture as a physical realizer entails the understanding of the manifestation of the physical states of culture. The notion of culture as physical realizer implies that the physical instantiation of culture as physical states shows the functional relation of cultural states as physical states and its physical state transitions in the world. The manifestation of culture as physical states demonstrates the plausibility of the understanding of the causal and functional basis of cultural states. The physical states of culture detail the causal--functional role of cultural states in the causal structure of the world. The postulation on culture as a physical realizer implies that the understanding of the plausibility of the physical instantiation of cultural states in the world.

The concept of culture as a biological computing machine broadens the understanding of culture as a physical realizer. Culture as a biological computing machine describes the information processing of biological mechanisms as a functional basis of cultural computation. The cultural computation of biological organisms shows the functional machinery that illustrates the role of cultural states as functional states of biological mechanisms. The adaptive machinery of biological organisms demonstrates the manifestation of cultural states as functional states of causal mechanisms. The cultural computation of biological organisms as a computing machine demonstrates the manifestation of culture as a physical realizer.

The patterns and regularities of culture as physical states of the natural world describe the lawlike principles of culture and computation. Cultural patterns and regularities detail the laws and principles of culture in causal structure of the natural world. The physical states of culture as biological computation shows the mechanistic basis of functional adaptation as a causal impetus of cultural computation. The manifestation of cultural patterns and regularities as physical states illustrates the actualities of culture as a physical realizer. The notion of culture as a physical realizer shows the multiple realizability of cultural computation. The cultural patterns and regularities of causal mechanisms shows the causation of cultural computation as physical states and their transitions in the natural world.

The cultural computation of the brain describes the physical states of culture as brain states. Cultural states as brain states demonstrate how culture manifests as brain states of biological organisms. The brain basis of culture shows how the causal mechanisms of brain function contribute to cultural capacities. Culture as a physical realizer encompasses the patterns of the cultural brain that hold. The cultural and brain states of biological organisms show the causal--functional role of cultural computation as processes and mechanisms of the functional machinery of biological organisms. The cultural computation of the brain illustrates the functional significance of the brain basis of culture.

The cultural brain as a physical realizer of culture demonstrates how the brain is a functional adaptation of culture. The cultural brain shows how the functional mapping of cultural and brain states contributes to the functional machinery of biological organisms. Cultural and brain states show a functional performance of culture that shows the functionality and adaptability of the cultural brain. The functionality and adaptability of the cultural brain illustrates the functional role of cultural and brain states as part and particulars of biological organisms. Culture as a physical realizer shows how the functional machinery of biological organisms is sufficient as a mechanistic basis of the functional adaptation of culture.

The functional machinery of the cultural brain implies the understanding of the causal--functional role of culture as adaptation. The brain basis of culture describes the causal--functional role of processes and mechanisms as functional adaptation. The cultural brain illustrates the causal and functional significance of causal mechanisms that show the cultural computation of the brain. The cultural brain is comprised of adaptive machinery that demonstrates the causal--functional role of culture as functional adaptation. The functional machinery of the cultural brain shows the functional processes and mechanisms of cultural and brain states.

The cultural brain as a physical realizer of cultural capacities demonstrates how brain states serve as a functional basis of cultural processing. The states of the cultural brain show functional significance on the cultural processing of brain function. Cultural states as brain states illuminate how the causal mechanisms of the cultural brain show the causation of cultural computation. The causal mechanisms of the cultural brain illustrate the causal role of brain states on the functional performance of culture. The cultural brain as a functional performance of culture as a biological computing machine facilitates the understanding of the causal mechanisms of the cultural computation of biological organisms. The cultural brain as machine computation shows the brain states of culture that demonstrate the naturalism of culture as a physical realizer.

Biological Organisms as a Cultural Species

The scholarly and scientific study of biological organisms as a cultural species postulates on the functional and mechanistic basis of cultural adaptation. The adaptive machinery of biological organisms comprises a functional basis of the multilevel adaptation of cultural species. From simple to complex organisms, the functional machinery of biological organisms illustrates the emergence of causal mechanisms that are fundamental to cultural adaptation. Functional machinery comprises a mechanistic basis for the causal–functional role of processes that contribute to the multilevel adaptation of cultural species. The emergence of functional machinery illustrates the preparedness and adaptiveness of biological organisms to respond to the cultural environment.

The functional basis of the multilevel adaptation of biological organisms as cultural species encompasses the processes and mechanisms that show a causal--functional role. The adaptiveness of biological organisms arises from the functional processes and basic mechanisms that facilitate the biological basis of cultural computation. The functional processes and basic mechanisms of biological organisms show the functional capacities that are essential to the performance of cultural capacities. The functional and mechanistic basis of the cultural computation of biological organisms provides a means for the functional performance of culture capacities. The characterization of the processes and mechanisms of adaptive machinery contributes to the functionality and adaptability of simple to complex organisms.

The adaptive machinery of biological organisms as a cultural computing machine illuminates on the functional significance of biological mechanisms as a mechanistic basis of cultural computation. Biological mechanisms demonstrate the functional capacities for the multilevel processing of adaptation. The functional significance of biological machinery as a causal mechanism of adaptation shows the importance of processes and mechanisms on the multilevel processes of cultural capacities. The biological basis of cultural capacities show the functional mechanisms that contribute to a mechanistic understanding of the adaptive machinery of biological organisms as cultural species.

The functional role of biological processes and mechanisms as a causal impetus of cultural computation shows the adaptiveness of biological organisms. The adaptive machinery of biological organisms is essential to the adaptive responsiveness of the cultural and social environment. The interaction of biological organisms with the cultural environment shows the functional importance of adaptive machinery as a mechanistic basis of adaptive responsiveness. Biological processes and mechanisms demonstrate the functional preparedness of biological organisms for the cultural and social environment. The functional performance of biological organisms as a cultural species demonstrates the preparedness and responsiveness of biological species to the cultural environment.

The biological mechanisms of cultural capacities serve as a higher-level process of multilevel adaptation. Cultural capacities demonstrate the functional performance of higher-level processing that is essential to the understanding of the cultural and social environment. Biological mechanisms comprise adaptive machinery that arises based on the adaptation of the higher-level function of biological organisms. The interaction of biological organisms with the cultural environment demonstrates the functional significance of adaptive responsiveness. The functional performance of the higher-level processing of biological organisms demonstrates the adaptive machinery that is of functional significance to multilevel adaptation.

The adaptive machinery of cultural capacities illuminates on the functional and mechanistic basis of cultural adaptation. Functional processes of cultural capacities describe the information processing that is essential to the functional components of adaptation. The functional processing of cultural capacities comprise information-processing mechanisms that serve as a functional basis of task performance. The biological mechanisms of cultural capacities illustrate a mechanistic basis of cultural processing. The biological basis of cultural capacities shows how biological mechanisms have functional significance on the cultural processing of biological organisms. The functional and mechanistic basis of cultural capacities demonstrate the adaptive machinery of biological organisms that is of importance to cultural adaptation.

Biological organisms as a cultural species demonstrates the feasibility of the functional machinery of biological organisms as a physical instantiation of cultural computation. The functional machinery of biological organisms shows how the cultural computation of multilevel mechanisms provides a mechanistic basis for adaptation. The cultural computation of multilevel mechanisms characterizes the wide range of functional machinery that demonstrates the functional performance of cultural capacities. The characterization of the multilevel mechanisms of the cultural computation of biological organisms shows the functional role of adaptive machinery on the multilevel processing of adaptation.

The understanding of biological organisms as cultural species illuminates on the importance of the functional machinery of cultural capacities. The demonstration of biological mechanisms as adaptive machinery shows the functional role of cultural processes and mechanisms on adaptation. The biological basis of cultural capacities entails the understanding of the causation of cultural processes and mechanisms as the functional machinery of biological organisms. The causal influence of cultural processes and mechanisms on the information processing of biological organisms shows the functional role of adaptive machinery on culture and adaptation. The causal--functional role of cultural processes and mechanisms demonstrates the causation of higher-level processes on the functional and mechanistic basis of adaptation.

The functional mechanisms of the cultural capacities of biological organisms show the higher-level processes of simple to complex organisms. The cultural causation of functional mechanisms illustrate the causal–-functional role of cultural processes on the mechanistic basis of cultural capacities. The functional basis of cultural capacities demonstrates the causal role of cultural processing on the basic mechanisms of biological organisms. The characterization of the cultural processes and mechanisms detail the higher-level processing of adaptive machinery that is of importance to multilevel adaptation. The functional mechanisms of cultural capacities demonstrate the adaptation of higher-level function of biological organisms as cultural species.

The Cultural Brain Hypothesis

The cultural brain hypothesis posits on the cultural processing of the brain. The brain basis of cultural capacities demonstrates the functional and mechanistic understanding of cultural processing. The study of the cultural brain describes the postulation on the functional specialization and functional integration of the cortical brain organization for cultural capacities. The cultural processing of the cortical brain organization shows the functional mechanisms that constitute a mechanistic basis of cultural capacities. The functional and mechanistic understanding of the cultural brain contributes to the characterization of the multilevel mechanisms of cultural capacities.

The postulation on the cultural brain hypothesis provides a conceptual basis for the empirical study of the brain basis of culture. The empirical study of the cultural brain consists of the formulation of theory building and hypothesis testing on the brain basis of cultural capacities. Theory building on the cultural brain hypothesis encompasses the development of scientific concepts and language that facilitate the systematic investigation of the cultural processing of the brain ("cultural brain hypothesis"). The development of scientific concepts and language on the cultural brain hypothesis shows the formulation of a conceptual basis for the study of the cultural brain. The conceptual models and frameworks of the cultural brain hypothesis contributes to the conceptual understanding of the causal and functional significance of the brain basis of cultural capacities.

Conceptual models of the cultural brain hypothesis consist of the postulation on the causal relations of the functional mechanisms of the cultural brain. The formulation of a cultural brain model details the causal role of the cortical brain regions that comprise the cultural processing of the cortical brain organization. The cultural brain model ascertains on the causal relations of the cortical brain regions that predict and explain the cultural processing of the brain. The conceptual modeling on the cultural brain shows the feasibility of the hypothesis testing on the causal relations of cultural processing of the brain. The detailed characterization of the functional mechanisms of cultural processing of the brain shows the causality of the brain basis of cultural capacities.

The cultural brain model describes a conceptual basis for the hypothesis testing on the cultural brain. The functional specification of the cultural brain model includes prediction on the causal significance of the cortical brain regions on the functional performance of cultural capacities. The causal interaction of the cortical brain regions illustrates the flow of information of the cortical pathways that demonstrates the functional performance of culture ("cultural brain mechanism"). The hypothesis testing of the cultural brain model consists of the modeling of the causal interaction of the cortical brain regions that predict the functional performance of cultural tasks. The conceptual modeling of the cortical pathways of cultural processing as causal interaction facilitates the characterization of the brain as a mechanistic basis of cultural capacities.

The causal modeling on the causal interaction of the cortical pathways of cultural processing illustrates a means for the hypothesis testing on the functional correspondence of cultural capacities. The identification of the causal relations of brain mechanisms and behavioral performance of cultural capacities demonstrates the magnitude and directionality of causality that characterize the cortical pathways of cultural processing ("cortical pathway of culture"). The characterization of the magnitude and directionality of causal relations of the cortical brain regions and cultural processing demonstrates the functional specificity of the cortical pathways that show the causal interaction of functional mechanisms and cultural processes. The causal modeling on the cortical pathways of cultural processing contributes to the identification of the causal mechanisms that describe the cultural processing of the brain.

The study of the cortical pathways of cultural processing entails the understanding of the causality of the cortical pathways and its functional role in cultural processing. The causal interaction of the cortical brain regions demonstrate the processing of cultural information that predicts the functional performance of cultural tasks. The flow of cultural information across the cortical brain regions shows the cortical pathways that are responsible for cultural processing. The cortical pathways of cultural processing illustrate the causal directionality of functional brain activity across the cortical brain regions that predict the functional performance of culture and behavior. The characterization of the cortical pathways of cultural processing contributes to the understanding of the causal interaction of the cortical brain regions and its functional relation to culture and behavior.

The conceptual modeling of the cortical brain processing as a cultural predictor demonstrates the functional role of patterns of cortical processing as a predictor of cultural processing. The characterization of the patterns of the cortical brain activity describes the cortical brain activity that is responsive to cultural processing ("cultural brain predictor"). The spatiotemporal patterns of the cortical processing of brain regions demonstrates the processing of cultural information. The functional correspondence of cortical processing as a predictor of cultural processing illustrates the functional role of the cortical brain as a cultural predictor. The conceptual modeling of the cortical brain as a cultural predictor shows the causal–functional role of the cortical brain activity on cultural processing.

The functional brain activity of the cortical pathways illustrates the importance of the cultural patterns of the cortical brain activity and behavior. The characterization of the cultural patterns of the cortical brain activity and behavior detail the functional correspondence of the brain and behavior relationships of culture. The fluctuations of the functional brain activity of the cortical pathways are of causal importance to the cultural patterns of behavior. The cultural patterns of the cortical pathways as predictor of cultural patterns of behavior show the functional relations of culture in brain and behavior. The functional brain activity of the cortical pathways serves as a mechanistic basis of cultural capacities.

The study of the brain and behavior relationships of culture facilitate the understanding of the functional role of the cortical brain processing on the functional performance of cultural capacities. A first hypothesis on the brain–behavior relationships on culture is that levels of cortical brain processing predict the functional processing of cultural capacities. The fluctuations of the cortical brain activity of cultural brain regions predict levels of functional performance on cultural tasks. The demonstration of the functional correspondence of levels of cortical processing and levels of functional performance on cultural tasks describes a causal role of cortical brain activity as a functional basis of cultural capacities.

A second hypothesis on the brain and behavior relationships of culture is that levels of the cortical brain processing predict types of cultural processing. The cultural patterns of the cortical brain activity of the brain predict effortful processing of cultural tasks. The cultural patterns of functional brain activity show functional correspondence with the levels of effort on cultural tasks. The functional brain activity of the cortical brain regions is of importance to the effortful processing on cultural tasks. The demonstration of the functional role of the cortical brain activity as a predictor of effortful processing on cultural tasks shows the importance of the levels of cortical brain processing on cultural processing that is effortful.

A third hypothesis on the study of brain and behavior relationships of culture is that the functional relationships of brain and behavior describe how the levels of cortical brain activity predict cultural processing that is intrinsic. The levels of cortical brain activity show functional patterns that are based on the intrinsic processing of the cultural brain. The cultural differences in the levels of cortical brain activity at rest show how the intrinsic processing of the brain is inherently cultural. The levels of cortical brain activity predict intrinsic cultural processing that is based on the levels of activity of cultural entrainment. The observation of the functional correspondence of functional brain activity and intrinsic cultural processing illustrates how cultural entrainment is sufficient to demonstrate the intrinsic cultural processing of brain function.

The conceptual modeling on the cultural brain hypothesis arises from a wide range of conceptual approaches to the study of the cultural processing of the brain. The conceptual models of the cultural brain provide scientific concepts and language for the formulation of theory building and hypothesis testing that is essential to the systematic investigation of the cultural brain. The development of conceptual language that describes the functional basis of the cultural processing of the brain that contributes to theory building and hypothesis testing. The conceptual language of the cultural brain hypothesis provides a conceptual basis for the understanding of the functional mechanisms that comprise cultural processing. The systematic investigation of the cultural brain hypothesis is essential to the understanding of the brain basis of culture.

The development of standard paradigms on the cultural brain hypothesis is fundamental to the functional operationalization of cultural tasks. Standard paradigms describe methodological approaches to the study of the cultural brain. The development of standard paradigms of the cultural brain hypothesis detail the functional operationalization of cultural tasks that show cultural processing. The design of cultural tasks illustrates a paradigmatic approach to the study of the cultural brain hypothesis. The formalization of the functional operationalization of cultural tasks provides a methodological basis on the empirical study of scientific theories and hypothesis on the cultural brain.

The systematic study of the cultural brain builds on the conceptual and methodological foundations of the cultural brain hypothesis. The formulation of scientific theories and hypothesis testing is central to the systematic investigation of the cultural brain. The conceptual and methodological foundations of empirical study contribute to the development of scientific concepts and language that are fundamental to the understanding of brain basis of culture. The elaboration of conceptual and methodological approaches to the study of the cultural brain is essential to the prediction and explanation of the cortical pathways of cultural processing. The cultural brain hypothesis is a fundamental conceptual approach to the empirical study of the brain basis of culture.

Computational Principles of Cortical Brain Organization

The study of the computational principles of cortical brain organization serves as a foundation for the understanding of the structural and functional principles of the cortical brain. Computational principles of the cortical brain organization describe the structural and dynamic principles of cortical processing across levels of processing. The characterization of cortical brain processing at the level of computation facilitates the understanding of the multilevel mechanisms of the cortical brain organization. The structural and functional properties of cortical processing detail the computational principles that are fundamental to the understanding of the cortical brain organization. Functional specialization describes the specialized and dedicated machinery of the cortical brain regions that show the functional performance of core capacities. The functional integration of the cortical brain organization shows the functional role of interconnectivity on the relaying of information processing across the cortical brain regions of neural networks. The understanding of the structural and functional properties of cortical processing is fundamental to the computational principles of the cortical brain organization.

The structural principles of the cortical brain organization describe the hierarchical structure of the cortical pathways of information processing. The hierarchical structure of the cortical brain organization describes the cortical layers of information processing that demonstrates the functional processes of the cortical brain function. The layers of cortical processing detail the types of neurons and their interconnection that comprise the structural properties of the cortical brain organization. The cortical layers of information processing show how the columnar organization of neuronal mechanisms demonstrates the types of information processing that comprises the structural basis of the cortical brain function. The hierarchical layers of processing across cortical areas show the transformation of information of specialized cortical pathways. The multilevel processing of neuronal mechanisms details the types of information processing that is describes the hierarchical structure of the cortical brain organization.

The functional principles of the cortical brain organization demonstrate the importance of the understanding of the functional properties that comprise cortical processing. The functional specialization and functional integration of the cortical brain organization describe the functional properties that are fundamental to the cortical processing of the brain. The functional specialization of the cortical brain regions details the domain specificity of cortical processing that is content-specific. The functional integration of the cortical brain regions consists of the cortical processing of brain function that is process-specific. The characterization of the multilevel processing of cortical mechanisms contributes to the understanding of the functional basis of information processing.

The functional specialization of the cortical brain regions describes the specialized and dedicated machinery of cortical mechanisms for the information processing of core capacities. The cortical brain regions show the functional responsivity of information processing that is specialized for information processing mechanisms. The functional specificity of information processing illustrates the neuronal mechanisms that show response that is domain-specific to types of information processing. The functional specialization of adaptive machinery details the patterns of neuronal response that are specialized for the domain specificity of information processing. The adaptive machinery of cortical mechanisms shows the functional specialization of the cortical brain regions to specific types of information processing.

The domain specificity of functional specialization shows how the adaptive machinery of the brain is specialized for information processing of the social and cultural environment. The functional specificity of information processing mechanisms details the response of neuronal mechanisms that is dedicated to social and cultural information processing. The preparedness of cortical brain mechanisms for the dedicated and specialized processing of social and cultural information illustrates the functional specificity of neuronal mechanisms. The functional specificity of the cortical brain regions to social and cultural information illustrates the information processing of core capacities. The domain specificity of functional specialization is of importance to the understanding of the adaptiveness of the cortical brain function.

The cortical brain regions show the aggregation of functional response of populations of neurons that show dedicated and specialized processing to distinct types of information of the environment. Neuronal mechanisms demonstrate the responsivity of cellular and molecular mechanisms to the information processing of the environment. The patterns of neuronal response show the representation and transformation of the sense data of the environment. Neurons serve as a cellular and molecular basis for the representation and transformation of information-processing mechanisms. The functional response of populations of neurons of the cortical brain regions demonstrates the dedicated and specialized processing of neuronal mechanisms.

The cellular and molecular basis of neuronal mechanisms describes the neurochemical signaling of neurons and their interconnection. The neurotransmission of cellular and molecular mechanisms describes the excitatory and inhibitory processing of neurons and the neurochemical signaling of neurons at the synapsis. Neurons as a cellular and molecular basis of neurotransmission demonstrate distinct types of information processing that are responsive to the environment. The distinct types of excitatory and inhibitory processing of neurons and their interconnection illustrates the neurotransmission of cellular and molecular mechanisms. The distinct types of neurotransmission describe the information-processing mechanisms that comprise the cellular and molecular mechanisms of the cortical brain regions. The neurochemical signaling of neurons and their interconnection describes a cellular and molecular basis of neuronal mechanisms.

The functional specialization of the cortical brain regions dedicated to social and cultural processing illustrates the distinct types of information processing that are responsive to the social and cultural environment. The neuronal mechanisms of the cortical brain regions show the dedicated and specialized processing of social and cultural information as a mechanistic basis of the response of biological organisms to the environment. The functional response of neuronal mechanisms illustrates the patterns of information processing to social and cultural information that is content-specific. The functional mechanisms of cultural and social patterns of information processing detail the functional specificity of cortical mechanisms of importance to social and cultural adaptation.

The large-scale distribution of information processing shows the distribution of representation across cortical areas of functional pathways. The layers of cortical representation of cortical pathways show the distribution of the functional processing of cortical mechanisms. The functional pathways of information processing illustrate the structural properties of representation across the cortical brain areas. The distribution of information processing as functional pathways demonstrate the dedicated, content-specific processing that comprises cortical areas. The functional pathways of information processing mechanisms show the types of content-specific processing that are based on the structural properties of cortical mechanisms. The large-scale distribution of information processing details how the distribution of representation across cortical areas facilitates the functional processing of cortical pathways.

The functional pathways of cortical processing illustrate the functional processing of cortical mechanisms that is process-specific. The distributed pathways of cortical processing show the functional pathways that demonstrate the relaying of information that is process-specific. The functional processing of cortical pathways shows the distinct types of processes that comprises the functional properties of the cortical brain regions. The cortical pathways of functional processes show how the cortical brain regions demonstrate the functional brain activity of information processing that is specific to types of processing. The distributed pathways of cortical processing illustrate how the functional properties of the cortical brain regions facilitate specific types of functional processing.

The functional integration of networks of neurons encompasses the structural and functional connectivity of the cortical brain regions and their function. The functional processing of networks of neurons describes the relaying of information across the cortical brain regions that is based on structural and functional properties of the cortical pathways. Structural properties of the cortical pathways are comprised of the cortical brain regions and their interconnection as relays of information processing. Structural connectivity of the cortical pathways describes the interconnection of the cortical brain regions that is based on the relaying of information of white matter tracts as networks of neurons. The functional properties of the cortical pathways describes the information processing across the cortical brain regions that comprise the patterns of functional brain activity. The functional connectivity of the cortical pathways demonstrates the information processing that is based on the functional patterns of the cortical brain activity. The functional processing of networks of neurons illustrates the structural and functional properties that comprise the functional integration of the cortical brain regions across levels of processing.

The interconnectivity of neural networks consists of the interconnection of networks of neurons that facilitate the relaying of information processing across the cortical brain regions. The interconnection of networks of neurons facilitates the functional processing of cortical mechanisms as the cortical brain regions. The functional processing of the cortical brain regions shows the patterns of functional brain activity that comprise the information processing of neural networks. The information processing of functional patterns of brain activity demonstrates the flow of information of activation patterns as matter and energy. The functional patterns of brain activity show the states of activation that comprise the activation patterns of neural networks. The functional brain activity of neural networks detail the patterns of activation that show functional correspondence with the information processing of cortical mechanisms.

The functional patterns of brain activity of cultural processing describes information processing and patterns of brain activation that comprise the cortical processing of cultural information. The cultural patterns of functional brain activity detail the flow of information as activation patterns of cultural neural networks. The cultural processing of functional brain activity as states of activation demonstrate the information processing of cultural activation patterns. Cultural patterns as activation states of functional brain activity depict the cortical processing of cultural information. The states of activation of cultural patterns comprise the cortical processing of cultural neural networks. The functional processing of cortical mechanisms demonstrates the activation patterns of neural networks that are specific to cultural processing.

The dynamic principles of the cortical brain processing demonstrate the computational basis of cortical mechanisms. The dynamical principles of the cortical processing illustrate the states of activation of cortical mechanisms that describe optimal states. The principle of multiple constraint satisfaction of cortical brain dynamics shows how patterns of functional brain activity entails levels of satisfaction of activation states of the cortical brain. Brain activation patterns describe the states of activation that comprise levels of energy that are optimal to the neural network. The functional brain activity of cortical processing shows the activation patterns that facilitate the satisfaction or pattern completion of neural networks. The principle of attractor dynamics comprises the attractor states of patterns of activation that demonstrate a state of activation of neural networks. Attractor dynamics characterize the states of activation that describe attractor states of the patterns of functional brain activity. The cortical brain dynamics of information processing entails the understanding of attractor states as patterns of functional brain activity.

Multiple constraint satisfaction as a computational principle shows how functional brain activity describes patterns of activation that are characteristic of neural networks. Functional brain activity as patterns of cortical processing consist of the flow of information and states of activation of neural networks. The patterns of functional brain activity show the activation patterns that demonstrate the satisfaction of multiple constraints of the environment. The activation patterns of neural networks describe states of activation of the cortical brain that are optimal to the cortical processing of neural networks. The functional brain activity of the cortical brain regions demonstrate the pattern completion of the activation states of neural networks. The activation patterns of neural networks show overall satisfaction that illustrates the flow of information and cortical processing of neural networks that is optimal.

The computational principle of attractor dynamics describes the dynamics of the cortical brain activity that are important to the information processing of brain function. Cortical brain dynamics describe states of activation that serve as attractor states of patterns of functional brain activity. Attractor states comprise states of activation that comprise a local or global minimum of activation as patterns of functional brain activity. Cortical brain dynamics show the states of activation that describe the overall levels of satisfaction of patterns of activation of functional brain activity. The dynamics of the cortical brain activity show the patterns of activation that comprise the information processing of the cortical brain function. The understanding of the attractor dynamics of the cortical brain activity show the states of activation that comprise cortical processing.

The cortical brain dynamics of cultural processing demonstrate the patterns of functional brain activity that comprise the activation states of cultural neural networks. The functional brain activity of cultural neural networks describes the states of activation that comprise cultural processing. The activation patterns of cultural brain activity show the states of activation that are of importance to the processing of cultural tasks. The patterns of functional brain activity of cultural processing consist of the flow of information and activation patterns that comprise cortical brain dynamics. The cultural processing of the cortical brain activity demonstrates the patterns of activation that show the functional correspondence of information flow and activation states of the cultural brain. The cortical brain dynamics of cultural processing illustrate the functional brain activity that is essential to the processing of cultural information.

The study of the computational principles of the cortical brain organization provide a foundation for the understanding of the physical instantiation of functional processes and basic mechanisms of cortical brain function. The computational basis of cortical brain function demonstrates the fundamental principles that are of importance to the understanding of the structural and functional properties of the cortical brain. Computational principles of the cortical brain organization are fundamental to the understanding of the cortical processing of multilevel mechanisms. The structural and functional principles of cortical brain organization describe the structural and functional properties that facilitate the information flow and functional brain activity of cortical processing. The interconnectivity of the cortical brain regions shows the functional pathways that comprise cortical processing. The dynamical principles of computation are fundamental to the understanding of the flow of information and activation patterns of functional brain activity. Cortical brain dynamics facilitate the understanding of the states of activation that comprise functional brain activity and its function in cortical processing. The computational level of the brain is fundamental to the understanding of the information processing of cortical brain function.

Cultural Neurocomputation

The study of cultural neurocomputation encompasses understanding of the cultural processing of the functional mechanisms of the nervous system. The structural and functional organization of the nervous system is comprised of core brain regions that are dedicated to cultural processing. The cultural processing of functional mechanisms consists of the representation and transformation of cultural input ("sense data") into output ("motor output"). The study of the cultural processing of the functional mechanisms of the cortical brain regions contributes to the understanding of the functional and mechanistic basis of cultural neurocomputation. The cultural computation of functional mechanisms demonstrates the biological plausibility of the brain as a functional basis of the cultural processing of biological organisms as cultural species.

Research on cultural neurocomputation investigates how cultural processing manifests from the neurons and networks of neurons that are physically instantiated as functional mechanisms across levels of processing. The cultural processing of neurons and networks of neurons illustrate the role of aggregation of response of populations of neurons as functional processes of culture. The automatic response of neurons to the processing of culture details the functional response of basic mechanisms that contribute to behavior and adaptation. The distinct response of neurons to cultural processing shows how neuronal response facilitates the representation of the cultural environment.

The encoding and decoding of culture across levels of processing illustrates the representation of cultural information processing mechanisms. The distinct response of neuronal mechanisms and their functional contributes to the cultural representation of information processing. The neuronal response of functional mechanisms details the representation of cultural processing that is of importance to functional adaptation. Neurons show the dedicated response of functional mechanisms that are illustrative of the encoding and decoding of cultural representation. Individual to multiple response patterns of neuronal mechanisms demonstrate the level of detailed representation that comprises cultural processing. From sparse to distributed representation, the functional response of neuronal mechanisms illustrates the encoding and decoding of cultural information processing mechanisms. The distinct types of encoding of neuronal mechanisms illustrate the encoding of cultural processes that is characteristic of functional mechanisms at the level of neurons.

The population-level activity of neurons shows the aggregation of functional response that is consistent with patterns of neuronal activity of the cortical brain regions. The functional activity of population of neurons that comprise brain regions shows the encoding of cultural representation as population-level neuronal activity. The cortical brain regions are comprised of thousands of neurons that show patterns of neuronal activity consistent across functional mechanisms. The cortical brain regions dedicated to cultural processing illustrate the patterns of neuronal activity of the cortical brain function that is consistent with the functional mechanisms of behavior and adaptation. The aggregation of functional response that is consistent with the patterns of neuronal activity shows the representation of cultural processing that arises based on population-level activity.

The cortical brain regions that show functional specialization in cultural processing illustrate a functional response that is of importance to multilevel adaptation. The functional specificity of the cortical brain regions demonstrates the dedicated, specialized processing of cortical mechanisms that are content-specific. The cortical brain regions detail the functional mechanisms that are essential to the processing of cultural information. The cultural processing of neuronal mechanisms illustrates the functional response to content that is culture-specific. The functional response of neuronal mechanisms of the cortical brain regions show the distinct types of representation that are of importance to functional and cultural adaptation.

The functional response of the cortical brain regions demonstrates types of information processing that comprises cultural adaptation. The functional response of specific brain regions show the feedforward response of specialized processing that is specific to culture. The automatic response of cultural brain regions details the flow of information processing that is essential to cultural patterns of behavior. The distinct types of response patterns of cultural brain regions show the functional specificity of cultural processing. The functional specificity of cultural brain activity detail the patterns of responsivity that is culture-specific.

The functional integration of the cultural brain activity of the cortical brain regions shows the role of interconnectivity on the integration of functional processes and information processing. The interconnectivity of the cortical brain regions shows the bidirectional processing of information flow that is of importance to cultural brain activity. The functional connectivity of the cultural brain regions details the feedforward and feedback processing of cultural information that is of importance to cortical processes. The functional integration of patterns of cortical brain activity demonstrates the feedforward and feedback processing that are illustrative of the functional processing of inhibition or control. The functional basis of cultural brain activity and its interconnectivity shows the bidirectional processing of functional mechanisms that contributes to cultural processing.

The functional patterns of the brain activity of the cultural brain regions detail the flow of information across distinct areas of the cortical brain organization. Cultural patterns of functional brain activity illustrate the cultural processing that corresponds with the interconnectivity of the cortical brain regions. The cultural processing across the cortical brain regions shows the interconnection of functional brain activity that is based on the effortful or task-based activity of mental function. The cultural brain activity of the cortical brain regions shows the specific functional patterns of cortical activity that correspond with effortful processing. The cultural patterning of functional brain activity demonstrates the functional pathways of information processing across cultural brain regions.

The functional brain activity of the cortical brain regions shows the distinct response of neural networks to the flow of cultural information that is process-specific. The cultural processing of the cortical brain regions demonstrates the functional specificity of cortical response to the task activity of cultural processes. The functional specificity of the cortical processing of brain regions shows the task-based activity that is of importance to distinct cultural processes. The cultural brain activity of the cortical brain regions demonstrates the processing of cultural information that is essential to the functional performance of cultural task activity. The functional brain activity of the cortical brain regions demonstrates the information processing of functional mechanisms that is process-specific to culture.

The cortical brain dynamics of cultural processing describe the flow of information and activation patterns that comprise the cultural brain. The brain dynamics of cultural processing show the activation patterns and the flow of cultural information that comprises the cultural brain activity of neural networks. The patterns of cultural brain activity consist of states of activation of the cortical brain regions that are of importance to cultural processing. The patterns of cultural brain activity show the states of activation that are consistent with the flow of cultural information that is essential to cultural processing. The cultural brain dynamics of functional patterns of activation show how cultural processing shows functional correspondence with activation patterns of the cultural brain. The dynamics of cortical brain activity facilitates the understanding of the cultural patterns of information flow and activation patterns of cultural brain function.

The functional patterns of activation of neural networks demonstrate the flow of cultural information and patterns of activation of cultural brain activity. The cultural brain dynamics of information processing show the flow of information that comprises cultural processing. The patterns of cortical brain dynamics detail the functional patterns of activation that are of importance to the information processing of cultural neural networks. The cortical brain dynamics of cultural processing illustrate the activation patterns of functional brain activity that are fundamental to the activation patterns of the cultural brain. The characterization of the functional brain activity of cultural neural networks is of importance to the understanding of the cultural information flow and the patterns of cultural brain activation that are essential to the study of the neural networks of the cultural brain.

The information processing of cultural neural networks shows the states of activation that are of importance to the functional performance of cultural tasks. The cultural processing of neural networks describes the activation patterns of functional brain activity that is essential to the functional performance of task-based cultural activity. The activation patterns of cultural neural networks describe the states of activation that are optimal to cultural processing. The states of activation of cultural neural networks are essential to the cultural processing of task-based activity. The activation patterns of functional brain activity contribute to the functional performance of cultural tasks.

Cultural neural networks show the functional patterns of cortical brain activation of cultural processing. The activation patterns of cortical brain activity detail the cultural information flow and patterns of activation that comprise levels of energy that are optimal. The patterns of cortical brain activity of cultural processing demonstrate the states of activation that are essential to the pattern completion of the cultural states of cortical brain activation. The functional brain activity of cultural processing as activation patterns of cortical processing show the states of activation that are optimal for cultural information flow. The information flow and cortical brain function of cultural processing comprise the states of activation of cultural neural networks.

The study of cultural neurocomputation contributes to the understanding of the computation of the functional mechanisms of the cultural brain. From neurons to networks of neurons, the cultural brain activity of functional mechanisms demonstrates the cultural computation of the adaptive machinery of the cultural brain. The functional response of neuronal mechanisms show the cultural brain activity of cortical brain regions that are of importance to the neurocomputation of culture. Cultural neural networks demonstrate the patterns of cortical brain activity that are of importance to cultural processing. The cultural neural networks of cortical brain activity contribute to the functional patterns of cultural processing.

Cultural Neural Networks

Cultural neural networks describe the neural networks that are dedicated to cultural information processing. Cultural neural networks are comprised of the distinct brain regions that are fundamental to cultural capacities. The cultural processing of distinct brain regions contributes to the interconnectivity of cortical brain regions that corresponds with functional patterns of brain activity. The interconnection of the structural and functional connectivity of the cortical brain regions demonstrate the functional basis of the cultural processing of neural networks. Cultural patterns of functional brain activity consist of the flow of information across distinct brain regions that are linked to cultural processing. The characterization of the distinct brain regions that subserve the information processing of cultural capacities illustrate the functional role of cultural neural networks in multilevel adaptation.

The cultural processing of distinct brain regions shows the functional significance of multiple brain regions that are linked to cultural capacities. The functional patterns of cultural brain activity show the information processing cortical brain regions for cultural tasks. Cultural patterns of functional brain activity that arises from cortical brain regions that correspond with cultural processing. The information flow of cultural processing across distinct brain regions shows the interconnectivity of functional processes that contribute to cultural brain activity. The functional patterns of cortical brain activity show how cultural processing arises from the functional and emergent properties of cultural neural networks.

Research on cultural neural networks identifies distinct neural networks that show functional processes that correspond with cultural processing. The study of cultural neural networks entails the identification of the core brain regions that comprise functional processes that are essential to cultural capacities ("cultural brain region"). The identification of the core brain regions consists of the distinct functional processes that are of importance to cultural processing. The detailed characterization of distinct brain regions and their functional role show the cultural processing that arises as functional and emergent properties. The cultural and functional processing of cultural neural networks illustrates the population-level activity of networks of neurons that contribute to cultural capacities.

The identification of the core brain regions show how functional processes comprise cultural capacities. Cultural neural networks consist of the core brain regions that are responsible for the cultural processing of core capacities. The functional processing of cultural capacities elicit functional patterns of cortical brain activity that show cultural processing. The detailed characterization of the bidirectional processing of cultural information across cultural neural networks shows the feedforward and feedback processing that comprises cultural processing. The bidirectional information processing of cultural neural networks shows the functional basis of cortical brain activity that contributes to core capacities.

The functional patterns of cortical brain activity of cultural neural networks describes the bidirectional processing of the cortical brain regions that is important to cultural capacities. The flow of cultural information across the cortical brain regions shows the functional patterns of brain activity that are of importance to cultural processing. The feedforward and feedback processing of cortical brain activity describes the causality and directionality of the functional patterns of the cortical brain regions that are important to cultural processing. The bidirectional processing of cortical brain activity shows how the flow of cultural information across the cortical brain regions is essential to cultural processing.

The interconnectivity of the cortical brain regions that are dedicated to cultural processing illustrates a functional basis of cultural neural networks. The interconnection of the cortical brain regions shows the functional role of the relay of information across the cortical brain regions that comprise cultural neural networks. The interconnectivity of neural network detail the structural and functional properties of cortical processing that correspond with cultural processing. The cortical processing of the neural network demonstrates the relay of cultural information that shows the causality and directionality of cultural processing. The interconnectivity of the cortical brain regions describes a functional mechanism of the cultural processing of neural networks.

The structural connectivity of cultural neural networks entail the characterization of the functional relation of the structural properties of neural networks and cultural processing. Structural connectivity describes the strength of interconnection of the cortical brain regions that is based on the density of white matter tracts. The magnitude of structural connectivity of cortical brain regions has functional significance for the cultural processing of neural networks. The strength of structural connectivity of neural networks is of importance to the facilitation and inhibition of cultural information processing. The structural connectivity of cultural neural networks demonstrates the functional role of structural properties on cultural processing.

The functional connectivity of cultural neural networks entails the role of interconnection on the functional processing of cultural information. The causality and directionality of functional connectivity shows a functional basis of cultural information processing. The strength of functional connectivity of neural networks demonstrates how the relay of information across the cortical brain regions is related to the interconnection of neural networks. The functional connectivity of cultural neural networks illustrates the flow of cultural information across the cortical brain regions that is specific to functional task. The functional connectivity of cultural neural networks contributes to the understanding of the functional relation of cortical and cultural processing.

The spatiotemporal dynamics of cultural neural networks facilitate the understanding of the spatiotemporal patterns of the cortical brain regions that contribute to cultural processing. The fluctuations of cortical brain activity that show functional correspondence with the flow of cultural information across the cortical brain regions demonstrates a functional basis of culture and behavior. The characterization of the spatiotemporal dynamics of the cultural processing of neural networks show how the patterns of information processing contribute to the functional basis of the cortical brain regions. The changes of spatiotemporal patterns of the cortical brain regions detail the information processing that is of relevance to culture and behavior.

Cultural Affective Neural Networks

Research on cultural affective neural networks shows the cultural and affective processing of the cortical brain regions. The cultural and affective processing of the cortical brain regions describes the specific functional processes that are essential to the cultural processing of emotion. The cultural processing of the cortical brain regions such as limbic structures and their interconnected regions shows the specific processing mechanisms that are responsible for the cultural processing of emotion. The cultural patterns of functional brain activity of neural networks contribute to the understanding of the functional basis of affective processing. The cultural processing of functional brain activity illuminates on the functional role of affective processing and its correspondence to cultural and emotional adaptation.

The functional brain activity of cultural affective neural networks demonstrates the functional processes and mechanisms that are of importance to emotional processing in cultural context. The functional brain activity of cortical and subcortical brain regions shows the levels of processing that are essential to the emotional processes of culture and behavior. The cultural patterns of functional brain activity illustrate the levels of activity that show functional correspondence with cortical and subcortical brain regions. The levels of functional activity of the cortical brain regions are of importance to the understanding of the functional relations of physiological and psychological activity that are of relevance to behavior in cultural context. The patterning of functional brain activity shows the functional responsivity that is illustrative of contextual influences.

The cultural patterns of functional brain activity depict the functional responsivity of cortical and subcortical regions during task-related processing. The functional brain activity of task-relevant processing shows the patterning of functional activity that is linked to contextual processing. The task-relevant processing of functional brain patterns demonstrates the levels of effortful activity that are of importance to the cultural patterning of functional activity. Effortful processing is essential to the cultural patterning of functional brain activity. Cultural patterns of functional brain activity display the functional correspondence of effortful and task-based processing that is linked to culture and behavior.

The automatic processing of functional brain activity of the cortical and subcortical brain regions shows the role of unconscious processing on emotion and behavior in cultural context. The automatic responsivity of functional activity of limbic and prefrontal brain regions demonstrates the role of subcortical and cortical activity as a functional circuit of emotion and behavior. The levels of activity of limbic and prefrontal brain regions such as the corticolimbic circuitry shows the specific brain regions that are of importance to the functional processes of emotion and behavior and the influence of contextual processing. The functional processes of emotion and behavior and its link to the cultural patterns of the functional activity of corticolimbic circuitry detail the role of emotion in multilevel processing across cultural contexts.

The functional patterns of brain activity of corticolimbic circuitry show the specificity of functional response to the contextual processing of emotion and behavior. The functional specificity of the amygdala and prefrontal cortex demonstrates the differential patterns of cortical and subcortical processing that are of importance to emotion and behavior. Differential functional patterns of cortical and subcortical processing show the levels of activation of the amygdala and prefrontal function that are linked to emotion and behavior. Cultural patterns of amygdala and prefrontal activity show the role of task-based emotion processing on behavior.

Cultural patterns of functional brain activity of the cortical and subcortical brain regions show the link of functional correspondence of effortful and task-based processing of emotion. The effortful and task-based processing of emotion shows the conscious levels of activity that are linked to functional activity. The conscious processing of emotion that is linked to functional brain activity details the role of control and regulation on emotion and functional adaptation. The interaction of emotion and cognition illustrates the use of cognitive or regulatory control on the adaptive processing of functional brain activity. The functional use of cognition as regulatory function is of importance to the adaptive processing of functional brain activity within cortical and subcortical brain regions.

The conscious processing of emotion that is of importance to task-based processes illustrates the role of cultural influences on the functional mechanisms of emotion and behavior. The conscious processing of culture and emotion shows the levels of conscious activity that are linked to emotional processes. The conscious recognition of facial expressions is linked to the functional activity of cortical and subcortical brain regions that comprise corticolimbic brain circuitry. The cultural patterning of the functional activity of cortical and subcortical brain regions that are linked to emotion processing shows the role of conscious levels of activity on the functional processes of behavior.

The conscious recognition of facial expression depicts the cultural influences on the functional brain activity of emotion processing. The cultural differences in functional brain activity of emotion processing illustrates the differential patterns of functional processes that are linked to emotion and behavior. Cultural differences in emotion processing show the differential levels of conscious activity that are linked to the functional activity of emotion and brain function. The conscious activity of emotion and brain function details differential patterns of functional processes that are linked to contextual influences. Cultural differences in emotion processing are of importance to the understanding of conscious processing and the multilevel mechanisms of emotion.

The functional patterns of brain activity that are linked to emotion and behavior show the role of cultural influences on the functional processing of emotion. The cultural influences on emotion processing depict the differential patterns of functional activity that are culture-specific. Cultural specificity in emotion processing at the level of functional activity illustrates the role of cultural influences on multilevel processing of emotion and brain function. The cultural influences on multilevel processing of emotion and brain function detail the differential levels of activity that are linked to culture-specific functional activity. Cultural influences on the patterns of functional brain activity show the functional specificity of multilevel processing that is culture-specific to emotion.

The functional specificity of the cultural patterns of brain activity show the differential levels of activity that is of importance to effortful processes. Cultural patterning of functional brain activity shows differential levels of activity that are based on functional roles. The functional role of contextual processing in effortful activity shows the cultural patterning of functional response that is specific to culture. Contextual processing of effortful activity details the functional specificity of cultural patterns that is linked to functional response. The functional responsivity of patterns of cortical brain activity show how effortful activity is of importance to the contextual processing of levels of activity.

The cortical brain dynamics of cultural affective neural networks describes the activation patterns of functional brain activity that are of importance to the cultural processing of emotion. The brain dynamics of cortical neural networks show the functional patterns of cortical brain activity that are essential to cultural and affective processing. The dynamical processing of neural networks demonstrates the states of activation that contribute to the cultural processing of emotional information. The demonstration of the cortical brain dynamics of neural networks facilitates the understanding of the information flow and activation patterns that comprise cultural affective neural networks.

The study of cultural affective neural networks demonstrates the neurocomputation of cultural and affective processing. The characterization of the multilevel mechanisms of cultural and affective processing illustrates the causal–functional role of neurons and networks of neurons as a functional and mechanistic basis of culture and emotion. The multilevel processing of cultural and affective information details the functional processes and mechanisms that facilitate the information processing on culture and emotion. From neurons to networks of neurons, the neurocomputational study of cultural and affective processing illustrates the functional mechanisms that show the functional significance of adaptive machinery. The study of the multilevel mechanisms of cultural and emotional capacities contributes to the broader understanding of the functional role of cortical brain mechanisms on the cultural processing of emotion.

Cultural Cognitive Neural Networks

Research on cultural cognitive neural networks shows the cultural processing of cognitive neural networks. The study of cultural cognitive neural networks depicts the functional processing of the cortical brain regions that are of importance to cognition. Functional processes of the cortical brain regions detail the cultural patterns of functional brain activity that are linked to cognition and higher-level processes. The cultural patterns of cognitive neural networks detail the role of contextual influences on the multilevel processing of cognition and behavior. Cultural influences on cognition and higher-level function detail the role of multilevel mechanism on cognition and behavior.

The study of cultural influences on cognition and behavior shows the role of contextual processes on cognition and higher-level function. Cultural influences on cognitive neural networks demonstrate the contextual processing of cognition and higher-level processes that are linked to behavior. From sensation and perception to cognition and higher-level function, the contextual processing of cognition and higher-level processes is comprised of the multilevel mechanisms that contribute to the reflexive and deliberate processing of cognition and behavior. The contextual processing of cognition and higher-level processes facilitates the representation and interpretation of cognition that is essential to the control and regulation of behavior.

The early processing of sensory and cognitive systems illustrates the role of contextual processing on the perception and cognition of culture and behavior. The early perceptual processing of sense data of the cultural environment illustrates the representation and transformation of sensory information into perception and behavior. The perceptual processing of objects and the cultural environment illustrates the role of sensory and cognitive systems in the perception and cognition of culture and behavior. Perceptual processing of objects and backgrounds of the cultural environment contributes to differential levels of activity that is task-based and culture-specific. The perceptual processing of objects and backgrounds of the cultural environment details the role of contextual influences on perception and cognition.

The perceptual processing of distinct categories of knowledge shows the functional responsivity to the cultural environment. Differential responsivity to category-specific knowledge in the functional patterns of brain activity shows levels of responsivity that are linked to the functional specificity of the cortical brain regions. Cultural differences in the functional patterns of cortical brain activity of perceptual and cognitive processes illustrates the functional specificity of the cortical brain regions that is responsive to the cultural environment. The functional specificity of levels of activity within the cortical brain regions shows responsivity-specific brain regions to category-specific knowledge. The functional selectivity of cortical response to category-specific knowledge facilitates the understanding of the functional specificity of multilevel mechanisms to the cultural environment.

The functional specificity of cognitive brain regions to contextual and environmental influences shows the responsivity of specific brain regions to category-specific knowledge. The functional selectivity of cognitive brain regions shows the cortical response to featural and configural information of objects and backgrounds that is essential to category-specific knowledge. The functional responsivity of cognitive brain regions to featural and configural processing depicts the functional activity that is of importance to the encoding of category-specific knowledge. The encoding of category-specific information within neuronal populations located in occipitotemporal cortices details the levels of processing that are specific to categorical processing within cortical layers of brain organization. The categorical processing of cognitive information within occipitotemporal cortices shows the role of functional processing on the multilevel mechanisms of the cortical brain organization.

The encoding of category-specific information within population-level activity of neurons shows the functional processing that is specific to distinct categories of information. Functional patterns of brain activity show the population-level activity of neurons located within occipitotemporal cortices are responsive to the cultural environment. Functional brain patterns of the population-level activity of cognitive neural networks details the multilevel processing of cognitive information that is specific to distinct categories. The subregions of occipitotemporal cortices such as fusiform, lingual and parahippocampal gyri show the differential levels of functional brain activity that is specific to face processing to the cultural environment. The functional activity of the cortical brain regions located within occipitotemporal cortices demonstrates the cognitive processing that is responsive to contextual and environmental influences.

Research on the functional brain activity of occipitotemporal cortices and face processing shows the differential patterns of cortical brain activity that are linked to the cultural environment. Levels of functional brain activity of cortical brain regions such as the fusiform and lingual gyri show the differential patterns of activity that are linked to culture-specific processing. Differential patterns of functional activity that are linked to culture-specific processing of faces illustrate the functional processes of multilevel mechanisms to category-specific knowledge that are culture-specific. The functional processing of category-specific knowledge shows the multilevel processing of cultural specificity that is of importance to sensory and perceptual systems.

The cultural specificity of perceptual and cognitive processes such as face recognition details the multilevel mechanisms that are essential to adaptation and behavior. The cultural specificity of perceptual and cognitive processing shows how the functional mechanisms of face recognition detail the responsivity of multilevel mechanisms to the cultural environment. The functional mechanisms of face recognition illustrate the featural and configural processing that is culture-specific. The functional processes of face recognition comprised of featural and configural processing demonstrate the levels of effortful processing that show the differential responses of importance to functional and cultural adaptation.

The functional response within the cortical brain regions of the neural network shows multilevel processing that is culture-specific. The neuronal activity of cognitive processing within the fusiform and lingual gyri shows the population-level activity of the cognitive brain regions that is specific to face processing and responsive to the cultural environment. The functional responsivity of cortical brain activity that is specific to culture illustrates the population-level activity of cognitive neural networks. The aggregation of population-level activity of the cognitive brain regions shows the differential cultural processing of population-level responsivity of cognitive neural networks. The differential patterns of cultural processing that describe cognitive neural networks shows the functional specificity of culture and face recognition.

The functional activity within interconnected brain regions shows the responsiveness of cultural neural networks to cognitive processes. The functional responsivity of cognitive brain regions and their interconnction show the mutlievel processing of cognitive processes that is culture-specific. The interconnection of cognitive brain regions such as the fusiform, lingual and parahippocampal gyri comprises a cultural neural network of the cognitive processing of races and faces. The cognitive processing of race and face recognition shows the multilevel processing of the cultural neural network from occipital to temporal brain regions that comprise the early processing of cognitive systems. The relay of cognitive information from perceptual to cognitive processes illustrates the multilevel processes that comprise the cultural processing of face recognition.

The cognitive processing of culture and face recognition shows the differential levels of activity that are linked to the encoding of face recognition. The cognitive processing of different race faces illustrates the levels of activity that show responsiveness to contextual influences. The differential responsivity of cultural neural networks to cognitive processing shows the levels of activity that are of importance to the encoding of the featural and configural processing of face recognition. Heightened levels of activity within cultural neural networks of cognition show the depth of encoding to featural and configural processing of faces and complex visual recognition based on expertise. The differential levels of activity of cultural processing details the levels of encoding that describe the processing of faces and other types of complex visual recognition.

The functional brain activity of cortical brain regions contributes to the functional processes of cognition and higher-level function. The cortical brain regions of cultural processing show the functional basis of cognition and higher-level processes. The interconnection of the cortical brain regions demonstrates the relaying of cognitive information that facilitates cultural processing. The cultural processing of cognitive information as patterns of functional brain activity shows the levels of processing that are based on cortical brain function. The cultural patterns of functional brain activity describe the information processing mechanisms that are essential to the functional brain activation of culture and cognition. The functional mechanisms of culture and cognition serve as a mechanistic basis of the cultural processing of cognitive information.

The brain dynamics of cultural cognitive neural networks demonstrate the functional processes of cultural and cognitive processing. The cortical brain dynamics of cultural and cognitive processing detail the information flow and activation states that comprise the cultural processing of cognitive information. The dynamics of cortical brain function show the flow of cultural information across cognitive brain regions and the multilevel processing of information processing mechanisms. The brain dynamics of culture and cognition depict the functional correspondence of cultural patterns of cognitive information and functional brain activation. The cortical brain dynamics of cultural and cognitive processing serve as a functional basis of the cultural patterning of cognitive processing.

The study of cultural cognitive neural networks is essential to the understanding of the multilevel mechanisms of the functional processes of cultural capacities. The cultural processing of cognitive information demonstrates the multilevel processing of cortical brain function that is of importance to culture and cognition. The functional processes of cultural capacities contribute to the information flow of neural networks that are essential to the functional performance of cognition and behavior. The cultural processing of cognitive information shows the functional processes of cortical brain mechanisms as a mechanistic basis of the cultural processing of cognition and higher-level function.

Cultural Social Neural Networks

Research on cultural social neural networks investigates the cultural patterns of neural networks that correspond with social processing. Cultural influences on the functional patterns of social processes detail the cortical brain regions that are essential to the cultural processing of social information. The cultural patterns of functional brain activity comprise the patterns of activation of neural networks that are of importance to cultural and social processing. The study of the neural networks of cultural and social processing shows the cortical processing of cultural information in social context. The functional brain activity of cultural neural networks demonstrate the cortical brain regions that contribute to the cultural processing of social information. Systematic investigation of cultural social neural networks is essential to the characterization of the multilevel mechanisms of cortical brain function that comprise cultural and social processing.

The neural networks of cultural social processing show the functional basis of the information processing that is of importance to culture and social behavior. Cortical brain activity of neural networks describe the cortical brain regions that show the functional role of information processing that is essential to cultural and social capacities. The cortical brain activity of neural networks illustrate the causal-functional role of functional mechanisms on the cultural processing of social information. The neural networks of cultural and social processing facilitate the functional role of cortical mechanisms that are essential to cultural processing in social context.

The cortical brain regions of cultural social neural networks demonstrate the cultural processing of social information. The functional processing of cultural and social information shows the functional role of the cortical brain regions as a basis for the systematic study of the causality and directionality of the information processing of cortical mechanisms. The cultural processing of social information describes the understanding of the causal–functional role of the cultural processing of social information and its functional basis. The cultural processing of social information relies on the functional brain activity of the cortical brain regions. The demonstration of the cultural processing of social information and its causal–functional role illustrates the functional processing of cultural social neural networks.

The cultural processing of social information of cortical brain function describes the information processing based on the cultural patterns of social processes. The cultural patterns of social processing detail how cortical mechanisms facilitate the functional processing of social information in context. The cultural and social processing of cortical mechanisms shows the casual–functional role of the functional brain activity of cortical brain regions on the processing of cultural and social information. The characterization of the cultural patterns of social processing facilitates the understanding of the functional brain activity of the cortical brain regions that are of importance to cultural and social processing.

The cortical brain regions of neural networks show the functional mechanisms of cultural and social information processing. The cortical processing of neural networks demonstrates the functional basis of the cultural processing of social information. The cortical brain regions serve as a functional mechanism of the understanding of other minds. The functional patterns of cortical brain activity demonstrate the causal–functional role of multilevel processing on the detection and recognition of other minds. The cortical processing of neural networks illustrates the functional processing of the cortical brain regions that facilitates the understanding of other minds. The functional properties of cortical neural networks are of importance to cultural and social processing.

Neural networks of cultural social processing illustrate the cortical brain mechanisms that are essential to cultural and social capacities. The cortical brain regions of occipitotemporal cortices show the cultural processing of social information. The functional role of occipitotemporal cortices on the processing of cultural and social information demonstrates the cortical brain mechanisms that are of importance to cultural social processing. The patterns of functional brain activity of occipitotemporal cortices show the activation states that correspond with the cultural processing of social perception. The functional activation patterns of the cortical brain regions that are located within occipitotemporal cortices show the functional significance of cortical mechanisms on the cultural processing of social information.

The functional brain activity of the occipitotemporal cortices shows the cultural patterns that contribute to social information processing. The cultural patterns of the cortical brain regions located within the occipitotemporal cortices shows the causal–functional role of cortical mechanisms on the detection and recognition of social information in context. The cultural processing of perceptual cues illustrates the interaction of top-down and bottom-up processing on social perception. Cultural processing of perceptual cues shows the functional role of the perceptual and motivational salience on the processing of social information. The functional brain activity of the cortical brain regions located within the occipitotemporal cortices show the cultural processing of social information.

Cultural social neural networks demonstrate the functional role of the cortical brain regions on the understanding of cultural and mental states. The cortical brain regions of the superior temporal gyrus are of importance to the cultural processing of mental state understanding. The cortical brain function of the superior temporal gyrus shows the functional role of cortical mechanisms on the understanding of mental states. The cultural patterns of the functional brain activity of the superior temporal gyrus show the patterns of activation that comprise the cultural processing of social information. The functional activation patterns of the cortical brain regions demonstrate the functional role of cortical mechanisms on the cultural processing of mental states. The neural networks of cultural social processing facilitate the patterns of activation that are essential to the cultural processing of mental state understanding.

The cultural processing of the cortical brain regions located within the superior temporal gyrus shows the functional processes that are of importance to the understanding of the cultural processing of mental state inference. The functional patterns of brain activity of the cortical brain regions shows the functional role of cortical mechanisms on the inferential reasoning on mental states. The cultural processing of mental state inference facilitates the levels of processing of the cortical brain regions that contribute to the inferential reasoning of social processes. The cultural patterns of functional brain activity of the superior temporal gyrus demonstrate the functional basis of the cultural processing on the mental states of others.

The cortical processing of the prefrontal cortex shows the functional role of cortical brain function on the cultural processing of social information. The functional subdivisions of the prefrontal cortex demonstrate the functional processes that facilitate cultural and social processing. The prefrontal cortex shows the functional basis of the information processing of cultural and social capacities. The prefrontal cortex serves as a functional basis of the neural representation of social and cultural information. The representational content of prefrontal brain function shows the neural representation of the functional processing of cultural and social capacities. The cortical processing of the prefrontal cortex demonstrates the functional states of cortical brain function that contribute to cultural and social information processing mechanisms.

The prefrontal cortex is a cortical brain region that is of importance to the functional processes of culture and social behavior. The cortical processing of the prefrontal cortex contributes to the cultural processing of social information. The functional patterns of cortical brain activity of the prefrontal cortex show the functional processing of culture and social information. The prefrontal cortex demonstrates the functional processes of importance to the control and regulation of cultural and social capacities. The functional patterns of prefrontal brain activity show the information processing of higher-level function. The control and regulation of cultural and social processing demonstrates the functional basis of cultural and social capacities. The functional brain activity of the prefrontal cortex shows the control and regulation of cultural and social capacities.

The cortical brain dynamics of cultural social neural networks show the functional role of the cortical brain regions on cultural and social processing. The brain dynamics of cultural and social information processing illustrate how the functional patterns of brain activity of the cortical brain regions contribute to information processing. The cortical brain dynamics of cultural and social processes demonstrate the flow of information and patterns of activation that comprise information processing. The functional processing of brain dynamics depict the cultural patterns of social information and activation states that are essential to functional performance. The cortical brain dynamics of neural networks illuminate on the functional role of the cortical brain regions on the functional processes of cultural and social information.

Cultural neural networks of social processing demonstrate the functional role of the interconnectivity of the cortical brain regions. The interconnectivity of neural networks shows the functional processes of cultural and social information mechanisms. The functional brain activity of neural networks illustrate the information-processing mechanisms of cultural and social capacities. The relaying of cultural and social information across the cortical brain regions shows the functional role of cortical processing on information flow. The cortical brain regions and their interconnection demonstrate the causality and directionality of information processing mechanisms that are essential to the functional performance of cultural and social capacities.

The Cortical Brain Dynamics of Culture

The cortical brain dynamics of culture demonstrate the dynamical principles of cortical brain function that underlies cultural information processing. The dynamical principles of cortical brain function contribute to the understanding of the cultural patterns of functional brain activity. The cortical brain dynamics of cultural processing show the information flow and activation states of cultural brain activity. The cultural processing of cortical brain dynamics illustrates the activation patterns that comprise cultural brain function. The cortical brain dynamics of cultural processing shows how the patterns of cultural brain function are essential to the processing of cultural information.

The interconnectivity of the cultural brain regions shows the interconnection of the cortical brain regions and relaying of information processing of cortical neural networks. The interconnection of the cortical brain regions serves as a functional basis of the flow of cultural information across interconnected brain regions. The interconnectivity of cultural brain regions demonstrates the relaying of information processing across cortical mechanisms. The cultural processing of the cortical brain regions illustrates the functional role of interconnectivity as a conduit of the processing of cultural information. The functional processes of cortical neural networks show the functional correspondence of cultural information flow and activation patterns of cultural brain activity.

The functional correspondence of cultural information flow and patterns of neural activation of cortical neural networks demonstrates the functional basis of cultural processing. The information processing of cortical brain function describes how the relaying of information across the cultural brain regions shows the bidirectional processing of cultural information ("cultural information flow"). The patterns of neural activation of cortical neural networks are essential to the functional processing of culture. The functional concordance of cultural information flow and patterns of neural activation detail the cultural processing that demonstrates the function of cultural activation patterns.

The cortical brain dynamics of cultural neural networks comprise the activation patterns that are essential to cultural processing. The processing of cultural information coincides with cultural states of activation that demonstrate the patterns of cultural brain function. The activation patterns of cultural brain activity show the local or global states of activation that comprise cultural information flow. The cultural patterns of neural activation of cultural neural networks show the flow of cultural information that is essential to cultural processing ("cultural activation"). The cultural activation patterns of functional brain activity show the dynamical processing of cultural information.

The functional brain activity of cultural processing illustrates the cortical brain dynamics of culture. The patterns of functional brain activity of cultural processing demonstrate the interactive processing of cultural information as a functional process of the cortical brain activity. The states of activation of cultural patterns as cultural information flow show the pattern completion of activation states of cultural neural networks. The set of states of cultural activation of neural networks are fundamental to the flow of cultural information. The functional relation of cultural brain activation and cultural information flow shows the importance of cortical brain dynamics on functional processes.

The pattern of activation states of cultural neural networks describes states of activation that are essential to cultural processing. The functional patterns of cultural neural networks show the levels of energy that are of importance to the completion of the activation states of cultural processing. The states of cultural activation show that the overall levels of energy at a local minimum of the cortical brain regions characterize the pattern completion of cultural brain activity. The states of cultural activation of the cortical brain regions that show overall levels of energy at a local or global minimum are the states of cultural information flow that are essential to cultural processing.

The cortical brain dynamics of cultural neural networks demonstrate the functional role of cultural brain activity. The dynamical principles of computation contribute to the understanding of the importance of cortical brain dynamics as a functional basis of cultural processing. The functional brain activity of the cultural brain shows how patterns of functional brain activity are of importance to the processing of cultural information. The cultural patterns of cortical brain dynamics illustrate how cultural states of the cortical brain regions comprise the flow of information that is of importance to the functional performance of culture. The cortical brain dynamics of culture are fundamental to the understanding of the cultural processing of neural networks.

Discussion

The identification of cultural brain regions shows the biological plausibility of the cortical brain organization that is specialized and dedicated to cultural processing. The biological mechanisms of cultural capacities of simple to complex organisms shows the computational machinery that is essential to the neurocomputation of culture. The cultural patterns of functional processes and mechanisms contributes to the understanding of the adaptive machinery that is essential to cultural capacities. Cultural processes that are physically instantiated within biological organisms show the functional processes and mechanisms that contribute to multilevel adaptation.

The cultural brain as a physical realizer of cultural computation shows the functional role of the adaptive machinery of biological organisms as cultural species. The functional processes and mechanisms of the cultural brain demonstrate how brain states are cultural states. The physical realization of the functional mechanisms of cultural processing detail how the functionalism of brain states as cultural states illustrates the feasibility of the computational level of adaptive machinery. The functional machinery of cultural brain processing contributes to the understanding of the causal–functional roles of functional mechanisms as physical realizers of cultural computation.

The characterization of the multilevel mechanisms of cultural processing is essential to the understanding of functional and cultural adaptation. Cultural neurocomputation describes the processes and mechanisms of cortical brain function that are fundamental to the neurocomputation of cultural processing. The biophysical mechanisms of cortical brain function detail the mechanistic basis of the information processing that comprises cultural adaptation. The multilevel mechanisms of cultural processing describe how neurons and networks of neurons show the multilevel processing of cultural information. The multilevel mechanisms of cultural processing demonstrate the mechanistic basis of functional and cultural adaptation.

Conclusion

Cultural neurocomputation provides characterization of functional mechanisms that are essential to cultural information processing. The structural and functional properties of the cortical brain organization contribute to the cultural processing that is essential to multilevel adaptation. The study of cultural neurocomputation contributes to the understanding of cultural information processing mechanisms and their functional role. The functional and mechanistic basis of cultural processing contributes to the detailed characterization of the neuronal mechanisms that comprise culture and behavior. The cultural patterns of neurons and networks of neurons illustrate the cultural processing that comprises cultural neurocomputation.

Cultural patterns of functional brain activity belie the adaptive response of the cortical brain regions. The identification of the cortical brain regions dedicated to cultural processing illustrates the functional mechanisms that contribute to behavioral and cultural adaptation. The functional specialization and functional integration of the cortical brain organization is fundamental to the cultural processing of functional mechanisms. The understanding of the computational principles that are central to the functional processes of the cortical brain organization is foundational to the study of the computational level of cultural brain processing.

The multitude of empirical approaches to the study of cultural neurocomputation illustrates the impact of empirical observation of the functional mechanisms of the nervous system on the understanding of cultural processing. The empirical study of the functional and mechanistic basis of cultural processing provides insight into the fundamental elements of the neurocomputational study of culture. The empirical study of cultural neurocomputation contributes to theory building on the causal and functional roles of processes and mechanisms of biological organisms as cultural species in the natural world. The causal role and functional significance of processes and mechanisms of biological organisms contributes to broader understanding of the cultural computation in the mind and in the world.

6

CULTURAL NEURAL NETWORKS

Introduction

Empiricism as scientific discovery demonstrates the interest of scholarly and scientific thought on the fundamental principles that contribute to the understanding of the biology and culture of behavior and its role in adaptation. The contemporary notions of evolution and adaptation show the scholarly interest on the rise of scientific empiricism and its role as a conceptual foundation of the scientific discovery of the biology and culture of behavior. The theoretical foundations of evolutionary biology postulate on the interaction of biology and culture and its functionality and causality as fundamentals of evolution and adaptation. The theory building on the conceptual foundations of the biology and culture of behavior posits the fundamental principles that guide the causal mechanisms that are of importance to the biological basis of culture and behavior.

The history of neuroscience of the 19th century shows the importance of scientific empiricism and the early scientific work in biology on the understanding of the structural and functional organization of the nervous system. The scientific discovery of the neuron as a cellular basis of the nervous system broadens scholarly interest on the structural and functional principles of the brain. The scientific work on the nervous system demonstrates the importance of neurons and networks of neurons as levels of processing. The characterization of the structural and functional properties of neurons and networks of neurons is of importance to the understanding of the higher-level processes of the brain. The characterization of neuronal networks imparts on the functional role of cortical brain regions

DOI: 10.4324/9781003384236-7

and the information processing of interconnected brain areas. The scientific discovery of the organization of the nervous system contributes to the understanding of the biology and culture of behavior. The empirical discovery of the fundamental elements of the organization of the nervous system contributes to the broader understanding of the importance of functional adaptation as a mechanistic basis of biological organisms as a cultural species.

The computational principles of cortical brain organization are fundamental to the understanding of the structural and functional properties of brain function. The fundamentals of computation serve as a conceptual foundation for the characterization of the structural and functional principles of cortical brain organization. From large-scale organization to interconnectivity, the structural and functional properties of cortical brain organization illustrate the importance of computational principles that guide the cortical brain function and its causal–functional significance. Across levels of processing, neurons and networks of neurons are fundamental as elemental units of the multilevel processing of cortical brain function. The cortical brain function of neural networks shows the functional role of interconnected cortical brain regions on the cognition and higher-level processes of brain function. The understanding of the multilevel mechanisms of cortical brain organization contributes to the characterization of the biological basis of culture and behavior.

The goal of the chapter is to provide a review of the study of cultural neural networks. Cultural neural networks describe the functional processes of cortical brain regions that demonstrate the functional basis of cultural processing. The characterization of cortical brain regions as nodes of neural networks shows the functional role of the interconnection of cortical brain regions. The interconnectivity of cultural neural networks demonstrates the functional significance of the cultural processing of core brain regions. The understanding of cultural neural networks provides a conceptual foundation for the exploration of cortical brain dynamics and the functionality and causality of neural networks as a mechanistic basis of cultural processing. Implications of research on cultural neural networks for health, medicine and related fields are discussed.

The Biology and Culture of Behavior

The study of the biology and culture of behavior contributes to the understanding of the fundamentals of evolution and adaptation. The scholarly interest on the biological basis of culture and behavior entails the understanding of the functional and mechanistic basis of the higher-level function of biological organisms as cultural species. The demonstration of the cultural capacities of biological organisms shows adaptive preparedness and responsiveness to the cultural environment. The adaptiveness of biological organisms to the cultural environment illuminates on the functionality and adaptability of functional machinery for behavioral and cultural adaptation. The functional adaptation of higher-level processes shows the presence of adaptive machinery for the processes and mechanisms of multilevel adaptation.

The biological basis of culture and behavior encompasses the characterization of the functional mechanisms that have causal–functional significance on the multilevel adaptation of biological organisms. The characterization of functional mechanisms describes the adaptive machinery that comprises the biological mechanisms that show a causal–functional role on culture and behavior. The biological machinery of cultural species demonstrates the functional processing that serves as a functional basis of culture and behavior. The functional processing of biological mechanisms describes the mechanistic basis of cultural capacities. The biological mechanisms of complex organisms detail the functional processing and causal mechanisms that comprise the functional and mechanistic basis of behavioral and cultural adaptation.

The adaptive machinery of biological organisms shows the functional preparedness and responsiveness to the cultural environment. The functional adaptation of higher-level processes of biological organisms shows the functional processing that is of importance to the multilevel mechanisms of culture and behavior. Cultural capacities illustrate the functional processing that is of importance to the multilevel adaptation of biological organisms as cultural species. The functional processes and mechanisms of adaptive machinery demonstrate the core capacities that are essential to the cultural processing of biological mechanisms and its relation to behavior. The functional adaptation of cultural processes and mechanisms is foundational to the adaptive machinery that facilitates the responsiveness of biological organisms to the cultural environment.

The interaction of biological organisms and the cultural environment is of importance to the understanding of the causal–functional role of adaptive machinery. The processes and mechanisms of adaptive machinery are essential to the adaptiveness and preparedness of biological organisms to the cultural environment. The responsiveness of complex organisms to the cultural environment shows the functional significance of adaptive machinery. The adaptive machinery of biological organisms facilitates the representation and transformation of sense data into motor output. The transformation of representational content of the cultural environment into cultural patterns of behavior illustrates the causal–functional role of adaptive machinery for the purpose of multilevel adaptation.

The biological evolution of cortical brain organization shows the functional role of the cortical brain as a functional mechanism of biological organisms. The cortical brain as a functional adaptation is of importance to the cognition and higher-level processing of complex organisms. The functional machinery of the cortical brain demonstrates the presence of functional mechanisms that are essential to the information processing of cortical brain function and its adaptive response to the cultural environment. The functional adaptation of the cortical brain as a functional and mechanistic basis of the core processing of cortical brain function shows the adaptive preparedness and responsiveness of biological organisms as a cultural species. The cortical brain organization of biological organisms shows the adaptive machinery that is essential to the functional responsiveness of adaptation.

The cortical brain as a functional adaptation illustrates the functional processes and mechanisms of higher-level processes. The cortical brain demonstrates the functional processing of cortical mechanisms that are of importance to the higher-level function of culture and behavior. The functional processes of cultural capacities show the causal–functional of cortical brain mechanisms as a mechanistic basis of behavior. The cultural processing of functional mechanisms shows the higher-level function of cultural capacities. The cultural capacities of the cortical brain details the higher-level function that underlies behavior. The cultural patterns of cortical brain mechanisms are essential to the cultural processing of behavior.

The functional processing of cortical mechanisms demonstrates the cultural patterns of information processing and its functional relation to behavior. The cortical processing of functional mechanisms facilitates the cultural processing of core capacities that are of importance to behavior. The cultural processing of core capacities shows the functional role of bidirectional processing on the functional mechanisms of behavior. The cultural patterns of information processing illustrate the functional processes that contribute to the production of behavior. The functional relation of cultural and cortical processing shows the importance of functional processes and mechanisms on the performance of behavior.

The cultural processing of core capacities illustrates the functional processes and mechanisms that are essential to the performance of behavior. From affect and cognition to social behavior, core capacities of functional mechanisms encompass the mental processes that facilitate behavior. The cultural capacities of functional mechanisms show the causal–functional role of cultural processing on the functional basis of core capacities. The cultural influences on emotion, cognition and social processes show the functional basis of the cultural processing on the higher-level processes of core capacities that underlie behavior. The functional basis of core capacities illustrates how cultural processing affects the functional components of mental processes and behavior.

The cultural influences on core capacities show the functional processes that are of importance to behavior. The cultural processing of emotional information contributes to the functional basis of cortical mechanisms and behavior. The cultural influences on emotional capacities show the changes of affect and emotion that are based on the cultural processing of emotional information. Cultural influences on affect and emotion demonstrate the functional role of cultural processing on the functional basis of emotion. The cultural processing of emotional information shows the functional mechanisms of cortical brain function that are responsive to the cultural environment. The cultural influences on emotional processes and its cortical brain mechanisms demonstrate the brain basis of culture and emotion and its role in behavior.

The brain basis of culture and emotion shows how cortical brain mechanisms serve as a mechanistic basis of the cultural processing of emotion. The brain mechanisms of culture and emotion detail the functional role of cortical brain function on the cultural processing of emotion. The response patterns of cortical brain mechanisms detail the functional processing that comprises culture and emotion. The functional relation of cortical brain mechanisms and behavior entails the causal–functional role of cultural processing on emotion and behavior. The response patterns of cortical brain mechanisms facilitate the functional basis of culture and behavior. The brain basis of culture and emotion is of importance to the understanding of the functional mechanisms that comprise culture and behavior.

The cortical brain function of culture and emotion details the functional processing of culture and emotion. The patterns of cortical brain function show the functional processes and mechanisms that comprise the cultural processing of emotion and behavior. Cultural patterns of cortical brain function depict the functional responsivity of functional mechanisms to the cultural processing of emotion. The functional responsivity of cortical brain function shows the causal–functional role of functional mechanisms to emotion processing that is culture-specific. The functional basis of cortical brain function that is culture-specific depicts the information processing of culture and emotion. The functional responsivity of cortical brain function facilitates the cultural processing of emotional information.

The cultural processing of cortical brain mechanisms illustrates the causal–functional role of cognition and higher-level function on culture and behavior. The functional processes of cognition and higher-level function demonstrate the perceptual and cognitive processing of the cortical brain function and its causal-functional significance on culture and behavior. The cultural processing of cognitive information is essential to the cortical brain function of culture and behavior. The cultural influences on cognition demonstrate the functional processing that is of importance to the cortical brain function that underlies culture and behavior.

The brain basis of culture and cognition demonstrates the functional processes and mechanisms that are of importance to the cultural processing of cognitive information. The brain mechanisms of cultural and cognitive processing show the functional mechanisms that facilitate behavioral performance. The functional relation of culture and cognition depicts the functional processes that underlie culture and behavior. The cultural processing of cognition details the functional mechanisms that contribute to the cortical brain functioning of cultural capacities. The brain basis of culture and cognition is fundamental to the functional processing of culture and behavior.

The cultural processing of cognition and brain function facilitates the functional basis of culture and behavior. The cortical brain function of culture and cognition illustrates the patterns of cortical mechanisms that comprise cultural and cognitive processing. The functional mechanisms of culture and cognition detail the patterns of cortical brain function that are of importance to the behavioral expression of culture. The functional relations of cortical brain function and behavior detail the role of cultural processing on the functional basis of cognitive processing. The cultural influences on cognition and brain function show the functional role of cognitive processing on culture, brain and behavior.

The functional adaptation of cortical brain function is of importance to the understanding of the cultural processing of social behavior. The functional mechanisms of biological organisms as cultural species illustrate the functional role of the cortical brain as the adaptive machinery that is essential to social behavior. The adaptive machinery of the cortical brain shows the preparedness and responsiveness of biological organisms in response to the cultural environment. The functional processes of social capacities contribute to the understanding of functional significance of the cultural processing of social information. The functional basis of social capacities illustrates the causal role of the cortical brain as a mechanistic basis of culture and social behavior.

The brain basis of culture and social behavior details the cortical brain mechanisms that serve as a mechanistic basis of cultural and social processing. The functional patterns of cortical brain mechanisms show the cultural processing of social information that comprises the functional processing of brain function. The functional mechanisms of cortical brain function show the importance of the cultural processing of social behavior. The adaptiveness and responsiveness of cortical brain function to the social information of the cultural environment facilitate the cultural influences on the social processes of behavior. The cultural patterns of cortical brain function contribute to the processing of social information. The brain basis of culture and social behavior demonstrates the causal–functional role of cortical brain mechanisms on the cultural and social processing of the brain and behavior.

The considerations of the biological basis of culture and behavior encompass the postulation on the evolution and adaptation of biological organisms as cultural species. The understanding of the importance of adaptation as a causal mechanism of the higher-level functioning of biological organisms contributes to the broader notion of the characterization of the brain basis of culture and behavior. The postulation on the biological basis of culture and behavior is of importance to the considerations of evolution and adaptation. The evolutionary adaptation of the brain demonstrates the functional and mechanistic basis of culture and behavior of complex organisms. The functional basis of the brain for cultural capacities shows the preparedness and responsiveness of biological organisms for cultural adaptation. The understanding of the mechanistic basis of the cultural brain is essential to the characterization of the multilevel mechanisms of culture and behavior.

Computational Principles of Cortical Brain Networks

The computational principles of cortical brain organization show the fundamental principles of the cortical brain that are of importance to the structural and functional properties of brain function. Structural and functional principles describe the structural and functional properties that describe cortical brain function. The structural principles of cortical brain organization detail the functional role of structural properties on the cortical brain. The functional principles of cortical brain organization describe the role of functional properties of cortical brain function. The functional specialization and functional integration of cortical brain function demonstrates the importance of dedicated and specialized processing of cortical brain regions and their interconnection. The interconnectivity of cortical brain networks shows the relay of processing that comprises brain function across cortical brain regions. The understanding of the fundamentals of cortical brain organization is essential to the characterization of the computational principles of cortical neural networks.

Networks of Neurons

Neuronal networks comprise a functional basis of the local networks of neurons that comprise cortical brain regions. Neuronal networks show the structure and function of cortical networks. The neuronal network is comprised of neurons and synapses that are located within cortical tissue. The interconnectivity of cortical networks demonstrates the aggregation of populations of neurons and their response properties. The local networks of neurons show levels of neuronal activity of populations of neurons that demonstrate the aggregation of neuronal activity. The cortical brain dynamics of populations of neurons show the local and global representation of the aggregation of population-level neuronal activity. The structural and functional properties of cortical brain networks comprise the functional basis for the information processing of interconnected brain regions.

The functional integration of interconnected brain regions shows the structural and functional properties of networks of neurons. The interconnectivity of cortical brain regions describes the structural and functional principles that characterize the information processing of neuronal networks. The structural and functional connectivity of interconnected brain regions describes the structural and functional properties of networks of neurons. The structural properties of neuronal networks refer to the structural interconnection of cortical brain regions that are based on the density of white matter tracts. The functional properties of neuronal networks reflect the functional connectivity of interconnected brain regions that are of importance to the functional processing of cortical brain mechanisms.

Large-scale Brain Organization

Large-scale brain organization is a computational principle that describes the structural and functional organization of the nervous system. The distribution of representation across cortical brain areas shows the hierarchical structure of the large-scale brain organization of information processing streams. The functional pathways of cortical processing demonstrate the large-scale distribution of information processing that comprises cortical brain organization. The functional pathways of cortical processing show the distribution of representation that comprises the dedicated, content-specific processing of information. The large-scale distribution of representation across multiple functional pathways shows the multitude of processing mechanisms that comprise the structural and functional properties of cortical brain organization. The large-scale brain organization of cortical brain function shows the structural and functional organization of the nervous system.

The distributed pathways of functional processing illustrate the cortical mechanisms that show distinct types of processing. The functional pathways of distributed representations demonstrate the relaying of information that is specific to types of processing. The functional processing of cortical mechanisms shows the information processing that is specific to distinct types of processing. The distributed processing pathways demonstrate the specific types of processing that comprise functional pathways. The distributed processing of functional pathways demonstrates the functional role of information processing that is process-specific. The distributed pathways of functional processing are of importance to the understanding of the distribution of representation across large-scale brain areas.

The specialized streams of functional pathways show the information processing that demonstrates the specialized information processing of distributed representations. The functional pathways of distributed representation show the specialized information processing streams that facilitate specific types of processing ("visual recognition", "spatial recognition"). The distribution of spatially invariant representation of the hierarchical structure of information processing streams shows the specific types of processing that comprise functional pathways. The specialized information processing streams of functional pathways detail the specific types of representation that are of importance to the large-scale organization of brain function.

Large-scale brain organization is of importance to the understanding of the distribution of representations that comprise functional pathways. The distribution of specialized information processing streams shows the large-scale distribution of representations. The hierarchical structure of large-scale brain organization demonstrates the cortical layers of representation that are of importance to the specialized processing streams of functional pathways. The distribution of types of processing streams across large-scale brain areas shows the hierarchical structure of cortical brain organization. The layers of cortical representation that comprise specialized processing streams facilitate the distribution of representation that are of importance to the functional processing of distributed pathways. The cortical layers of distributed representation are essential to the characterization of specialized functional pathways that underlie large-scale distribution of cortical brain areas.

Interconnectivity of Neuronal Networks

The interconnectivity of the cortical brain demonstrates a computational principle of cortical brain organization that is based on the structural and functional connectivity of the brain. The interconnection of cortical brain regions that comprise cortical brain organization shows the functional significance of the relaying of information across cortical brain regions. The interconnectivity of the cortical brain details the functional pathways of information processing streams that are of importance to the relaying of information across cortical brain regions. The bidirectional processing of cortical brain regions shows the relaying of information that is of importance to the structural and functional connectivity of the brain.

The interconnectivity of networks of neurons that comprise cortical brain regions shows the flow of information that is essential to the functional processing of the brain. The interconnection of networks of neurons illustrates the relaying of information that comprises the bidirectional pathways of functional processing of cortical brain regions. The interconnectivity of networks of neurons demonstrates the processing of bidirectional pathways that is of importance to the feedforward and feedback processing of information flow. The bidirectional pathways of functional processing facilitate the relaying of information that comprises the interconnection of neuronal networks. The interconnectivity of networks of neurons shows how cortical brain regions demonstrate information processing that is of importance to functional mechanisms.

Networks of neurons comprise the cortical layers of functional architecture that perform information processing. Cortical layers of neuronal networks show the functional processing of distinct layers of information flow that are of importance to the transformation of information processing from input to hidden to output layers. The input layer of cortical processing shows the unidirectional pathways of information flow that facilitate the relaying of information from sensory processing to motor output. The hidden layer of neuronal networks describes the cortical layers of information processing that are of importance to the transformation of representational content into motor output. The hidden layer of cortical processing shows the transformations of representational content that is based on the interpretation of information flow. The output layer shows the cortical layer that relays information flow to the hidden layer to motor output. The bidirectionality of information flow is essential to the relaying of information across cortical layers of functional architecture.

Neuronal networks demonstrate computational principles that describe the functional significance of cortical layers of information processing across multiple nodes of neural networks. The information flow of neuronal networks describes the unidirectional and bidirectional processing of information that facilitates the functional basis of input to output layers. Unidirectional processing demonstrates the excitatory processing that is based on the interconnection of excitatory neurons. The interconnection of excitatory processing facilitates the amplification of information processing that facilitates the transformation of representational content. The interaction of excitatory neurons facilitates the feedforward processing of information flow that contributes to the representation and transformation of sensory information into motor output.

The unidirectional processing of cortical neural networks shows the processing of information flow that is of importance to the representation of distributed cortical areas. The information processing of neuronal networks shows the interconnection of excitatory neurons as the functional basis of the local and distributed representation of information that comprises cortical brain regions. The local representation of information flow shows the detection of content-specific information that contributes to specialized information processing. The local representation of cortical brain regions facilitates the content-specific processing of specific types of information that are of importance to specialized functional mechanisms. The distributed representation of cortical brain regions shows the response of cortical networks to multiple types of representations. The local and distributed representation of information is of importance to the unidirectional processing of cortical neural networks.

The bidirectional processing of neuronal networks shows the information processing mechanisms that comprise the functional properties of the cortical brain regions. The functional properties of cortical brain regions show the information flow of feedforward and feedback processing that facilitate the inhibition of cortical neural networks. The feedforward and feedback processing of neuronal networks arises from the interaction of excitatory and inhibitory neurons that show the mutual influence of activation patterns that comprise bidirectional processing. The interaction of excitatory and inhibitory processing contributes to the control or inhibition of the feedforward processing of information flow. The bidirectional processing of cortical networks also shows the amplification of activation patterns that comprise the information flow across cortical brain regions. The functional properties of cortical brain networks show the interaction of excitatory and inhibitory processing that is of importance to the control or inhibition of information processing across interconnected brain regions.

The bidirectional processing of cortical brain networks shows the pattern completion of activation patterns that is based on feedforward processing of cortical brain regions. The pattern completion of activation patterns shows the importance of inhibitory processing on the completion of activation patterns of functional processing. The interaction of excitatory and inhibitory processing shows the pattern completion of partial activation that is based on the mutual activation of cortical brain regions. The pattern completion of cortical neural networks demonstrates the functional processing of cortical brain regions that is essential to activation patterns. The completion of activation patterns shows the functional processing of information flow that is essential to the feedforward and inhibitory processing of cortical brain networks.

The structural and functional connectivity of the cortical brain serves as a foundation to the understanding of cortical brain organization. The structural and functional properties of the cortical brain show the functional architecture of cortical brain organization that is of importance to the information processing of functional mechanisms. The structural properties of cortical brain organization show the neuroanatomical basis of the structural connectivity of cortical brain structures. The functional properties of cortical brain organization demonstrate the functional basis of the functional connectivity of interconnected brain regions. The structural and functional principles of cortical brain organization are essential to the understanding of computational principles of cortical brain function.

Structural Connectivity

The structural connectivity of the cortical brain encompasses the neuro-anatomical structures that comprise the structural properties of cortical brain regions. The strength of the interconnection of cortical brain regions based on white matter tracts describes the functional basis of structural connectivity. Structural connectivity comprises the magnitude of interconnection that arises from the density of white matter tracts that show the strength of interconnection as the structural properties of cortical brain regions. The structural properties of interconnected brain regions describe the neuroanatomical basis of the structural connectivity of cortical brain regions. The structural properties of cortical brain networks facilitate the information processing of interconnected brain regions. The structural connectivity of cortical brain regions is of importance to the magnitude of information processing of cortical brain networks.

Functional Connectivity

The functional connectivity of cortical brain networks describes the flow of information across interconnected brain regions. The information processing of interconnected brain regions consists of the distributed processing of cortical brain regions. The distributed representation of information processing across interconnected brain regions comprises the functional processes of cortical neural networks. The functional connectivity of interconnected brain regions shows the information flow of distributed processing streams across cortical brain regions. The specialized processing of cortical brain regions that facilitates the distribution of representation across multiple brain regions illustrates the functional processes of cortical neural networks.

Functional connectivity of cortical brain networks demonstrates the functional processing that is based on the interconnection of cortical brain regions. The interconnection of cortical brain region facilitates the flow of information processing across multiple cortical brain regions. The distribution of information processing across interconnected brain regions facilitates the information processing of functional pathways. The functional processing of cortical brain networks shows the processing of information flow across the interconnection of cortical brain areas. The information processing of cortical brain areas facilitates the distribution of processing that facilitates the transformation of distributed representation into motor output.

The information flow of cortical brain networks is of importance to the representation and transformation of information processing across interconnected brain regions. The bidirectional processing of interconnected brain regions facilitates the functional processing of cortical brain regions. The excitatory and inhibitory processing of interconnected brain areas facilitates the control or regulation of feedforward processing that is of importance to the functional processes of cortical brain regions. The information flow of cortical brain areas demonstrates the local and distributed representation of information processing that comprises functional properties of cortical brain organization. The functional properties of cortical brain organization illustrate a functional basis of the information-processing mechanisms of cortical brain function.

The functional connectivity of cortical brain networks contributes to the information processing of cortical brain function. The bidirectional processing of information flow across interconnected brain regions shows the causal role of functional properties on cortical brain function. The excitatory and inhibitory processing of populations of neurons of cortical brain regions shows the functional properties of the interconnectivity of cortical brain networks. The feedforward and feedback processing of information flow across interconnected brain regions details the casual–functional significance of functional connectivity. The causal interaction of cortical brain regions shows how the excitatory and inhibitory processing of neuronal populations facilitates cortical brain function. The feedback processing of information flow contributes to the inhibition of functional processing that is of importance to the functional activation states of cortical brain networks. The detailed characterization of the functionality and causality of cortical brain function is of importance to the understanding of the functional processing of the cortical brain.

Attractor Dynamics

The bidirectional processing of cortical brain networks describes the patterns of activation states that comprise attractor dynamics. The dynamics of cortical brain function of interconnected brain regions show the information flow of processing that comprises the activation states of cortical brain regions. The fluctuations of physiological signaling across interconnected brain regions describes the bidirectional processing of cortical brain regions that facilitate the pattern completion of activation states. The pattern completion of activation states of cortical brain networks demonstrates the multiple constraint satisfaction of information-processing mechanisms of cortical brain networks.

The information-processing mechanisms of cortical brain networks show the attractor dynamics of interconnected brain regions. The feedforward and feedback processing of information flow describes the brain dynamics of activation states that serve as attractor states of cortical brain networks. The attractor states of cortical brain dynamics describe the patterns of activation states that facilitate the pattern completion of information-processing mechanisms. The information flow of processing streams across cortical layers of brain organization shows the bidirectional processing that comprises the dynamical patterns of activation states. The information processing across cortical layers toward attractor states demonstrates the completion of patterns of activation that are based on the processing mechanisms of cortical brain dynamics.

Attractor states of cortical brain dynamics comprise the states of activation that are optimal for the functional processing of cortical brain regions. The attractor dynamics of cortical brain networks show that activation states and information flow comprise the information processing of cortical brain regions that is essential to the pattern completion of activation states. The activation states of attractor dynamics show levels of energy that are optimal for the neural network. The pattern completion of activation states of cortical brain networks details the optimal states that facilitate the overall satisfaction of multiple constraints. The attractor dynamics of cortical brain networks demonstrate the patterns of activation that are essential to the functional processing of interconnected brain regions.

The hierarchical structure of cortical brain organization facilitates the cortical layers of brain function that contribute to the pattern completion of attractor dynamics. The feedforward processing of information streams from input to output layers relies on the information flow across hidden or context layers of information processing streams. The hidden or context layer of the information processing stream facilitates the completion of patterns of activation based on the dynamical mechanisms of attractor states. Attractor states show the distribution of representation as patterns of activation within the hidden or context layer that facilitates the interpretation of the cortical input layer. The activation patterns of attractor states comprise the states of activation of cortical layers that facilitate the completion of input to output processing. The attractor state dynamics of cortical brain networks are important to the understanding of the pattern completion of activation states of that which comprises the functional processing of cortical brain regions.

Multiple Constraint Satisfaction

The computational principle of multiple constraint satisfaction details the importance of the functional processing of cortical brain networks that contributes to the pattern completion of information-processing mechanisms. The functional processing of interconnected brain regions shows the information flow and activation patterns that comprise the bidirectional processing of functional mechanisms. The feedforward and feedback processing facilitates the functional processes of cortical brain regions. Feedforward processing shows the functional processing that is based on the local dynamics of cortical brain regions; feedback processing illustrates the control of feedforward processing that comprises the inhibition of functional processes. The bidirectional processing of cortical brain regions demonstrates the multiple constraint satisfaction of cortical brain networks.

The pattern completion of information processing across cortical layers of brain organization is based on the satisfaction of multiple constraints. The feedforward processing of the cortical input layer relies on the pattern completion of activation states to the cortical output layer. The functional processing of information flow across cortical input to output layers shows the multiple constraint satisfaction of activation states that arises based on the pattern completion of cortical layers. The pattern completion of activation states from cortical input to output layers shows the dynamical mechanisms of information-processing mechanisms that are essential to cortical brain networks.

Computational principles of cortical brain organization are essential to the understanding of the computational foundations of cultural neural networks. The delineation of computational principles that describe the structural and functional properties of cultural neural networks shows how brain computation is essential to cultural processing. The detailed characterization of the computational principles of cortical neural networks contributes to the understanding of the structural and functional principles that describe the dynamical basis of cortical brain networks and cultural processing. The structural and functional properties of the cultural processing of neural networks are of importance to the understanding of the structural and functional principles of cortical brain organization.

Cultural Neural Networks

Cultural neural networks describe the cultural processing of information that comprises the functional patterns of interconnected brain regions. The cultural processing of cortical brain networks encompasses the information flow of interconnected brain regions that are of importance to cultural capacities. The processing of cultural information of neural networks shows the patterns of activation that comprise the functional processing of cultural brain regions. The cultural processing of interconnected brain regions comprises the functional basis of cultural neural networks. The understanding of the information processing mechanisms of cultural neural networks is essential to the characterization of the multilevel mechanisms of cultural adaptation.

The cultural processing of cortical brain regions is of importance to the understanding of the structural and functional properties of cultural neural networks. The structural and functional connectivity of interconnected brain regions facilitates the understanding of the structural and functional basis of the cultural processing of neural networks. The interconnection of cortical brain regions that show dedicated and specialized processing of cultural information shows the structural and functional basis of cultural neural networks. The information flow and activation patterns of cultural neural networks describe the cultural processing that is essential to the functional and emergent properties of cultural neural networks.

The structural and functional connectivity of cultural neural networks shows the functional significance of the structural and functional properties of cortical brain organization. The structural properties of cultural neural networks comprise the structural connectivity of cortical brain regions that is based on the density of white matter tracts. The structural connectivity of cortical brain regions that is linked to cultural processing shows the functional significance of the structural properties of cultural neural networks. The notion of the functional relation of structural connectivity that is linked to the depth of cultural processing facilitates the understanding of the functional significance of structural properties of cortical brain organization.

The functional properties of cultural neural networks are of importance to the understanding of the functional role of the interconnectivity of cortical brain regions. The functional basis of interconnectivity of cultural neural networks shows the functional significance of cultural processing. The functional connectivity of cortical brain regions demonstrates the functional properties that arise based on the interconnection of cortical brain regions. The cultural processing of interconnected brain regions shows the flow of cultural information that comprises the functional basis of cultural neural networks. The information flow of cultural processing as patterns of activation of interconnected brain regions demonstrates the functional processing of cortical brain regions that are of importance to cultural capacities.

The functional role of cultural processing as patterns of activity shows the functional processes that are of importance to the levels of processing of cultural capacities. Functional patterns of cortical brain activity that comprise the cultural processing of neural networks show the patterns of activation that correspond with levels of cultural task activity. The cultural patterns of cortical brain activity that are linked to levels of cultural task activity comprise the functional basis of cultural capacities. The cultural processing of neural networks shows how the information flow as patterns of activation facilitates the levels of processing of task-based activity that is cultural. The processing of cultural information as patterns of activation demonstrates the functional significance of information flow and activation patterns as the functional processing of cultural neural networks.

The functional patterns of cortical brain activity of cultural neural networks demonstrate the patterns of activation that show a causal influence on functional brain activity. Cultural neural networks comprise patterns of activation states that are of importance to the levels of processing of cultural capacities. The activation patterns of cultural neural networks facilitate the causal role of activation states as the processing of cultural information. The causal–functional role of patterns of activation that comprises cultural processing illustrates the functional significance of patterns of activation on the functional processing of cortical brain regions. The cultural patterns of activation of cortical brain activity show the causal–functional role of activation states on the cultural processing of interconnected brain regions.

The cultural patterns of functional brain activity of interconnected brain regions show a causal role on the functional performance of cultural capacities. The cultural patterns of task-based activity illustrate the functional role of effortful processing on the functional performance of cultural tasks. The functional brain activity of cultural neural networks is linked to the functional role of effortful processing on the functional performance of cultural tasks. Cultural task-based activity that relies on effortful processing shows the level of processing that comprises the functional performance of cultural tasks. The cultural patterns of functional brain activity show a causal–functional role on the levels of effortful processing that are of importance to cultural task performance. The cultural patterns of effortful processing that are linked to cultural patterns of functional brain activity show the functional role of cortical brain processing on the performance of cultural tasks.

The cultural processing of neural networks demonstrates the intrinsic and extrinsic properties of cultural neural networks. The intrinsic properties of cultural neural networks describe the cultural processing of cortical brain regions that is based on the maintenance of active representations of cortical layers that facilitate functional processes that are cultural. The intrinsic activity of neural networks that is based on the entrainment of cultural processing shows the spontaneous fluctuations of cortical processing that facilitates the functional processing of cortical brain regions. The intrinsic properties of cultural neural networks demonstrate the intrinsic activity of cultural brain regions that facilitates the entrainment of neural networks.

The intrinsic activity of cultural brain regions contributes to the automatic processing of cultural information of cultural neural networks. The feedforward processing of information flow of cultural neural networks demonstrates the cortical basis of the automatic processing of cultural information. The automatic detection of cultural information facilitates the feedforward processing of cortical brain regions that contributes to cultural processing. The information flow and activation patterns of cultural neural networks shows how activation patterns of cortical brain regions show functional relation of the automatic detection of cultural information. The cultural patterns of interconnected brain regions are of importance to the automatic processing of cultural neural networks.

The extrinsic properties of cultural neural networks describe the functional processing of interconnected brain regions that is based on the effortful processing of cultural information of cortical brain regions. The extrinsic activity of cortical brain regions shows the functional significance of effortful processing on the cultural information flow of neural networks. The effortful processing of cultural information details the task-based activity of cultural capacities and its functional relation to the cultural processing of interconnected brain regions. The extrinsic activity of interconnected brain regions shows the functional responsivity of cortical neural networks based on the task-based activity of cultural processing and its functional and emergent properties. The extrinsic properties of cultural neural networks demonstrate the functional and emergent properties of cultural processing.

The cultural patterns of cortical brain activity describe the functional and emergent properties of cultural processing. The levels of cortical brain activity of cultural processing are linked to the emergence of cultural task activity that is based on effortful processing. The emergent properties of cultural neural networks describe the functional processing of interconnected brain regions that are linked to cultural processing. The effortful processing of cultural task activity coincides with the cultural patterns of functional brain activity. The functional correspondence of cultural patterns of functional brain activity with cultural task activity shows the functional processing of cultural capacities. The functional and emergent properties of cultural neural networks demonstrate the link of cultural patterns of task-based activity and cortical brain activity.

The functional and emergent properties of cultural neural networks describe the cultural patterns of information processing that are emergent based on the functional performance of cultural capacities. The cultural processing of cortical brain activity that describes the control or inhibition of cultural processing facilitates functional and emergent properties that comprise cultural neural networks. The control or inhibition of cultural processing shows controlled processing that is of importance to the regulation of cultural capacities. The functional brain activity that is linked to controlled processing of cultural information shows the functional and emergent properties that are linked to the control or inhibition of information flow. Cultural processing of functional brain activity shows the functional role of control or inhibition on the functional performance of cultural capacities.

The functional basis of cultural capacities facilitates the conscious processing of cultural information that is intentional. The conscious processing of cultural information demonstrates the functional role of cognition and higher-level function on the functional basis of cultural capacities. The conscious processing of cultural information contributes to the patterns of functional brain activity that are of importance to the understanding of the functional role of cortical brain regions. The patterns of functional brain activity of cultural neural networks describe the conscious processing of cultural information. The functional patterns of cortical brain activity show the cultural processing of information flow that is of importance to the facilitation of intentional action.

The cultural processing of action understanding demonstrates the functional role of interconnected brain regions on the perception and action of intentionality. The cultural patterns of cortical brain activity show how cortical representation is of importance to the perceptual and cognitive processing of intentionality and action. The perceptual and cognitive processing of the attentional and motivational salience of sensory information contributes to the understanding of intentional action. The functional processing of cultural brain activity shows the functional significance of perception and cognition on the understanding of intentional action. The cultural processing of action understanding demonstrates the functional significance of the cortical brain regions of cultural neural networks.

The functional processing of cultural neural networks shows the importance of higher-level processing on the functional performance of cultural capacities. The cognition and higher-level function of cultural processing demonstrates the functional role of higher-level processing on the culture and social behavior. The cultural processing of social information depicts the functional patterns of brain activity of cortical neural networks. The cultural patterns of social processing demonstrate the functional significance of cortical processing on the activation patterns of cultural brain function. The higher-level processing of cultural brain function shows the functional significance of cognition and higher-level function on the functional performance of cultural capacities.

The cultural patterns of information processing mechanisms contribute to the functional relation of cortical brain activity and behavior. The functional role of cultural processing on brain and behavior demonstrates the functional relation of cortical brain activity and behavior. The cultural patterns of cortical brain activity depict the causal role of cortical brain activity as a functional basis of cultural patterns of behavior. The culture of brain–behavior relationships shows the causal–functional significance of the cultural processing of cortical brain function on behavior. The functional basis of culture and brain–behavior relationships details the casual influence and functional role of cultural processing on cortical brain function and its functional relation to behavioral expression.

The cortical brain dynamics of cultural neural networks explore the dynamical mechanisms of cultural processing and cortical brain function. The brain dynamics of cultural processing demonstrate the functional basis of intrinsic and extrinsic processing of cortical brain function. The cortical brain dynamics of cultural neural networks show the information flow and activation patterns that comprise cultural processing. The intrinsic processing of cortical brain function demonstrates the spontaneous fluctuations of cortical brain activity that comprise the autonomous brain dynamics of cultural processing. The extrinsic processing of cortical brain activity shows the dynamical properties of cortical mechanisms that facilitate the task-based activity of cortical brain function and its functional role on cultural capacities.

The brain dynamics of cultural neural networks is of importance to the understanding of the spatiotemporal processing of cortical brain regions that show a functional role in cultural processing. The spatiotemporal processing of cultural information entails the fluctuations of physiological signaling of cortical brain regions that show functional correspondence with the functional performance of cultural tasks. The spatiotemporal dynamics of cortical brain regions detail the functional role of fluctuations of physiological signaling and its functional role on the processing of cultural information. The characterization of the brain dynamics of cultural neural networks facilitate the understanding of the spatiotemporal dynamics of cortical brain regions that are responsible for cultural capacities.

The study of cultural neural networks encompasses a wide realm of functional processes and mechanisms that comprise the cultural processing of cortical brain networks. The characterization of the multilevel mechanisms of cultural processing demonstrates the functional importance of cortical brain networks on the information processing of cultural capacities. The consideration of the causal–functional significance of cultural neural networks as a functional basis of cultural capacities illuminates the depth of processing that comprises cultural brain activity. The detailed characterization of the core brain regions that comprise cultural neural networks is of importance to the understanding of the mechanistic basis of cultural capacities.

Cultural Affective Neural Networks

The study of cultural affective neural networks entails the characterization of the cultural processing of emotional information that is based on the interconnected brain regions of neural networks. The cultural processing of emotional information comprises functional processes and mechanisms that are characteristic of cultural neural networks. The functional processing of cultural information facilitates emotion processing that is of importance to cultural capacities. The interaction of cultural and emotional processes demonstrates the functional and causality of the information processing of cultural affective neural networks. The causal influence of cortical brain regions on the cultural processing of emotional information shows the functional role of neural networks as a mechanistic basis of culture and emotion.

The characterization of the core brain regions that comprise cultural affective neural networks is of importance to the understanding of the cultural processing of emotional information. The neuroanatomical structures of subcortical and cortical brain regions its interconnected brain regions comprise a functional basis for the cultural processing of emotional information. The interconnectivity of limbic and prefrontal cortices demonstrates the functional processing of cortical brain regions that are of importance to culture and emotion. Limbic and prefrontal cortices show the interconnection of cortical and subcortical brain structures that are of importance to the automatic and controlled processing of culture and emotion. The corticolimbic circuitry of limbic and prefrontal cortices details a functional basis for the cultural processing of emotion.

Cultural influences of emotion processing illustrate a causal–functional role of culture on the processing of emotional information. The cultural processing of emotion demonstrates the functional role of bidirectional processing on the functional mechanisms of cultural capacities. The feedforward processing of cultural processing facilitates the automatic detection of cultural and emotional information. The cultural processing of emotional information shows the functional significance of top-down influences on the amplification of information that comprise emotion processing. Culture serves as a top-down influence on the information flow of emotion processing. Cultural influences on emotion processing facilitate the amplification of emotional information that is of importance to the functional processing of cultural capacities.

The cultural patterns of cortical brain activity show the functional role of information processing mechanisms on culture and emotion. The functional brain activity of corticolimbic circuitry demonstrates the functional basis of information processing mechanisms that are of importance to the cultural processing of emotion. Cultural influences on emotion processing show functional correspondence with the cultural patterns of functional brain activity. The cultural patterns of functional brain activity that is located within corticolimbic circuitry demonstrates the cultural processing of emotion. The functional processing of cultural and emotional information contributes to the amplification of information flow that is essential to the cortical brain activity of cultural capacities.

The cultural processing of emotional information contributes to the understanding of the motivational and social significance of culture and emotion. The cultural influences on emotion processing illustrate the functional role of motivational and social salience on the understanding of the functional significance of emotion in context. The contextual processing of emotion contributes to the interpretation of emotion processing that is of importance to the functional basis of cultural capacities. The processing of emotional information that is culture-specific demonstrates the functional role of motivational and social salience on the interpretation of emotional processing. The understanding of the motivational and social significance of emotion that is based on cultural context contributes to the cultural processing of emotion.

The distributed processing of cultural information shows the functional significance of higher-level association areas that comprise cortical brain function. The higher-level association areas of cortical brain regions demonstrate the functional basis of the semantic comprehension of conceptual knowledge that is of importance to the cognition and higher-level function of the brain. The cultural processing of higher-level association areas demonstrates the functional role of semantic processing as a conceptual basis of cultural knowledge. The semantic processing of cultural information contributes to the understanding of the shared meaning or significance of cultural processes. The functional role of higher-level association areas on the semantic comprehension of cultural processing is of importance to the understanding of cortical brain function.

The semantic processing of cortical brain regions shows the functional role of frontoparietal regions on the conceptual processing of cultural knowledge. The cultural processing of emotion relies on the understanding of the shared meaning or significance of emotional information. The understanding of the shared meaning of emotional information demonstrates the contextualizing of emotion that is of importance to the motivational and social significance of culture and emotion. The frontoparietal regions show the functional significance of functional brain activity on the semantic comprehension of cultural information. The functional processing of the conceptual basis of cultural knowledge contributes to the understanding of the functional role of frontoparietal regions on the conceptual processing of culture and emotion.

The culture and language of emotion serves as a conceptual basis of the cultural processing of emotion. The functional brain activity of cortical brain regions that are located within frontoparietal brain areas shows the functional significance of conceptual processing on the understanding of the emotion. The cultural processing of emotion that is based on the functional processes of culture and language facilitates the cortical brain activity of semantic processing. The cultural patterns of cortical brain activity that are linked to conceptual processing show the importance of semantic understanding on the cultural processing of emotion. The cultural and linguistic expression of emotion demonstrates the functional significance of the functional processing of cortical brain areas that are involved in higher-level association.

The control or regulation of cultural capacities demonstrates the functional role of interconnected brain regions on cultural and affective processing. The cultural patterns of functional brain activity show the functional processing of corticolimbic brain regions and their causal interaction that is of importance to the control or regulation of the cultural processing of emotion. The functional processing of interconnected brain regions facilitates the feedback or inhibitory processing that is of importance to the control or regulation of the cultural processing of emotional information. The inhibitory processing of cultural capacities depicts the control or inhibition of the automatic processing of cultural and emotional information. The cultural patterns of functional brain activity of corticolimbic circuitry shows the functional significance of interconnected brain regions on the control and regulation of cultural and affective processing.

The cortical brain dynamics of cultural affective neural networks is of importance to the understanding of the dynamical mechanisms of cultural and emotion processing. The brain dynamics of cultural processing show the fluctuations of cortical activation that show functional correspondence with the information flow of cultural processing of emotion. The functional relation of physiological signaling of cortical brain regions with the cultural processing of emotion demonstrates a functional basis of the brain dynamics of culture and emotion. The cortical brain dynamics of cultural processing show the functional significance of the fluctuations of physiological signaling on the functional processes of culture and emotion.

The brain dynamics of cultural processing show the functional role of interconnected brain regions on the flow of information and activation patterns of cultural neural networks. The flow of cultural information across interconnected brain regions demonstrates the patterns of activation that comprise the cultural processing of emotion. The cultural processing of emotion information facilitates information flow across interconnected brain regions that are of importance to culture and emotion. The information flow and activation patterns of cultural neural networks demonstrate a functional basis of the cultural processing of emotion. The activation patterns of functional brain activity and its link to cultural and affective processing show the functional significance of the cortical brain dynamics of culture and emotion.

The patterns of activation of cultural affective neural networks demonstrate the states of activation that comprise the pattern completion of cultural and affective processing. The functional brain activity of cultural neural networks shows the activation states that comprise the optimal states of the cultural processing of emotion. The pattern completion of activation states of core brain regions is of importance to the activation patterns of cortical brain regions that are involved in cultural and affective processing. The functional brain activation of cultural and affective processing details the states of activation that comprise the cultural processing of emotion.

The spatiotemporal dynamics of cortical brain function demonstrate the functional processing of culture and emotion. The fluctuations of physiological signaling of cortical brain regions show the dynamical processing of cultural and emotional information. The characterization of the spatiotemporal dynamics of cortical brain function facilitate the understanding of the information processing that underlies cultural capacities. The spatiotemporal dynamics of cultural brain activity detail the functional relation of physiological signaling and the mental activity of culture and affective processing. The functional correspondence of physiological signaling with cultural and affective processing demonstrates the importance of the characterization of the spatiotemporal dynamics of cultural neural networks.

Cultural Cognitive Neural Networks

Research on cultural cognitive neural networks investigates the cultural and cognitive processing of neural networks. The cultural processing of cognition and higher-level function are based on the functional brain activity of cultural neural networks. The cultural patterns of functional brain activity show the information flow of cultural processing that comprises the cortical brain function of interconnected brain regions. The cultural processing of cognition details the functional basis of the higher-level function of cortical brain activity. The study of cultural cognitive neural networks contributes to the understanding of the functional role of cognition and higher-level function on the activation patterns of interconnected brain regions. The characterization of the interconnected brain regions of cognition and higher-level function is of importance to the study of the culture and cognition of brain function.

The functional processing of cultural cognitive neural networks demonstrates the causal role of cognition and higher-level function on the cultural processing of cortical brain regions. The cultural processing of cognitive information shows the information flow and activation patterns that are characteristic of the cognition and higher-level function of cortical brain regions. The causal–functional role of cultural processing on cognition shows the functional significance of patterns of cortical brain activity on culture and cognition. The functional patterns of cortical brain activity of cultural and cognitive processing show the functional significance of interconnected brain regions on the multilevel processing of the cognition and brain function of cultural capacities.

The cortical brain activity of cultural cognitive neural networks describes the functional activation of functional processes that comprise the conscious or deliberate processing of cultural information. The functional patterns of cortical brain activity show the functional role of culture on the conscious processing of cognitive information. The perceptual and cognitive processing of cultural information shows the functional role of cortical brain function on the culture and cognition of brain function. The conscious processing of cultural and cognitive information shows the functional basis of conscious or deliberate processing and its link to cortical brain activity. The cultural patterns of functional brain activity show the importance of cultural and cognitive processing on brain function.

The functional patterns of cultural brain activity of core brain regions demonstrate the functional role of cognition and higher-level processes on the brain function of cultural capacities. The core brain regions of cultural and cognitive neural networks located within occipitotemporal, parietal and frontal cortices show the neuroanatomical structures that comprise the functional processing of the cognition and brain function of culture. The functional brain activity of core brain regions details the perceptual and cognitive processing of cultural information that is linked to the activation patterns of interconnected brain regions. The cultural and cognitive processing of neural networks depicts the functional role of core brain regions on the higher-level processing of cortical brain function. The cultural patterns of functional brain activity are of importance to the understanding of the cognition and brain function of cultural capacities.

The functional processing of occipitotemporal brain regions shows the functional role of perception and cognition on cultural processing. The functional brain activity of occipitotemporal brain regions illustrates the perceptual and cognitive processing of cultural information. The perceptual and cognitive processing of cultural information depicts the functional patterns of cortical brain activity that are located within occipitotemporal cortices. The functional specificity of cortical brain activity of cortical brain regions facilitates the information processing of cultural capacities. The perceptual and cognitive processing of occipitotemporal cortices describes a functional and mechanistic basis for the cultural processing of cognitive information. Cultural patterns of cortical brain activity located within occipitotemporal lobes show the functional basis of the cortical brain function of perceptual and cognitive processing.

The cortical brain function of parietal lobes shows the functional role of cortical brain activity on the semantic processing of cultural information. The functional brain activity of cortical brain regions of the parietal lobe demonstrates a functional basis of the higher-level association of cortical areas. The functional brain activation of cortical areas located within the parietal lobe illustrates the functional integration of information-processing streams into the higher-level processing of cognition and brain function. The cortical brain activity of cortical brain regions show how higher-level processing facilitates the semantic comprehension of conceptual knowledge. The cultural processing of cognitive information within the cortical areas of the parietal lobe contributes to the semantic processing of conceptual knowledge that is cultural.

The functional patterns of brain activation of higher-level association areas facilitate the understanding of the shared meaning or significance of the conceptual processing of cultural information. The cultural patterns of functional brain activity located within the parietal lobe detail the activation states that comprise cultural and cognitive processing. The functional integration of information-processing streams within cortical areas facilitates the conceptual processing of cultural and cognitive information. The functional brain activity of cortical brain regions shows the causal–functional role of higher-level function on cultural and cognitive processing. The functional brain activity of cortical brain regions details the functional significance of higher-level association areas for cultural and cognitive processing.

The prefrontal cortex is a cortical brain region that is of importance to the cultural and cognitive processing of brain function. The functional role of prefrontal cortex in cultural and cognitive processing shows the functional basis of the control or regulation of cultural capacities. The functional brain activity of prefrontal cortices contributes to the control or inhibition of cultural processing. The cultural patterns of functional brain activity detail the activation patterns that are of importance to the control or inhibition of cultural and cognitive processing. The cortical brain function of prefrontal cortex demonstrates how cultural patterns of functional brain activity contribute to the controlled processing of cultural and cognitive information.

The cortical brain dynamics of cultural cognitive neural networks show the dynamical mechanisms of the cognition and higher-level function of cultural processing. The brain dynamics of culture and cognitive neural networks detail the information flow and activation patterns that comprise cultural and cognitive processing. The relaying of information across interconnected brain regions of cultural and cognitive neural networks shows the causal role of functional connectivity on the dynamical mechanisms of cortical brain dynamics. The information flow of cultural and cognitive processing illustrate the patterns of activation that are of importance to the cognition and higher-level function of cultural processing. The cortical brain dynamics of cultural cognitive neural networks depict the dynamical mechanisms of the higher-level processes of culture and cognition.

The brain dynamics of cultural and cognitive neural networks show the information flow and patterns of activation that comprise cultural processing. The patterns of activation states of cultural and cognitive processing demonstrate the global and local dynamics of interconnected brain regions that comprise neural networks. The activation patterns of cultural brain activity show the functional processing of cognition and higher-level function that is optimal as activation states of cultural and cognitive neural networks. The flow of information of cultural and cognitive processing is linked to the cultural brain activity of cognitive neural networks. The brain dynamics of cultural and cognitive neural networks detail the information processing and activation states that are of importance to cultural and cognitive processing.

The spatiotemporal dynamics of cultural and cognitive neural networks demonstrate the functional relation of physiological signaling and mental activity that comprises cultural and cognitive processing. The spatiotemporal brain dynamics of cortical brain regions show the fluctuations of physiological signaling that correspond with the levels of processing of mental activity that facilitate the performance of cultural tasks. The cultural patterns of dynamical processing of cortical brain regions facilitate the cultural and cognitive processing of cortical and mental activity. The characterization of the spatiotemporal dynamics of cultural and cognitive processing contributes to the understanding of the multilevel processing of brain function.

Research on cultural cognitive neural networks is of importance to the understanding of the cultural and cognitive processing of brain function. The information flow and activation patterns of cultural and cognitive neural networks show the functional processing of culture and cognition that are essential to the higher-level processes of brain function. The cortical brain dynamics of cultural and cognitive processing details the information flow and patterns of activation that are of importance to the functional processes of interconnected brain regions. The characterization of the multilevel mechanisms of cultural and cognitive processing contributes to the understanding of the functional basis of cultural cognitive neural networks.

Cultural Social Neural Networks

The study of the cultural social neural networks details the cultural processing of social information that is based on the cortical brain function of interconnected brain regions. The functional basis of cultural and social processing shows the functional role of cortical brain activity on the interconnection of cortical brain regions. The cultural and social processing of interconnected brain regions shows the causal–functional significance of cortical brain function on the social processing of cultural capacities. The functional processing of cultural and social information is linked to the functional patterns of cultural brain activity. The cortical brain dynamics of cultural and social neural networks contribute to the understanding of the information flow and activation patterns of cultural and social processing.

The cultural and social processing of neural networks describes the flow of information across interconnected brain regions that comprise cultural capacities. The information flow of cortical brain regions details the information processing that is specific to cultural and social processing. The functional specificity of cultural and social processing details the patterns of functional brain activity that are of causal–functional significance to cultural capacities. The cultural patterns of functional brain activity are linked to cultural and social processing. The understanding of the functional patterns of cortical brain activity that are essential to the processing of cultural and social information is of importance to the characterization of the multilevel mechanisms of cultural social neural networks.

The study of the neural networks of cultural social processing illustrates the functional basis of the processing of cultural information in social context. The functional processing of cultural information in context facilitates the cultural and social processing of cortical brain function. The cultural brain activity of social processing shows the patterns of functional activation that are linked to cultural and social processing. The cultural processing of social information contributes to the cortical brain activity of neural networks that demonstrates the functional activation patterns that comprise the social processing of cultural capacities. The cultural patterns of functional brain activity show the functional role of social processing as a functional basis of cultural task performance.

The cultural and social processing of cortical brain networks shows the functional role of cortical brain regions as a mechanistic basis of cultural capacities. Cortical brain regions located within the occipitotemporal and prefrontal cortices and their interconnection demonstrate the functional processing of social and cultural capacities. The functional patterns of activation of interconnected brain regions show the functional significance of cultural and social processing. The causal–functional significance of cultural and social processing contributes to the understanding of the functional basis of cortical brain function. The functional processing of social and cultural capacities illustrates the link of the functional brain activity of cortical brain networks and the cultural processing of social information.

The functional processes of occipitotemporal cortices show the causal–functional role of cortical brain regions and their interconnection on cultural and social processing. Brain regions located within occipitotemporal lobes such as the fusiform gyrus, lingual gyrus and hippocampus show the functional role of cortical brain regions on cultural and social processing. The functional patterns of cortical brain activity located within the neural network show the functional significance of the cultural processing of social information. The cultural patterns of functional brain activation located within occipitotemporal lobes are linked to the detection and recognition of social information in context. The amplification of cultural and social processing of cortical brain networks depicts the causal–functional significance of cultural patterns on functional processes. The functional brain activity of occipitotemporal regions demonstrates a causal mechanism of the processing of cultural and social information.

The cortical brain activity of cultural social neural networks demonstrates the detection and recognition of social information and the role of contextual influences. The cortical brain function of core brain regions and their interconnection show the functional significance of cultural and social processing on the functional performance of core capacities. The cultural task activity of social processing illustrates the functional relation of cultural patterns of cortical brain activity. The functional correspondence of levels of processing and levels of brain activity of cultural and social processing shows the causal–functional role of cortical brain regions on core capacities. The cultural processing of social information demonstrates task-based activity that is of importance to the functional processing of cortical neural networks.

The interconnection of cortical brain regions that comprise cultural and social processing show the functional relation of cortical brain activity and task-based processing. The functional patterns of activation of interconnected brain regions show the facilitation of task-based processing that is based on the mutual activation patterns of nodes of the cortical brain network. The mutual activation patterns of interconnected brain regions show the amplification of information processing that comprises the task-based activity of cultural and social processing. The functional brain activity of interconnected brain regions shows how the mutual activation patterns of interconnected brain regions facilitate the functional processing of cultural social processing.

The functional brain activity of cortical brain regions such as the superior temporal gyrus demonstrates the functional specificity of cultural and social processing. The cortical brain activity of the superior temporal gyrus is of importance to the cultural processing of social information. The functional patterns of activity of the cortical brain and its interconnection with other cortical brain areas demonstrate the causal–functional significance of the cortical brain region on the cultural processing of mental state understanding. The cultural processing of understanding the mental states facilitates the patterns of functional activation of cortical brain regions that is cultural. The cultural patterns of functional brain activity located within the superior temporal gyrus demonstrate the contextual influence of culture on social processing. The functional brain activity of the superior temporal gyrus details the functional role of the cortical brain region on cultural and social processing.

The functional processing of cultural neural networks shows the functional role of interconnected brain regions on the understanding of mental states. The cultural processing of cortical brain activity that is located within the superior temporal gyrus details the functional responsivity of the cortical brain region to the perceptual processing of social information. The cultural patterns of functional brain activation detail the social processing of perceptual cues that is of importance to mental state inference. The cultural processing of mental state understanding shows functional correspondence with the functional patterns of cortical brain activity. The cultural processing of social information facilitates the information flow and activation patterns of interconnected brain regions.

The cultural processing of social information is of importance to the functional processing of prefrontal cortex. The functional brain activity of prefrontal cortex contributes to the control and regulation of cultural and social capacities. The functional role of prefrontal cortex in the control or inhibition of cultural and social processing facilitates the regulation of cultural capacities. The prefrontal cortex is a cortical brain region that demonstrates the functional processing of control and inhibition on higher-level processes such as cultural and social processing. The cultural patterns of functional brain activity located within prefrontal cortex shows the functional role of the cortical brain region on the control and regulation of cultural capacities.

The cortical brain activity of the prefrontal brain function is of importance to the control and regulation of cultural capacities. The control and regulation of cultural and social processing are essential to the functional basis of cultural capacities. The cortical brain activity of prefrontal cortex and its interconnected brain regions demonstrate a causal mechanism of the control and regulation of social and cultural processing. The control or regulation of cultural and social processing is of importance to the facilitation and inhibition of functional processing that is fundamental to cultural capacities. The inhibitory control of cultural and social processing contributes to the cognitive representation of cultural and social information. The cultural patterns of functional brain activity located within the prefrontal cortex details the causal–functional role of activation patterns on the control and regulation of cultural capacities. The cultural patterns of prefrontal brain activity show the functional mechanisms of cultural and social processing.

The causal interaction of prefrontal cortex with interconnected brain regions illustrates the cortical brain dynamics of cultural processing. The causal–functional role of prefrontal cortex function with interconnected brain regions details the flow of information processing that comprises cultural capacities. The causal interaction of interconnected brain regions details the directionality and magnitude of the flow of information and activation patterns that comprise the dynamics of cortical brain function. The functional processing of prefrontal cortex facilitates the cognitive representation of cultural information that demonstrates the active maintenance of the representational content of cultural capacities. The active maintenance of the representation of cultural information facilitates the higher-level processing of cultural capacities. The cultural processing of prefrontal brain function shows the causal–functional significance of control and regulation on cultural capacities.

The cortical brain dynamics of cultural social neural networks depict the dynamical mechanisms of cultural and social processing. The dynamical basis of cortical brain function shows how information flow and patterns of activation of cortical neural networks are essential to the cultural processing of social information. The brain dynamics of cultural and social neural networks show the activation states that facilitate the pattern completion of cultural social processing. The information processing of cortical layers that comprise cultural social neural networks demonstrate the relaying of information from cultural input to cultural output. The cultural processing on social information illustrates the functional role of cortical brain dynamics on the understanding of the optimal states of activation that comprise cultural and social capacities.

The cortical brain dynamics of cultural social neural networks are of importance to the understanding of the functional relations of dynamical mechanisms and cultural social processing. The fluctuations of cortical brain activity that correspond with the levels of processing of cultural task activity show the functional relation of the physiological and mental activity of cultural social processing. The dynamical mechanisms of cortical brain function depict the functional relation of physiological and mental activity as a mechanistic basis of cortical brain dynamics and its functional significance. The brain dynamics of cultural and social processing show the functional significance of dynamical mechanisms on the functional processes of cultural capacities.

The spatiotemporal dynamics of cortical brain regions that are responsible for cultural and social processing show the functional role of the dynamical mechanisms of cortical brain regions on the processing of cultural and social information. The spatiotemporal dynamics of interconnected brain regions show the flow of information processing that comprises cultural capacities. The fluctuations of cortical brain processing and the mental activity of cultural and social tasks show the dynamical mechanisms of cortical brain function. The understanding of the spatiotemporal dynamics of cortical brain activity is of importance to the understanding of the cortical brain dynamics of cultural social neural networks.

The study of cultural social neural networks comprises the characterization of the multilevel mechanisms of cultural and social adaptation. The functional patterns of cortical brain activity comprise the mechanistic basis of the cultural and social processing of biological organisms as cultural species. The understanding of the multilevel processing of cortical brain activity shows the cultural patterns of social processing that are of importance to cortical brain networks. The cultural and social processing of cortical neural networks shows the functionality and causality of information processing mechanisms that are essential to cultural capacities.

Culture and Cortical Brain Dynamics

Cortical brain dynamics are foundational to the understanding of the cultural processing of cortical brain function. The brain dynamics of cortical neural networks show the global and local dynamics that comprise the cultural processing of brain function. The global and local dynamics of the cortical brain describe the flow of cultural information of cortical brain regions that comprise the dynamics of cortical brain function. The dynamical mechanisms of cortical brain function show the relaying of information processing that facilitates the functional processes of cortical brain regions and their functional significance. The cortical brain dynamics of cultural processing encompas the understanding of the dynamical mechanisms that contribute to the global and local dynamics of cortical brain function.

The dynamical mechanisms of neuronal populations comprise the mechanistic basis of the functional processing and dynamics of cortical brain networks. The changes of intrinsic and extrinsic processing of neuronal populations that comprise cortical brain regions show the dynamic changes of cortical brain function that are essential to cultural processing. The dynamical mechanisms of neuronal populations reflect on the changes of intrinsic and extrinsic processing that are of importance to the functional and emergent properties of cortical brain function. The dynamics of cortical brain networks show the aggregation of dynamic changes of neuronal populations that comprise the functional processing of networks of neurons.

Global brain dynamics of cortical neural networks describe the global representation of distributed information processing across a multitude of cortical brain areas. The global dynamics of cultural neural networks demonstrate the distributed processing of cultural information as multiple pathways of cortical brain regions. The distributed processing of cultural information shows the functional pathways of cortical brain regions that show the functional processing of cultural information. The global representation of the dynamical changes of cortical brain regions that comprise the relaying of cultural information depicts the overall aggregation of functional responsivity of populations of neurons that are based on the distributed representation of cultural processing.

The global dynamics of cultural neural networks describe the dynamics of distributed information processing across interconnected brain regions. The global representation of brain dynamics illustrates the self-organizing principle of cortical brain function. The dynamics of distributed information processing show the changes of cortical brain activity that are based on the structural and functional properties of interconnected brain regions. The interconnection of cortical brain regions facilitates the flow of information processing that contributes to the dynamics of neuronal populations. The global dynamics of cultural neural networks comprise the dynamics of cultural information processing that is distributed across cortical brain regions and their interconnection.

Local brain dynamics of cultural neural networks describe the local representation of cultural processing that is based on the specialized processing of specific cortical brain regions. The local representation of cultural processing encompasses the specialized information processing of cortical brain regions. The local representation of cortical brain regions depicts the feedforward processing of cortical brain function that shows the functional basis of cultural processing. The levels of activation of cortical brain regions describe the aggregation of population-level neuronal activity that encompasses the cortical brain area. The local brain dynamics of cortical brain networks detail the local dynamics of the cortical brain that contribute to the processing of cultural information.

The global and local brain dynamics of cultural neural networks depict the dynamical mechanisms of cortical brain activity that are linked to task-based activity. The cortical brain dynamics of neural networks demonstrate the dynamical fluctuations of cortical brain activity that arise based on the effortful processing of task-based activity. The patterns of activation of cortical brain dynamics show the optimal states of activation that comprise the pattern completion of the functional processing of cultural neural networks. The brain dynamics of cortical brain function show the global and local representation of the neuronal activity of populations of neurons that comprise the distributed and local processing of cultural information. The characterization of the cortical brain dynamics of cultural neural networks is of importance to the understanding of the functional processing of cultural task activity.

The attractor dynamics of cultural neural networks detail the attractor states that demonstrate the multiple constraint satisfaction of dynamical mechanisms. The attractor states of cultural neural networks describe the activation states that comprise the pattern completion of cultural processing. The representation and transformation of cultural information from the cortical input layer to the cortical output layer illustrates the states of functional activation that are of importance to the information flow of cultural processing. The information processing of cortical layers shows the functional role of activation states on the pattern completion of functional mechanisms. The information processing of cultural activation states details the flow of information towards the attractors states and activation patterns that comprise the optimal states of cultural processing. The cortical brain dynamics of cultural neural networks depict how dynamical mechanisms facilitate the information flow and activation patterns of cultural mechanisms.

The attractor states of cortical brain dynamics depict the patterns of activation that are optimal of cultural processing. The activation patterns of cultural neural networks describe the levels of energy that comprise the pattern completion of cultural processing. The attractors states of brain dynamics detail the overall levels of energy at a local minimum that characterize the optimal states of information processing. The processing of cultural information of cortical brain dynamics toward attractor states shows the functional specificity of activation patterns that facilitate cultural processing. The attractor dynamics of cortical brain networks show the functional brain activation that is of importance to the optimization of the levels of processing of cultural capacities.

The autonomous brain dynamics of cultural neural networks describe the spontaneous fluctuations of cortical brain activity that are based on the structural properties of brain function. The autonomous brain dynamics of cultural neural networks describe the endogenous brain activity that arises based on the resting-state activity of the brain. The autonomous dynamics of cortical brain activity of resting state networks comprise the fluctuations of physiological signaling that arises from maintenance of the active and itinerant representations of information processing that comprise cortical brain function at rest. The physiological signaling of autonomous brain dynamics shows the spontaneous brain activity that comprises resting state function. The maintenance of active and itinerant representations facilitates the spontaneous brain dynamics that comprise the entrainment of brain activity that is autonomous. The brain dynamics of cultural neural networks that are autonomous demonstrate the functional role of spontaneous brain activity on the entrainment of cultural brain function.

The intrinsic activity of cultural neural networks is comprised of the information processing that facilitates the endogenous dynamics of cortical brain function. The spontaneous and intrinsic activity of cortical neural networks show the endogenous brain activity that is based on the maintenance of active and itinerant representations of the cultural environment. The maintenance of active and itinerant representation of cultural information contributes to the patterns of activation states of cortical layers that comprise the cultural processing of the brain at rest. The intrinsic properties of cortical brain dynamics demonstrate the functional role on spontaneous brain activity on the cultural entrainment of cortical brain function. The intrinsic activity of cortical brain function shows the spontaneous fluctuations of brain activity that comprise the autonomous dynamics of cortical brain activity. The spontaneous and intrinsic activity of cortical brain dynamics shows the functional significance of cultural entrainment on the maintenance of the active representations of resting state activity. The endogenous brain activity of cortical brain function details the autonomous and self-organizing principle of cortical brain dynamics.

The causal interaction of cortical brain dynamics describes the information processing of interconnected brain regions that characterize brain function. The dynamical mechanisms of cortical brain function show the causal interaction of functional brain activation across cortical brain regions that comprise the intrinsic and extrinsic activity of populations of neurons. The causal dynamics of the brain activity of neuronal populations entail the functionality and directionality of information flow of cortical brain networks. The causal role of information flow and activation patterns on the dynamical mechanisms of cortical brain function show the functional significance of brain dynamics on the understanding of the intrinsic and extrinsic processing of the cortical brain.

The study of culture and cortical brain dynamics are of importance to the understanding of the dynamical mechanisms that are fundamental to cultural processing. From intrinsic to extrinsic processing, cortical brain dynamics reveal the functional significance of dynamical mechanisms and the causal–functional role of information processing on the intrinsic and extrinsic properties of cortical brain function. The causal interaction of dynamical processing of the cortical brain shows the functional importance of the understanding of brain dynamics that characterize the cultural processing of interconnected brain regions. The characterization of the functionality and causality of cortical neural networks is essential to the understanding of the causal role of the simple to complex dynamics of the information processing of cultural capacities.

Discussion

The study of cultural neural networks is fundamental to the understanding of the cultural processing of cortical brain networks. The computational principles of cortical brain networks describe the structural and functional principles that comprise cortical brain organization across levels of processing. The cultural processing of cortical brain networks shows the information flow that comprises the activation patterns of cortical brain regions. The characterization of the cultural information flow of cortical brain regions and its activation patterns is of importance to cultural information processing. The information processing of cortical brain networks demonstrates the functional basis of the cultural processing of brain and behavior. The structural and functional principles of cortical brain organization are fundamental to the understanding of the cultural processing of cortical brain networks.

The interconnectivity of cultural neural networks shows the role of structural and functional connectivity on the cultural processing of cortical neural networks. The interconnection of cortical brain regions shows the information flow that comprises the functional processing of culture. The flow of cultural information across cortical brain regions illustrates the role of interconnectivity on the functional processing of cultural capacities. The bidirectional processing of cultural information flow is of importance to the understanding of the control or inhibition of cultural processing that is based on the interconnection of cortical neural networks. The interconnectivity of cultural neural networks is of importance to the characterization of the multilevel mechanisms of cultural adaptation.

The cortical brain dynamics of cultural processing details the functional significance of dynamical changes of cortical brain function on the processing of cultural information. The attractor dynamics of cortical brain function show the activation patterns that comprise the pattern completion of cultural information-processing mechanisms. The global and local dynamics of cortical brain function demonstrate the activation patterns that are essential to the functional processing of cultural capacities. The autonomous brain dynamics of cultural neural networks facilitate the functional processing of cortical brain regions that are of importance to the intrinsic processing of cultural entrainment. The understanding of the cortical brain dynamics of cultural processing shows the importance of dynamical changes of functional activation patterns that are essential to the understanding of the functional and emergent properties of cultural capacities.

Conclusion

The study of the cultural neural networks contributes to the empirical discovery of the computational principles of cortical brain networks and its role on cultural processing. The scholarly and scientific interest on the computational foundations of cultural neural networks demonstrates the inherent naturalism of the cortical brain and its role on multilevel adaptation. Empirical research on cultural neural networks facilitates the understanding of the intrinsic and extrinsic processing of cultural information that comprises cortical brain function. The functional capacity of cortical brain function for the autonomous and task-based processing of cultural information contributes to the functionality and adaptability of the cortical brain. Research on cultural neural networks is of importance to the broader understanding of the functional significance of the evolution and adaptation of the brain and its importance for cultural adaptation.

The scholarly and scientific interest on the computational approaches to the study of cultural neural networks shows the importance of computational tools and technologies on the scientific understanding of the functional processes and mechanisms of cortical brain organization. The empirical approaches to the study of cultural neural networks show the functional role of interconnectivity on the cultural information processing of the brain. The structural and functional properties of cortical neural networks demonstrate the functional processing that is essential to cortical brain organization. The understanding of the structural and functional properties of cultural neural networks illustrates the representations and transformations of cultural information that are essential to the functionality and adaptability of cortical brain function.

The evolution and adaptation of biological organisms as cultural species brings forth a compendium of considerations regarding its causal–functional significance. The cultural capacities of biological organisms as complex species show the intricacies of the cultural processing of cortical brain function and its functional role. The demonstration of the functional role of cortical neural networks as a level of processing of cultural capacities brings forth a wide realm of interest on the manifestation of the representation of cultural information as networks of neurons. The computational discovery of the functional significance of cultural neural networks provides a foundation for scholarly and scientific inquiry on the cultural brain and its underlying computational principles.

Implications

Research on cultural neural networks has broad implications for the development of computational approaches to the study of the information-processing mechanisms of cultural neural networks. Computational tools and technologies are beneficial to the understanding of the functionality and causality of cultural neural networks. The development of computational tools and technologies in health and medicine is of importance to the scientific and technological innovation on culture and health in neuroscience. The development of computational approaches facilitates the development of scientific models that facilitate the prediction and explanation of the brain basis of cultural capacities. The scientific and technological innovation of research capabilities in health and medicine is beneficial to the understanding of the casual interaction of cortical brain networks and their functional significance. The building of the research capabilities and scientific infrastructure for the development of computational tools and technologies is foundational to the generation of scientific knowledge that facilitates the innovation of science and technology on culture in health and medicine.

7

CULTURAL BRAIN DYNAMICS

Introduction

The rise of scientific empiricism encompasses the advances of scholarly and scientific interest on the importance of observation and scientific discovery as a means of understanding the fundamental laws and principles of the causal structure of the natural world. The scholarly and scientific interest on the discernment of the role of parts and particulars of causal mechanisms are of importance to the characterization of the patterns and regularities of simple to complex systems. The lawlike principles of patterns and regularities of simple to complex systems illuminate on the importance of basic mechanisms as a causal impetus of functional and emergent properties. The characterization of the lawlike principles of basic mechanisms facilitates broader understanding of the functionality and causality of the inherent naturalism of the world.

The scientific realism of the causal impetus of basic mechanisms as a fundamental basis of functional and emergent properties illuminates on the causal interaction of the patterns and regularities of simple to complex systems. The causal interaction of mechanisms and their causal effects illustrates the notion of the importance of patterns and regularities of aggregate systems. The scientific observation and inherent naturalism of the functional principles of causal mechanisms facilitates the understanding of patterns and regularities of aggregate systems that hold. The characterization of the causation of basic mechanisms as fundamental elements of patterns and regularities illustrates the importance of the causal effects of functional and emergent properties.

DOI: 10.4324/9781003384236-8

The scientific discovery of the 19th century of the neuron as the fundamental unit of the nervous system contributes to the understanding of the cellular and molecular basis of information processing of the cortical brain organization. The cellular and molecular basis of the nervous system demonstrates the functional role of neurons and networks of neurons on the information processing of the cognition and higher-level function of brain function. The structure and function of neurons as populations of cells located within the core brain regions illustrate the cellular and molecular basis of neuroanatomical structures. The causal interaction of cortical brain regions and their causal effects detail the structural and dynamical principles of cortical brain organization. The structural and functional properties of cortical brain organization detail the functional role of neurons and neuronal networks as a causal mechanism of cortical brain function across levels of processing. The structural and functional principles of cortical brain organization detail the understanding of the structural and dynamical basis of cortical brain function.

The goal of the chapter is to provide a comprehensive review of cultural brain dynamics. The study of the structural and dynamical basis of cortical brain function is of importance to the characterization of the cortical brain dynamics of cultural processing. The structural and functional properties of large-scale cultural brain networks describe the causal interaction of the core brain regions that facilitate the processing of cultural information. The brain dynamics of cultural brain networks describe the functional role of dynamical mechanisms and its causal effects on the characterization of the cultural patterns of cortical brain function. The understanding of the structural and dynamical principles of cultural brain dynamics is essential to the detailed characterization of the computational principles of the nervous system.

Functional Architecture of the Cultural Brain

The functional architecture of the cultural brain describes the structural and functional properties of the nervous system that are of importance to the cultural processing of core capacities. The structural and functional properties of cortical brain organization describe the basic mechanisms of cortical processing and its causal effects as an aggregate system of neurons and networks of neurons. The characterization of the multilevel mechanisms of cultural processing encompasses the causal mechanisms of large-scale cultural brain networks that are of importance to the functional processing of cultural capacities. The study of the structural and functional properties of cortical brain organization contributes to the characterization of the basic mechanisms of cortical processing and its causal effects as cultural patterns of cortical brain function. The structural and dynamical principles of cortical brain function are essential to the understanding of the functional architecture of the cultural brain.

Neurons and Networks of Neurons

Neurons are the cellular and molecular basis of the structural and functional organization of the nervous system. The structure and function of neurons and its response properties detail the electrophysiological processing that comprises the information processing of brain function. The neuroanatomical basis of cortical tissue of the nervous system is based on the structure and function of neurons. The neuroanatomical structure of the core brain regions is comprised of the anatomical and physiological properties of distinct types of neurons and networks of neurons. The structural and functional properties of neurons comprise the fundamental unit of neural computation of cortical brain regions and its causal interaction.

The anatomical and physiological structure of neurons detail the structural properties that comprise the function of neuronal activation. The electrical signaling of cellular components of neurons shows the neural communication that formulates the functional basis of the excitatory and inhibitory neurotransmission of neuronal activation. The synaptic transmission of neurons comprises the excitatory and inhibitory interaction of neurotransmission describes the neurochemical signaling of specific types of neurons. The electrophysiological signaling of neurons as a cellular mechanism describes the neural information processing based on excitatory and inhibitory neurotransmission. The synaptic dynamics of excitatory and inhibitory processing show the electrophysiological signaling that describe patterns of neuronal activation. The patterns of neuronal activity of neurons describe the electrophysiological response of neurons and its response properties that are of importance to the encoding and decoding of sensory information of the environment.

Distinct types of neurons show patterns of activation that are comprised of distinct patterns of neuronal activity and its response properties. The functional properties of neurons are comprised of distinct patterns of neuronal activity that are intrinsic and deterministic to the type of neuron. Distinct types of neurons show neurochemical properties that comprise distinct patterns of neuronal activity based on neuroanatomical location, structural and functional interaction and type of neurotransmission. The spontaneous and intrinsic activity of neurons detail the patterns of neuronal activation that are based on the electrophysiological signaling and neural transmission of cellular mechanisms. Patterns of activation of neuronal mechanisms show the intrinsic properties of spontaneous activity that comprises neuronal activation and its response properties. The multitude of patterns of intrinsic and extrinsic activity of neuronal activation illustrate the functional properties of distinct types of neurons located within the neuroanatomical structures.

The neural transmission of neurons and networks of neurons describe the multilevel processing of cortical organization. The synaptic transmission of networks of neurons details the aggregate levels of neuronal activity of populations of cells that comprise cortical brain regions. The synaptic connection of networks of neurons describes the neural communication of cortical brain regions based on excitatory and inhibitory neurotransmission and its synaptic efficiency. The neuronal activity of neuronal networks demonstrates the neural transmission of neuronal populations and the causal effects of its synaptic connection. The patterns of neuronal activity of networks of neurons demonstrate the activation patterns that are of importance to cortical brain regions and their function.

Neuroanatomical Structures

The neuroanatomical structures of cortical brain organization show the structural and functional properties of the nervous system. The structural and functional properties of cortical brain organization detail the structural components and hierarchical structure of the cortical brain. The structural neuroanatomy of cortical brain organization is comprised of cortical and subcortical brain areas that comprise major cortical structures. The cortical and subcortical areas of neuroanatomical structures such as the occipital, temporal, parietal and frontal lobes constitute the cortical brain structures that are of importance to the structural and functional properties of cortical brain organization. The cortical areas of neuroanatomical structures describe the cortical brain regions and their structural and functional properties that are essential to the information processing of cortical brain function.

The hierarchical structure of cortical brain organization details the large-scale organization of the cortical brain. The large-scale organization of the cortical brain describes the distributed representation of information processing that encompasses the structural and functional components of cortical brain function. The distributed representation of information processing details the functional role of cortical mechanisms as a functional basis of cortical pathways. The distribution of representation as comprised of a multitude of core brain regions demonstrates the distributed processing of information that comprises the cortical processing of brain function. The large-scale organization of the cortical brain describes the structural and functional components of cortical brain organization.

Core Brain Regions

The core brain regions describe the functional processing of cortical mechanisms that shows the specialized information processing of cortical brain function. The core brain regions that comprise neuroanatomical structures detail the specialized information processing that is based on the neuronal activity of populations of neurons located within specific neuroanatomical structures. The core brain regions show the patterns of neuronal activity that demonstrate the functional specificity of information processing mechanisms. The identification of core brain regions shows the functional basis of information processing that demonstrates the dedicated and specialized mechanisms of cortical brain function.

The structural and functional properties of the core brain regions detail the functional processing of cortical brain function that is dedicated and specialized to information-processing mechanisms. The specialized information processing of the core brain regions and their interconnection comprise a functional basis for the functional specificity of large-scale cortical brain networks. The distribution of information processing across the core brain regions shows the distribution of types of representation that comprise the functional processing of cortical brain networks. The functional processing of interconnected brain regions shows the information processing that is dedicated and specialized to the structural and functional properties of large-scale cortical brain networks.

The core brain regions of large-scale cultural brain networks show the functional processing of cultural information that is dedicated and specialized to the cortical brain regions and their interconnection. The cultural processing of the core brain regions demonstrates a functional basis of the information-processing mechanisms that show functional specificity to interconnected brain regions. The processing of cultural information that is based on specialized information-processing mechanisms details the distributed processing of cultural representation that comprises the interconnectivity of cultural brain networks. The functional processing of cultural information depicts the functional specificity of specialized mechanisms and their causal effects on the information processing of cortical brain function.

The interconnection of the core brain regions depicts the structural and functional properties that comprise the information processing of large-scale cortical brain networks. The information processing of interconnected core brain regions shows the distribution of representation across multiple brain areas. The structural and functional connectivity of the core brain regions facilitates the relaying of information processing across the core brain regions. The structural connectivity of the core brain regions demonstrates the role of synaptic connection based on white matter tracts as a structural basis of cortical brain function. The functional connectivity of the core brain regions depicts the distributed processing of cortical brain regions and their interconnection. The structural and functional basis of the core brain regions details the information-processing mechanisms that comprise large-scale cortical brain networks.

The structural and functional properties of the core brain regions contribute to the interconnectivity of large-scale cultural brain networks. The structural and functional connectivity of cortical brain function shows the cultural processing of information that is dedicated and specialized to the core brain regions. The cultural processing of affective, cognitive and social information details the functional role of specialized information processing mechanisms that are located within interconnected brain regions. The distribution of representation across specialized functional pathways shows the functional basis of cultural processing of that comprises large-scale cultural brain networks.

Computational Principles of Cortical Brain Dynamics

The computational principles of cortical brain dynamics show the structural and functional principles that comprise the dynamical mechanisms of cortical brain function. The structural and functional properties of cortical brain organization are essential to the understanding of the dynamical aspects of cortical brain organization. The structural and functional properties of cortical brain function show the causal mechanisms that facilitate the dynamical processing of cortical brain function. The computational foundations of cortical brain dynamics are of importance to the characterization of the structural and dynamical principles of cortical brain function.

Large-scale Cortical Brain Networks

Large-scale cortical brain networks detail the distributed processing of the core brain regions that comprise cortical brain networks. The large-scale cortical organization of cortical brain function describes the importance of the distribution of representation across a multitude of core brain regions. The functional pathways of information-processing streams show the functional significance of the distributed processing of cortical brain regions and their interconnection. The distributed processing of information of large-scale cortical brain networks details how the relaying of information across the core brain regions is essential to functional processing. The distribution of processing streams across interconnected brain regions facilitates the functional processing that is of importance to the large-scale organization of cortical brain networks.

The distributed processing of cortical brain networks entails the characterization of specialized information-processing streams and their functional relation to the dynamical aspects of cortical brain networks. The specialized information processing of the core brain regions that comprise cortical brain networks shows the causal relations of the core brain regions that demonstrate the relaying of information processing mechanisms. The bidirectional processing of the core brain regions shows the directionality and causality of information flow that comprises the functional processing of cortical brain networks. The feedforward and feedback processing of the core brain regions details the information processing that is specialized to the core brain regions of cortical brain networks. The dynamical aspects of cortical brain networks show the specialized processing of the core brain regions that facilitates the functional basis of large-scale brain organization.

The structural and functional properties of large-scale cortical brain networks detail the bidirectional processing of the core brain regions and its causal effects. The bidirectional processing of the core brain regions shows the relaying of information that is essential to the functional processing of cortical brain function. The excitatory and inhibitory processing of cortical brain regions depicts the functional processing that facilitates the causal interaction of the core brain regions. The bidirectional processing of the core brain regions depicts the causal influence of information processing on the functional activation of the core brain regions. The functional processing of the core brain regions shows the bidirectional processing of the core brain regions that is of importance to the brain function of cortical brain networks.

The causal interaction of the core brain regions that comprise large-scale cortical brain networks shows the directionality and causality of functional processing of interconnected brain regions. The causal interaction of the core brain regions shows the flow of information processing that is of importance to the patterns of activation that comprise cortical brain networks. The bidirectional processing of information flow facilitates the causal role of the core brain regions as a functional basis of patterns of cortical brain activation. The directionality and causality of information flow and activation patterns depict the functional processing that comprises large-scale cortical brain networks.

Structural and Functional Connectivity

The functional integration of cortical brain regions shows the role of interconnection on the flow of information processing across distributed cortical brain areas. The functional integration of cortical brain regions demonstrates the distributed processing of information based on the information flow of a multitude of cortical brain areas. The distributed processing of functional mechanisms that comprise cortical brain regions details the distribution of representations of interconnected brain regions. The interconnection of cortical brain regions depicts the causal role of synaptic connection as a functional basis of cortical brain dynamics. The information flow of processing mechanisms facilitates the distribution of representation that comprises the functional integration of cortical brain function.

The structural and functional connectivity of cortical brain networks shows the role of interconnection on the bidirectional processing of cortical brain function. The structural and functional properties of cortical brain networks depicts the characteristics that comprise the functional basis of information processing. The interconnection of cortical brain regions that is based on structural and functional properties shows how neural transmission serves as a conduit of the information processing mechanisms of cortical brain function. The structural and functional connectivity of cortical brain regions facilitate the bidirectional information processing that is essential to the structural and functional principles of cortical brain organization. The understanding of the structural and functional principles of cortical brain organization is of importance to the computational foundations of cortical brain networks.

The structural connectivity of cortical brain regions consists of the structural properties of cortical brain function. The interconnection of cortical brain regions is comprised of the synaptic connection of neuronal populations based on the white matter tracts that show the interconnection of cortical brain regions. The synaptic density of white matter tracts depicts the strength or magnitude of structural connectivity that characterizes the structural basis of interconnected brain regions. The strength or magnitude of structural connectivity shows causal relation to the functional processing of cortical brain function. The structural connectivity of cortical brain regions is of importance to the characterization of the structural properties of cortical brain function.

Functional connectivity of cortical brain networks entails the functional properties of cortical brain regions. The functional connectivity of interconnected brain regions details the relaying of information processing across cortical brain regions. The interconnectivity of cortical brain regions consists of the unidirectional and bidirectional processing that describes the information flow of cortical brain networks. The feedforward and feedback processing of information contributes to the pattern of activation states that comprise cortical brain networks. The functional connectivity of cortical brain regions shows how the unidirectional and bidirectional processing of information flow is sufficient as a functional basis of interconnectivity.

The unidirectional processing of information flow describes the local dynamics of cortical brain regions that show specialized information processing. The feedforward processing of information flow depicts the specialized information processing of cortical brain regions. The unidirectional processing of information flow demonstrates the amplification of functional properties of information processing that are of importance to the pattern completion of activation states. The feedforward processing of cortical brain regions details the local dynamics of information flow that are of importance to specialized processing mechanisms. The unidirectional processing of cortical brain networks shows the causality of information flow that has functional significance to the pattern completion of activation states that comprise cortical brain function.

Attractor Dynamics

The attractor dynamics of cortical brain networks demonstrate the attractor states and patterns of activation that comprises the cortical processing of brain function. The bidirectional processing of neuronal populations shows the dynamical aspects of cortical brain function that comprise the patterns of activation of cortical brain networks. The excitatory and inhibitory processing of networks of neurons depict the information flow and activation patterns that facilitate the functional processing of interconnected brain regions. The attractor states of cortical brain networks show the bidirectional processing of activation patterns that are of importance to the information flow of cortical layers.

The local attractor network is comprised of neuronal populations that comprise the excitatory and inhibitory processing of cortical brain function. Local attractor dynamics describe the dynamical processes of excitatory and inhibitory processing of populations of neuronal activity that facilitates the aggregation of neuronal response located within the core brain regions. The local representation of the dynamical basis of functional processing shows the fluctuations of population-level neuronal activity that comprise cortical brain function. The local attractor dynamics facilitate the aggregation of neuronal response for the representation and transformation of input to output. The dynamical processing of attractor states of the core brain regions shows the bidirectional processing of information flow that is of importance to cortical brain function.

The attractor dynamics of cortical brain networks show the importance of attractor states as patterns of activation that facilitate cortical processing. The information processing of cortical brain networks aims for the dynamical processing of information flow across interconnected brain regions towards patterns of activation states that show functional processing. The information flow and activation patterns of cortical brain dynamics demonstrate the pattern completion towards attractor states that comprise the optimal states of activation patterns. The brain dynamics of cortical brain networks detail the information processing of cortical layers that comprise states of activation patterns that are essential to bidirectional processing.

The patterns of activation of cortical brain networks detail the information processing of cortical layers that facilitates the excitatory and inhibitory processing of cortical brain regions. The activation patterns of cortical layers describe the states of activation patterns that facilitate the representation and interpretation of information processing. The relaying of information of interconnected brain regions from cortical input layer to cortical hidden layer to cortical output layer shows the activation states that comprise the representation and transformation of sensory data into motor output. The information flow and patterns of activation states of cortical brain networks demonstrate the bidirectional processing of cortical layers that is essential to functional processing.

Attractor states describe the activation states that comprise the optimization of activation patterns for the pattern completion of bidirectional processing. Attractor dynamics show how activation patterns of functional processing facilitate the completion of activation states toward optimal states of activation based on levels of energy. The activation pattern consists of attractor states that demonstrate the global or local minimum levels of energy. The attractor states of brain dynamics detail the states of activation that illustrate the pattern completion of transformations of sensory input to motor output. The cortical brain dynamics of brain networks illustrate attractor states that comprise the activation patterns of bidirectional processing.

The brain dynamics of cortical brain networks show how the representation and transformation of sense data relies on the sets of activation states that comprise functional processing. The bidirectional processing of sense data into the cortical input layer represents the activation of patterns that relies on the completion of functional processing. The information flow from cortical input layer to cortical output layer facilitates the representation and interpretation of information processing across the hidden or context layer of cortical brain function. The bidirectional processing of information flow from cortical input layer to cortical output layer depicts how the representation and transformation of information processing are essential to the pattern completion of activation states of cortical brain function.

The demonstration of attractor dynamics of cortical brain networks is of importance to the identification of the states of activation that comprise the satisfaction of multiple constraints. The attractor states as fixed points or states of activation of information processing mechanisms shows the maintenance of active representations of bidirectional processing that facilitate the information flow and activation patterns that comprise functional processing. The activation states of attractor dynamics detail the activation patterns and information flow that is of importance to the cortical brain function. The brain dynamics of the cortical brain regions show the activation patterns that are essential to the pattern completion of functional processing and the satisfaction of multiple constraints.

Multiple Constraint Satisfaction

Multiple constraint satisfaction entails the functional processing of cortical brain networks that is essential to the activation patterns of information processing mechanisms. The functional processing of cortical brain networks comprises the patterns of activation that are of importance to the pattern completion of activation patterns. The bidirectional processing of cortical brain networks shows the satisfaction of multiple constraints that is of importance to the functional basis of cortical brain activity. The feedforward and feedback processing of information flow entails the patterns of activation that facilitate the functional role of input–output relations. The feedforward and inhibitory processing of information flow demonstrates the functional processing and activation patterns that are essential to cortical brain networks.

The pattern completion of activation patterns of cortical brain networks shows the information flow of functional processing across cortical layers of brain organization. The information flow and cultural patterns of cortical brain networks detail the information processing across cortical input layer to cortical output layer. The activation patterns of cortical brain networks describe the flow of information processing that facilitates the transformation of representation into motor output. The partial activation of cortical brain networks of sensory information undergoes pattern completion as information processing streams from cortical input layer to the cortical output layer. The states of activation patterns that show pattern completion of information processing streams illustrates the functional processing that is characteristic of cortical brain networks.

The computational principle of multiple constraint satisfaction details the functional significance of the activation states that show the overall satisfaction of multiple constraints of cortical brain networks. The functional significance of activation patterns details the importance of states of activation that comprise the pattern completion of activation of cortical brain networks. The functional role of multiple constraint satisfaction as a computational principle of cortical brain dynamics facilitates the demonstration of the pattern completion of activation that is essential to functional processing.

Computational Approaches to Cultural Brain Dynamics

Computational approaches to the study of cultural brain dynamics abound of conceptual and methodological considerations that are of importance to the understanding of the dynamical principles of the cultural brain. The understanding of the dynamical principles of the cultural brain relies on the conceptualization of the functional relation of cultural processing and cortical brain dynamics and its causal effects. The cultural processing of cortical brain dynamics shows the structural and functional properties of cultural brain networks that are of importance to the dynamical aspects of cultural brain function. The characterization of the causal mechanisms of cultural brain dynamics is of importance to the functional significance of the cultural and cortical processing of cultural brain networks.

Conceptual approaches to the study of cultural brain dynamics consider a wide realm of scientific approaches. The development of scientific concepts and language on cultural brain dynamics facilitates the theoretical and methodological development of computational tools and technologies that facilitate the characterization of the dynamical mechanisms of the cultural brain activity. The development of scientific concepts and language of cultural brain dynamics considers the fundamental basis of the causal interaction of cortical brain regions that comprise cultural brain networks. The characterization of the causal interaction of dynamical mechanisms that underlie cultural brain networks is essential to the understanding of cultural brain dynamics.

Methodological development of computational tools and technologies is beneficial to the understanding of the computational modeling of the causal interaction of cortical brain regions that comprise cultural brain networks and its causal effects. The computational modeling of cultural brain dynamics entails the formalization of conceptual models that describe the directionality and causality of the interaction of cortical brain regions and its functional significance. The development of computational tools and technologies that facilitate the theory building and hypothesis testing of conceptual models is of importance to the study of cultural brain dynamics. Conceptual models that hypothesize the causal interaction of cortical brain regions that comprise cultural brain networks are of importance to the theory building and hypothesis testing of the computational modeling of cultural brain dynamics.

Computational modeling of cultural brain dynamics postulates on several considerations regarding the causal interaction of cortical brain regions of cultural brain networks. First, computational approaches consider the identification of the core brain regions that are of importance as a brain basis of cultural processing. The identification of the core brain regions that are responsible for cultural processing is essential to the understanding of cultural brain networks. Second, detailed characterization of the structural and functional connectivity of cultural brain networks is of importance to ascertain the structural and functional properties of cultural processing. The structural and functional connectivity of cultural brain networks is essential to the characterization of the structural and functional properties of the cultural brain activity. Third, the computational modeling of the causal interaction of cortical brain regions that comprise cultural brain networks is fundamental to the hypothesis testing of the dynamical aspects of the cultural brain activity. Finally, consideration of the autonomous brain dynamics of cultural brain networks is of importance to the understanding of the intrinsic cultural processing of the cultural brain activity. The study of the autonomous brain dynamics of endogenous brain activity contributes to the characterization of the patterns of activation that are linked to intrinsic cultural processing.

The empirical study of cultural brain dynamics relies on the conceptual and methodological development of scientific concepts and language that provide a foundation for the theory building and hypothesis testing on the dynamical principles of the cultural brain. The identification of the dynamical mechanisms that comprise the cultural patterns of cortical brain activity is essential to the characterization of the causal mechanisms of cultural brain dynamics. The development of conceptual and methodological approaches for the study of cultural brain dynamics contributes to the scientific and scholarly inquiry on the computational principles of the cultural brain.

Large-scale Cultural Brain Networks

The detailed characterization of large-scale cultural brain networks is of importance to the understanding of cultural brain dynamics. Large-scale cultural brain networks describe the interconnection of cortical brain regions that are responsible for cultural processing. The interconnectivity of cortical brain regions that comprise a functional basis of cultural processing facilitates the understanding of the intrinsic and extrinsic properties of cultural brain networks. The distributed processing of representation across interconnected brain regions facilitates cultural and cortical processing. The information flow and activation patterns of cultural brain networks illustrate the dynamical aspects of cultural processing that are fundamental to cultural brain dynamics.

Large-scale cultural brain networks describe the core brain regions that comprise the interconnected brain regions of cortical brain networks. The distributed representation of cultural information across the core brain regions located across cortical brain areas shows the functional significance of interconnection on cultural processing. The relaying of information across interconnected brain regions shows the functional significance of the functional processing of cultural brain networks. The directionality and causality of information processing across the core brain regions depicts the causal interaction of interconnected brain regions that is of importance to the understanding of the dynamical basis of cultural processing.

The unidirectional or feedforward processing of cultural information describes the relaying of information across interconnected brain regions toward the pattern completion of activation that comprise cultural processing. The feedforward processing of cultural information facilitates the amplification of features of cultural processing that are of importance to the activation patterns of cultural brain dynamics. The information-processing stream entails the flow of information across cortical layers of distributed representation. The information flow of cultural processing from cortical input to cultural output layers shows the activation pattern of the feedforward processing of cultural brain networks.

The information flow and activation patterns of cultural brain networks describe the cultural processing that is based on cortical brain mechanisms. The excitatory and inhibitory processing of neurons and networks of neurons illustrates the activation patterns that show functional relation to cultural processing. The feedforward processing of cultural information demonstrates the flow of information that is based on the excitatory processing of interconnected brain regions. The flow of cultural information of cortical layers of distributed representation entails the fluctuations of cortical brain activity that facilitate the feedforward processing of cultural brain networks.

The pattern completion of activation states shows the cultural processing that facilitates the cortical brain activity of cultural neural networks. The information flow and activation patterns of cultural processing depict the feedback and inhibitory processing of cortical brain networks. The feedback or inhibitory processing of cultural information facilitates the control or inhibition of the cultural processing of cortical brain activity. The bidirectional processing of cultural activation patterns contributes to the pattern completion of activation states that are of importance to cultural brain networks. The cultural processing of cortical brain activity illustrates the relaying of information of interconnected brain regions that is of importance to the characterization of the functionality and causality of cultural brain dynamics.

Causal Interaction

The study of the causal interaction of cultural brain regions is essential to the understanding of the dynamical basis of cultural and cortical processing. The interaction of the core brain regions and its causal effects facilitates the information flow and activation patterns that comprise the brain dynamics of cultural brain networks. The characterization of the causal interaction of cultural brain regions details the unidirectional and bidirectional processing of cultural information that is essential to cultural brain dynamics. The dynamical basis of cultural and cortical processing depicts the causal role of cultural patterns on the information and activation patterns of cultural brain networks.

The computational modeling of the causal interaction of the core brain regions that are responsible for the cultural processing is of importance to the understanding of the structural and dynamical properties of cultural brain networks. The structural properties of cultural brain networks describe the structural basis of the interconnectivity of core brain regions. The structural basis of the interconnectivity of cultural brain networks encompasses the neuroanatomical structures and their synaptic interconnection. The structural properties of cultural brain networks consider characteristics such as the magnitude and directionality of synaptic inter-connection that describes the structural connectivity of core brain regions. The study of the structural properties of cultural brain networks contribute to the characterization of the structural and dynamical properties of cultural processing.

The computational modeling of cultural brain dynamics includes the consideration of the functional properties of the cultural processing of cultural brain networks. The functional connectivity of interconnected brain regions describes the functional basis of the relaying of information of core brain regions of cultural brain networks. The functional properties of cultural brain networks demonstrate the distributed representation of cultural information across core brain regions. The functional processing of core brain regions illustrates the cortical brain activity that shows functional relation with cultural task activity that is stimulus driven. The characterization of the functional relation of the cultural processing of cortical brain activity and mental function is of importance to the functional properties of cultural processing. The computational modeling of cultural brain dynamics details the understanding of the causal interaction of core brain regions of cultural brain networks and its causal effects.

The computational modeling of cultural brain dynamics also considers the functional role of cultural brain function as a predictor of cultural task performance. The dynamic causal modeling of the causal interaction of core brain regions of cultural brain networks further considers the functional role of dynamical processes on the functional performance of cultural task activity. The study of the functional properties of brain–behavior relation-ships of cultural processing is of importance to the broader understanding of the functional significance of cultural brain dynamics. The computational modeling of cultural brain dynamics is fundamental to the characterization of cultural brain networks and their causal–functional significance.

Autonomous Brain Dynamics

Computational approaches to the study of autonomous brain dynamics consider the spontaneous and intrinsic brain activity of cultural brain networks. The computational modeling of cultural brain dynamics based on resting-state activity provides a conceptual basis for the systematic investigation of the structural and dynamical properties of the cultural brain activity that is endogenous. The consideration of the study of the structural and intrinsic properties of endogenous brain activity relies on the observation and meas-urement of the cultural brain activity at resting state. The characterization of the endogenous brain activity of cultural brain networks facilitates the understanding of how spontaneous fluctuations of cortical brain activity comprise the intrinsic cultural processing of core brain regions.

The cultural processing of cortical brain function that is based on resting-state activity details the cultural brain dynamics of endogenous brain activity. The spontaneous fluctuations of cortical brain activity show the intrinsic properties of cultural processing. The fluctuations of cortical brain activity based on resting-state activity illustrate the intrinsic properties of endogenous brain activity. The multitude of functional roles of endogenous brain activity shows the functional significance of intrinsic cultural processing. The cortical brain activity of core brain regions at rest illustrates the patterns of activation of intrinsic processing that is of importance to the encoding of cultural information and the prediction of the cultural brain activity.

The entrainment of the cortical brain shows the functional role of spontaneous and intrinsic properties on the cultural processing of cultural brain networks. The endogenous brain activity of cultural brain networks shows the spontaneous fluctuations of cortical brain activity that are based on resting-state activity. The spontaneous fluctuations of cortical brain activity contribute to the intrinsic cultural processing of resting-state function. The cortical brain activity of intrinsic processing that is cultural shows the active maintenance of representations that are of importance to the activation patterns of endogenous brain activity. The functional significance of activation patterns of endogenous brain activity illustrate the causal role of intrinsic properties on the encoding of cultural processing that is stimulus-independent.

Autonomous brain dynamics describe the dynamical basis of the intrinsic processing of cultural brain networks. The intrinsic properties of cortical brain activity show the autonomous and self-organizing principle of cultural brain dynamics. The dynamical basis of cultural and cortical processing shows the causal role of endogenous brain activity on cultural activation patterns and its causal–functional significance. The structural and intrinsic properties of cultural brain networks detail the causal effects of endogenous brain activity on the cultural processing of core brain regions. The intrinsic processing of cultural information of core brain regions shows the functional basis of autonomous brain dynamics that comprise cultural brain networks.

Cultural Brain Dynamics

The study of cultural brain dynamics encompasses a wide realm of considerations of the structural and dynamical basis of cultural and cortical processing. The characterization of the causal interaction of core brain regions of cultural brain networks describes the structural and functional properties that are of importance to the understanding of cultural brain dynamics. The structural and functional properties of the interconnection of cortical brain regions show the dynamical aspects of the cultural processing of cultural brain networks. From synaptic dynamics to large-scale cortical network dynamics, the structural and dynamical basis of neuronal activity shows the functional role of the causal interaction of core brain regions on cultural processing. The dynamical mechanisms of neurons and networks of neurons show the multilevel processing of cortical brain activity that comprises the functional basis of cultural brain dynamics.

The scholarly study of cultural brain dynamics is fundamental to the structural and dynamical principles of cultural brain networks. The characterization of the structural and dynamical principles of cultural brain networks is of importance to the understanding of the cultural processing of core brain regions and their interconnection. The structural principles of cultural brain networks describe the structural basis of dynamical aspects of cultural brain function. The functional principles of cultural brain networks detail the functional basis of the dynamical processing of cultural information of the core brain regions. The cultural processing of core brain regions illustrates the distributed processing that comprises the structural and dynamical basis of large-scale cultural brain networks.

The characterization of core brain regions that are responsible for cultural processing is of importance to the computational modeling of cultural brain dynamics. The neuroanatomical structures of main lobules including the occipitotemporal, limbic, parietal and frontal cortices comprise the structural basis of core brain regions that underlie cultural brain networks. The core brain regions located within the cortical and subcortical areas and their interconnection show the neuroanatomical basis of the structural and functional properties of cultural brain networks. The functional specialization of the core brain regions located within the cortical areas detail the functional processing of interconnected brain regions that are responsible for cultural processing. The functional integration of core brain regions that include the occipitotemporal, limbic, parietal and prefrontal cortices demonstrate the functional basis of the interconnectivity of the cultural brain network. The neuroanatomical structures of the main lobules of cortical areas comprise a structural and functional basis of the functional processing of the cultural brain network.

The structural and functional connectivity of core brain regions that comprise the cultural brain networks demonstrate the hierarchical structure of cortical brain organization. The structural connectivity of core brain regions depicts the synaptic connection of cortical and subcortical structures based on the interconnection of white matter tracts. The interconnection of cortical and subcortical brain structures illustrates a structural basis of the intrinsic and extrinsic connectivity of the core brain regions. The functional connectivity of core brain regions includes the information flow and activation patterns of the core brain regions that comprise the cultural brain network. The cultural patterns of functional brain activity of core brain regions show the functional properties of cortical brain function that comprise the cultural brain network. The structural and functional properties of core brain regions illuminate on the hierarchical structure of cortical brain function.

The cultural brain dynamics of core brain regions show the functional processing of core brain regions that comprise cultural capacities. Cultural brain dynamics detail the functional basis of cortical brain activity that corresponds with the levels of processing of cultural capacities. The dynamical aspects of cultural and cortical processing demonstrate the patterns of cortical brain activity that show the functional relation of the culture of brain–behavior relationships. The functional basis of cortical brain activity of core brain regions shows that the flow of information across interconnected brain regions facilitates the functional correspondence of cortical and cultural processing. The cultural brain dynamics of cultural processing illustrate the dynamical aspects of cortical brain activity that facilitate cultural mental function.

The synaptic dynamics of neurons and networks of neurons describe the fluctuations of neuronal activity and its response properties that correspond with cultural processing. The synaptic transmission of neurons and networks of neurons shows the patterns of neuronal activity and response properties that are of importance to the higher-level function of cultural processing. The excitatory and inhibitory processing of populations of neurons that comprise cortical brain regions illustrate the bidirectional processing of cultural information that is based on the synaptic dynamics of cellular and molecular mechanisms. The interconnection of neural transmission of cortical brain areas shows the functional role of electrophysiological signaling on the functional processing of interconnected brain regions that comprise cultural brain networks.

The global brain dynamics of cultural brain networks show the distribution of representation of cultural processing that comprises the cortical brain activity of core brain regions. The global dynamics of core brain regions describe the aggregation of population-level neuronal activity that is located within interconnected brain regions of the cultural brain network. The global dynamics of cultural brain networks demonstrate the overall satisfaction of activation patterns that comprise the pattern completion of cultural processing. The cortical brain function of core brain regions demonstrates the cultural patterns of functional brain activity that are responsible for cultural processing. The characterization of global brain dynamics facilitates the distributed processing of cultural information that is of importance to cultural brain networks.

The interconnectivity of core brain regions that comprise the cultural brain network describes the pattern of activation states that comprise global brain dynamics. The global dynamics of brain function depict the optimal states of activation as the global minimum of activation states across interconnected brain regions. The satisfaction of activation patterns illustrates the activation states of pattern completion that comprise the global dynamics of cultural brain networks. The global representation of the cortical brain activity of interconnected brain regions facilitates the characterization of the level of cultural processing of core brain regions that comprise the cultural brain network.

The local brain dynamics of cultural brain networks describe the cultural processing of cortical brain activity that comprise cultural brain networks. The local representation of cultural brain dynamics demonstrates the specialized information processing of the core brain regions of the cultural brain activity. The local dynamics of cortical brain activity show the neuronal activity of cortical brain regions that show the specialized processing of functional mechanism. The local brain dynamics detail the cultural patterns of neuronal activity that contribute to the cultural processing of cortical brain regions. The functional basis of cultural processing facilitates the local brain dynamics of cultural brain networks.

The simple to complex dynamics of cultural brain networks illustrate the functional role of cortical brain dynamics on the levels of processing of cultural capacities. The spatiotemporal dynamics of cultural brain networks details the fluctuations of cortical brain activity that coincide with levels of processing on cultural task activity. Cultural patterns of spatiotemporal dynamics of cortical brain function detail the cultural processing that is based on the fluctuations of cortical brain activity and cultural task activity. The functional concordance of cortical brain activity and cultural task activity of core brain regions depicts the cultural patterns of the spatiotemporal dynamics that characterize cultural brain dynamics.

The spatiotemporal processing of core brain regions further demonstrates the fluctuations of cortical brain activity that comprise cultural processing. The spatiotemporal dynamics of core brain regions show the cultural patterns of cortical brain activity that comprise levels of processing. The fluctuations of spatiotemporal dynamics comprise the functional basis of the cortical and cultural processing of cultural brain networks. The cultural patterns of cortical brain function show the levels of processing that affect the spatiotemporal processing of cultural brain networks. The study of the spatiotemporal dynamics of core brain regions facilitates the understanding of the cultural processing of interconnected brain regions.

The attractor dynamics of the cortical brain activity of cultural brain networks detail the patterns of activation and information flow that comprise cultural processing. The brain dynamics of cultural processing illustrate the attractor states that are based on the cultural brain activity of the core brain regions. The cultural brain activity of the core brain regions depicts patterns of activation that are of importance to the pattern completion of the activation states of cultural brain networks. The cultural processing of cortical brain activity shows the flow of information processing across core brain region toward activation patterns that facilitate the cultural brain activity. The information flow and activation patterns of cultural brain networks depict the attractor states that comprise cultural processing.

The cultural patterns of cortical brain activity of cultural brain networks detail patterns of activation that correspond with the flow of cultural information. The activation patterns of cortical brain activity show the activation states or attractor states of core brain regions that facilitate the completion of cultural processing. The fluctuations of cultural brain activity across cortical layers of processing illustrate the flow of information from cortical input layer to cortical output layer. The cultural patterns of activation that comprise cortical brain activity show the pattern completion of cultural processing. The patterns of activation states of cultural brain networks are of importance to the understanding of the information flow and functional brain activation of cultural brain networks.

The attractor states of cultural brain dynamics depict the states of activation that are optimal as cultural patterns of functional brain activity. The cultural patterns of functional brain activity consist of the activation states that are optimal as attractor states of activation patterns. The information flow of cultural processing of core brain regions shows the cultural processing of interconnected brain regions toward attractor states of cultural brain dynamics. The cultural processing of core brain regions shows the functional significance of interconnected brain regions that facilitate the optimal states of activation. The attractor dynamics of the cultural brain activity show the information flow and activation patterns that are optimal for cultural processing.

The cultural processing of cortical brain networks shows the bidirectional processing of information that comprises cultural patterns of functional brain activity. The excitatory and inhibitory processing of neurons and networks of neurons facilitate the feedforward and inhibitory processing of cultural information across core brain regions. The bidirectional processing of cultural information illustrates the representation and transformation of sensory data into motor output across cortical layers of brain organization. The fluctuations of cortical processing across the cortical input layer to the cortical hidden layer to the cortical output layer. The cortical processing of the hidden or context layer comprises the sets of activation patterns that facilitate the functional relations of cultural input–output. The information processing of the hidden or context layer contributes to the representation and interpretation of cultural input to cultural output. The information flow of cultural brain networks across cortical layers of processing facilitates the cultural patterns of functional brain activity.

The attractor states of cultural brain networks describe the states of activation that comprise the optimal states of activation. Attractors states serve as fixed points of activation that detail the levels of energy that demonstrate the satisfaction of multiple constraints. The activation states of the cultural brain activity that show the global or local minimum of levels of energy comprise the attractor states that are essential to the pattern completion of cultural processing. The attractor dynamics of cultural brain networks show the activation patterns that facilitate the attractor states of the cultural brain activity. The cultural brain dynamics of the core brain regions show the information flow and activation patterns of cultural brain networks.

The autonomous brain dynamics of the cultural brain describe the spontaneous and intrinsic processing of endogenous brain activity. The intrinsic properties of cortical brain activity show the cultural processing of endogenous brain activity that is of importance to the autonomous brain dynamics of the cortical brain at rest. The endogenous brain activity of cultural brain networks shows the fluctuations of electrophysiological signaling that comprise the cortical brain activity of the core brain regions at rest. The intrinsic processing of cultural brain activity demonstrates the autonomous and self-organizing principles of the intrinsic and extrinsic processing of cortical brain dynamics. The autonomous brain dynamics of cultural brain networks show the intrinsic properties of cultural processing.

The intrinsic properties of cultural processing facilitate the entrainment of cultural brain function. The autonomous and self-organizing cortical brain activity of cultural brain networks show the spontaneous and intrinsic processing that is based on the functional processing of the core brain regions that is stimulus-independent. The functional processing of the core brain regions shows that the intrinsic cultural processing of cortical brain activity is of importance to the encoding of cultural information and the prediction of cultural patterns that comprise the entrainment of cultural brain function. The autonomous brain dynamics of endogenous brain activity show the intrinsic processing of cultural information that is of importance to the entrainment of the cultural brain.

The cultural brain dynamics of the core brain regions show the causal interaction of interconnected brain regions and its functional significance. The causal interaction of interconnected brain regions depicts the flow of information and patterns of activation that serve a causal–functional role on the cultural processing of cortical brain activity. The cultural processing of interconnected brain regions shows the interaction of the core brain regions and its causal effects. The cultural brain activity of core brain regions demonstrates the causal interaction of the core brain regions and the functional significance of cultural patterns of brain activity.

Cultural Brain Dynamics of Emotion

Research on the cultural brain dynamics of emotion investigates the structural and dynamical basis of the cultural processing of emotion information. The structural and dynamical properties of cultural processing detail the intrinsic and extrinsic processing of the core brain regions that facilitate cultural and affective processing. The structural properties of cultural brain dynamics detail the structural basis of the neuroanatomical structure that facilitates the interconnection of cultural brain networks of emotion. The dynamical properties of cultural brain dynamics demonstrate the functional basis of the information flow and activation patterns of the relaying of information across core brain regions. The intrinsic and extrinsic processing of cultural information of core brain regions shows the structural and functional properties of cultural brain dynamics of emotion.

The characterization of the core brain regions of cultural and affective processing contributes to the understanding of the cortical brain dynamics of culture and emotion. The cultural and affective processing of core brain regions that are located within the cortical and subcortical brain areas of limbic and prefrontal cortices show the cortical brain dynamics of cultural and affective processing. The core brain regions of corticolimbic circuitry detail the information flow and activation patterns that facilitate the structural and dynamical basis of the culture brain networks of emotion. The core brain regions of corticolimbic circuitry comprise the neuroanatomical basis of the structural and functional properties of the cultural brain network of emotion.

The cultural brain dynamics of emotion encompass the interconnection of core brain regions and its functional relation to cultural processing. The brain dynamics of cultural and affective processing detail the intrinsic and extrinsic properties that entail the functional processing of cultural information. The intrinsic processing of cultural brain networks reflects the cortical brain dynamics of endogenous brain activity that is based within corticolimbic structures and its interconnected brain regions. The extrinsic processing of large-scale cultural brain networks shows the brain dynamics of the cultural patterns of cortical brain activity that is stimulus-driven and its causal effects. The cultural patterns of cortical brain activity reflect the functional relation of cortical brain activity and cultural task activity. The cultural brain dynamics of emotion demonstrate the cultural processing of interconnected brain regions.

The cultural processing of cortical brain activity shows functional relation with cultural task activity. The functional concordance of the fluctuations of electrophysiological signaling and cultural task activity shows the functional role of task-based activity on the cultural patterns of cortical brain activity. The functional processing of the cultural brain activity illustrates the functional basis of the culture of brain–behavior relation-ships. The functional relation of cultural brain activity and cultural task activity shows the functional relation of cultural brain function and men-tal activity that is of importance to the understanding of the dynamical basis of functional processing. The cultural processing of functional brain activity contributes to the characterization of the cultural brain network of emotion.

The bidirectional processing of cultural brain dynamics shows the infor-mation flow and activation patterns that comprise the control or regula-tion of the cultural processing of emotion. The feedforward and inhibitory processing of cultural brain networks shows the causal mechanism for the control and regulation of cultural and affective processing. The inhibitory processing of the core brain regions shows the functional basis of the con-trol and regulation of culture and emotion. The bidirectional processing of cultural brain networks facilitates understanding of the cultural patterns of emotion processing that are linked to cortical brain activity.

The computational modeling of the cultural brain dynamics of emo-tion entails the study of the causal interaction of the core brain regions of cultural brain networks that are of importance to the cultural processing of emotion. The understanding of the directionality and causality of the causal interaction of the patterns of activation of core brain regions is essen-tial to the characterization of the cultural brain dynamics of emotion. The causal influence of cultural processing on cortical brain regions shows the top-down and bottom-up influences of information processing that con-tribute to the causation of the cultural brain activity of core brain regions. The functional basis of core brain regions that are of importance to cultural and affective processing facilitate the causal interaction of cultural brain networks that comprise the cultural brain dynamics of emotion.

The spatiotemporal dynamics of cultural brain networks of emotion show the cultural patterns of brain dynamics that comprise cultural and affective processing. The spatiotemporal dynamics of cultural and affec-tive processing show the fluctuations of electrophysiological signaling that is linked to the cultural processing of emotion. The electrophysiological signaling of neurons and network of neurons detail a functional basis of the cultural processing of emotion. The relaying of information flow across core brain regions shows the functional significance of cultural brain net-works of emotion.

The cultural brain dynamics of cortical brain networks depict the mul-titude of functional relations that characterize the causal interaction of the core brain regions of cultural and affective processing. The causal interac-tion of the core brain region illustrates the relaying of information across cortical layers of representation of interconnected brain regions. The cul-tural brain dynamics of cortical brain activity depict the functional relation of culture and brain–behavior relationships. The functional correspondence of cultural brain activity and cultural task activity detail the functional basis of the cultural patterns of cortical brain function. The study of the cultural brain dynamics of emotion shows the casual interaction of cultural and affective processing and its causal effects.

Cultural Brain Dynamics of Cognition

The study of the cultural brain dynamics of cognition details the structural and dynamical basis of the cortical brain activity that is responsible for cultural and cognitive processing. The brain dynamics of cultural brain networks demonstrate the information flow and activation patterns that comprise the functional basis of cognition and higher-level processes of cultural capacities. The dynamical aspects of cultural and cognitive processing show the cortical brain activity that is of importance to the functional relation of spatiotemporal brain dynamics and mental function. Cultural influences on cognitive processing and its neural bases illustrate the causal interaction of the core brain regions of cultural brain networks. The computational modeling of causal interaction of core brain regions that are responsible for cultural and cognitive processing contributes to the understanding of the cultural brain dynamics of cognition.

The structural and dynamical basis of cultural brain networks illustrate the functional and emergent properties of the dynamical mechanisms of the cortical brain activity of culture and cognition. The dynamical aspects of cultural brain dynamics depict the core brain regions that are of importance to cultural and cognitive processing. The core brain regions located within the occipitotemporal, parietal and prefrontal cortices show the cortical brain activity of large-scale cultural brain networks that is responsible for the cultural processing of cognitive information. The functional processes of the core brain regions show the functionality and causality of information flow across interconnected brain regions of cultural brain networks. The causal interaction of core brain regions illustrates the information processing of cortical layers that demonstrate the cortical brain activity of culture and cognition. The cultural patterns of functional brain activity contribute to the functional relation of the cultural and cognitive processing and its casual effects. The functional basis of core brain regions shows the causal–functional significance of cortical brain activity and its role in the understanding of the cortical brain dynamics of culture and cognition.

The interconnection of the core brain regions that comprise cultural brain networks show the functional basis of cortical brain activity that is of importance to culture and cognition. The cultural processing of cognitive information entails the cortical brain activity of interconnected brain regions that facilitate the bidirectional processing of information flow. The structural and functional connectivity of the core brain regions of cultural brain networks shows the structural and functional properties that are of importance to cultural and cognitive processing. The structural and functional basis of the bidirectional processing of information flow shows the cultural patterns of cortical brain activity that comprise cortical brain dynamics. The cultural and cognitive processing of interconnected brain regions demonstrates the functional basis of cultural brain networks.

The structural connectivity of cultural brain networks shows the interconnection of core brain regions that are of importance to the structural basis of intrinsic cultural processing and its underlying cultural brain dynamics. The structural basis of the interconnectivity of interconnected brain regions is based on the interconnection of networks of neurons that comprise core brain regions of the cultural brain network. The structural connectivity of interconnected brain regions shows the anatomical connectivity of core brain regions that facilitates the strength or magnitude of interconnection of white matter tracts. The strength or magnitude of structural connectivity of core brain regions that comprise cultural brain networks shows the structural properties of neuronal networks that are of importance to the functional processes of culture and cognition.

The structural basis of cultural brain networks demonstrates the anatomical connectivity that is of importance to the intrinsic cultural processing of endogenous brain activity. The endogenous brain activity of cortical brain regions that contribute to functional processing illustrates the spontaneous and intrinsic properties of cortical brain activity. The spontaneous and intrinsic activity of cortical brain regions detail the patterns of cortical brain activity that facilitate intrinsic processing. The endogenous brain activity of cortical brain regions that are involved in cultural and cognitive processing demonstrates the spontaneous and intrinsic processing of functional brain activity at rest. The functional brain activity of cortical brain regions during resting state activity shows the cortical brain activity that is based on intrinsic processing. The structural properties of neuroanatomical structures located within cortical brain areas of importance to culture and cognition show the intrinsic cultural processing that is based on endogenous brain activity.

The functional connectivity of cultural brain networks consists of functional and emergent properties that describe the interconnectivity of the core brain regions that are responsible for the cultural processing of cognition. The functional properties of the core brain regions show the cultural patterns of cortical brain activity that correspond with cultural task activity. The cortical processing of interconnected brain regions demonstrates the information flow and activation patterns that comprise the functional relation of cultural brain activity and its relation to behavior. The functional processing of the core brain regions details the patterns of functional brain activation that are of importance to cultural task activity. The cortical brain activity of the core brain regions facilitates the functional and emergent properties of the functional processing of cultural capacities. The functional connectivity of the cultural brain networks shows the functional and emergent properties of interconnected brain regions that are of importance to culture and cognition.

The functional processing of the core brain regions of cultural brain networks shows the cultural patterns of functional brain activity and its functional significance. The functional correspondence of cultural patterns of functional brain activation and cultural task activity shows the causal–functional role of cultural processing on cortical brain function. The patterns of functional activation of cultural brain activity correspond with levels of processing of cultural task activity. The functional relation of cultural patterns of cortical brain activity and cognitive activity illustrates the concordance of the culture of brain–behavior relationships. The cultural influences on cognitive processing and cortical brain activity illustrate the functional role of task-based activity on the functional processes of cultural brain activity.

The unidirectional processing of cultural information of cultural brain networks details the functional significance of information processing mechanisms on the cultural and cognitive processing of functional brain activity. The unidirectional processing of cultural information shows the functional basis of the cortical mechanisms of cognition and higher-level function. The feedforward processing of cultural information of core brain regions depicts the functional processes of cortical areas of importance to the dedicated and specialized processing of cognition. The activation patterns of functional brain activity of core brain regions illustrate the cultural and cognitive processing that is based on task-based activity. The feedforward processing of cultural information shows the functional basis of the cortical brain mechanisms of culture and cognition.

The specialized processing mechanisms of cultural brain networks demonstrate the functional role of content-specific processes and its underlying cortical mechanisms. The core brain regions of cultural brain networks comprise functional mechanisms that facilitate the dedicated and specialized processing of cognition. The cultural processing of cognition depicts the information flow and activation patterns of core brain regions that is of importance to functional processes that is content-specific. The cultural patterns of functional brain activity of core brain regions illustrate the content-specific processing of cultural brain networks. The specialized and dedicated processing of core brain regions demonstrates the functional basis of the cultural processing of cognition.

The bidirectional processing of functional brain activity details the cultural patterns of activation that comprise the functional processes of cultural brain networks. The feedforward and feedback processing of cultural brain networks depict the relaying of information across cortical layers of representation. The bidirectional processing of cultural information from cortical input to hidden to cortical output layers facilitates the representation and transformation of sensory data into motor output. The information processing of the hidden or context layer describes the states of activation patterns that facilitate the interpretation of cortical input. The cultural patterns of functional activation of cortical brain regions illustrate the of optimization of the activation states of functional processing. The storage and maintenance of cognitive representations as activation patterns of the hidden or context layer illustrates the activation states that facilitate the cognition and higher-level function of cultural brain activity. The information processing of core brain regions of cultural brain networks shows the functional basis of culture and cognition.

The distributed representation of cultural and cognitive processing across a multitude of core brain regions of large-scale cultural brain networks illustrates the flow of information that comprises cortical brain activity. The distributed processing of information across interconnected brain regions details the types of functional processes of cultural and cognitive processing. The cultural patterns of functional brain activation demonstrate cognitive processing that is process-specific. The distributed representation of information across core brain regions of cultural brain networks illustrates the functional processes of cognition. The distribution of representation of culture and cognition contributes to the understanding of the cultural brain dynamics of large-scale cultural brain networks.

The cultural patterns of functional brain activity of cognition illustrate the control or inhibition of cultural capacities. The cultural processing of cognition demonstrates the functional role of feedback or inhibitory processing on the control and regulation of culture and cognition. The cultural patterns of functional brain activity of cognition illustrate the inhibitory processing that facilitates the control and regulation of cultural and cognitive processing. The inhibitory processing of cortical brain activity of the core brain regions facilitates the control or regulation of cognition in context. The functional patterns of activation of cultural and cognitive processing detail the functional processes that are of importance to the control and regulation of cultural capacities.

The interconnectivity of cultural brain activity shows the functional basis of the core brain regions of cultural brain networks. The structural and functional properties of large-scale cultural brain networks illustrate the structural and functional basis of the information flow and activation patterns of culture and cognition. The interconnection of the core brain regions demonstrates the structural and functional basis of the relaying of information processing of the cultural brain network. The intrinsic and extrinsic processing of cultural information of cortical brain regions shows the functional significance of the functional and emergent properties of cultural brain activity. The interconnectivity of the core brain regions is fundamental to the intrinsic and extrinsic connectivity of cultural brain networks.

The computational modeling of the causal interaction of core brain regions of cultural brain networks details the functionality and causality of the information processing mechanisms of culture and cognition. The computational modeling of the dynamics of causal interaction of cultural brain activity aims to characterize the information flow and activation patterns that comprise cultural and cognitive processing. The cultural brain dynamics of the core brain regions illustrate the causal influence of the core brain regions and the relaying of information that comprises the functional basis of culture and cognition. The dynamical modeling of the causality of core brain regions and its functional processes contributes to the understanding of the causation of functional mechanisms that comprise cultural brain networks.

The understanding of the spatiotemporal processing of cultural brain dynamics is of importance to the characterization of the electrophysiological signaling of cultural and cognitive processing. The fluctuations of electrophysiological signaling of cortical brain regions and its functional relation to mental activity during cultural tasks illustrate the spatiotemporal processing of culture and cognition. The cultural patterns of cortical brain activity that correspond with the cultural task activity of mental function show the functional role of brain-behavior relationships. The spatiotemporal dynamics of cultural brain networks illustrate the functional role of fluctuations of cultural brain activity and mental activity on cultural task performance.

The autonomous brain dynamics of cultural brain activity demonstrate the cultural patterns of functional activation that comprise the intrinsic processing of the core brain regions. The functional brain activation of the core brain regions during resting state illustrates the spontaneous and intrinsic activity of cultural brain activity. The spontaneous and intrinsic properties of cultural brain networks detail the active maintenance of cortical representations that are of importance to the functional processing of cortical brain regions. The intrinsic cultural processing of cultural brain networks shows the cortical brain activity that facilitates the cortical representations of cultural and cognitive processing that is stimulus-independent. The functional brain activation of cultural patterns of cortical brain function shows the intrinsic processing that is essential to the autonomous brain dynamics of the cultural brain.

The intrinsic processing of cultural brain activity shows the functional processing of cortical brain regions that facilitate the entrainment of cultural brain function. The functional processing of cultural brain regions shows the endogenous brain activity of cultural brain function during resting state. The endogenous brain activity of cultural brain function illustrates the intrinsic properties of functional brain activity that is based on the spontaneous fluctuations of cortical brain regions. The endogenous brain activity of cultural brain function depicts the spontaneous activity of cortical brain mechanisms that comprise the intrinsic properties of the cultural brain. The intrinsic properties of functional brain activity contribute to the functional processing that is of importance to the entrainment of cultural brain function.

The autonomous brain dynamics of cultural brain function further demonstrate the functional processes that contribute to the functional significance of cultural brain activity. The functional processes of cultural brain activity facilitate the patterns of activation that contribute to the prediction of action patterns that is of importance to cultural and cognitive processing. The cultural patterns of functional brain activity depict functional processing that is of importance to the functional activation of culture and cognition. The functional brain activity of intrinsic cultural processing contributes to the activation states that facilitate the cultural processing of cognitive information. The autonomous and self-organizing principles of cultural brain dynamics detail the functional processing of cultural brain activity that is of importance to intrinsic processing.

Research on the cultural brain dynamics of cognition is of importance to the understanding of the structural and dynamical basis of the cultural processing of cortical brain function. The computational modeling of the core brain regions of cultural brain networks is of importance to the understanding of the functional basis of cultural and cognitive processing. The study of the causal interaction of the core brain regions of large-scale cultural brain networks details the directionality and causality of cultural patterns of functional activation and its relation to cognition and higher-level function. The structural and dynamical properties of cultural brain networks facilitate the cultural processing of cognitive information. The characterization of the cultural brain dynamics of cognition is fundamental to the understanding of the structural and dynamical basis of large-scale cultural brain networks.

Cultural Brain Dynamics of Social Processing

The systematic study of cultural brain dynamics of social processing encompasses the characterization of the cultural and social processing of large-scale cultural brain networks. The structural and dynamical properties of cultural brain networks entail the structural and functional basis of the cultural processing of social information. The structural and functional basis of cultural brain activity depicts the dynamical mechanisms of the cortical brain activity of cultural and social processing. The functional mechanisms of the brain basis of cultural and social processing illustrates the dynamical basis of cortical brain activity and cultural capacities. The cultural brain dynamics of social processing contribute to the understanding of the dynamical mechanisms of large-scale cultural brain networks.

The structural and dynamical principles of cultural brain networks demonstrate the intrinsic and extrinsic processing that comprises the cultural brain activity of the core brain regions. The structural and functional properties of cultural brain activity detail the intrinsic and extrinsic activity of cortical brain regions. The intrinsic and extrinsic activity of cortical brain mechanisms and its response properties are of importance to the functional basis of cultural and social processing. The intrinsic and extrinsic properties of the cortical brain activity of functional mechanisms show the structural and functional basis of cultural and social processing.

The structural and dynamical basis of cultural brain activity shows the functional significance of the cortical brain mechanism of cultural and social processing. The distinct types of representations of functional mechanisms illustrate the intrinsic and extrinsic properties that are of importance to the cultural processing of social information. The patterns of cortical brain response of functional mechanisms depict the types of representation that comprise the processing of cultural and social information. The intrinsic and extrinsic properties of functional mechanisms illustrate the structural and functional basis of information-processing mechanisms. The structural and dynamical basis of cultural brain activity illustrates the functional significance of functional mechanism on the processing of cultural and social information.

The neuroanatomical structures of large-scale cultural brain networks detail the structural and dynamical basis of the functional processes of cultural and social processing. The neuroanatomical structures of main cortical lobules such as occipitotemporal, limbic, parietal and frontal cortices show the core brain regions of cultural brain networks. The anatomical connectivity of cultural brain networks serves as a structural basis of the interconnection of the core brain regions. The core brain regions of cultural brain networks illustrate the functional mechanism of the social information processing of cultural capacities. The structural properties of the neuroanatomical basis of cultural brain networks demonstrate the interconnection of cortical brain regions that facilitate the cultural processing of social information.

The structural connectivity of cultural brain networks depicts the structural properties of core brain regions that facilitate the interconnection of cortical brain areas. The structural properties of the core brain regions are comprised of the synaptic connection of cortical brain mechanisms located within the core brain regions. The interconnection of the core brain regions based on white matter tracts contributes to the information processing of cultural and social capacities. The structural properties of cortical brain mechanisms detail the intrinsic activity of cortical brain regions and its interconnection. The structural basis of cultural brain networks facilitates the understanding of the intrinsic properties of structural connectivity and its causal effects.

The functional connectivity of large-scale cultural brain networks illustrates the functional basis of cultural and social processing. The functional and emergent properties of cultural brain regions depict the functional processes that are based on cortical brain activity. The functional processing of core brain regions details the intrinsic and extrinsic properties that facilitate the cultural processing of social information. The functional brain activity of core brain regions illustrates the patterns of activation that comprise the processing of cultural and social information. The intrinsic and extrinsic processing of cortical brain regions facilitates the functional basis of information processing mechanisms. The intrinsic and extrinsic properties of functional connectivity describe the functional basis of cultural and social processing.

The functional properties of cultural brain networks show the patterns of activation that comprise the cultural processing of social information. The cultural patterns of functional brain activity illustrate the functional specificity of cultural and social processing. The cultural processing of social information depicts the functional brain activity of cortical brain mechanisms. The functional specificity of the response patterns of cortical brain mechanisms shows the dedicated and specialized processing of core brain regions. The functional brain activity of cortical brain regions demonstrates the patterns of activation that are responsible for cultural and social processing.

The functional processing of cultural brain networks describes the cultural patterns of activation that are linked to cultural and social processing. The cultural patterns of functional activation of cortical brain regions show the functional relation of the cortical brain activity and cultural task activity of social processes. The functional correspondence of cultural brain activity and cultural task activity demonstrates the cultural patterns of brain function and behavior. The cultural basis of social processing illustrates how cortical brain activity predicts the cultural task activity of social behavior. The cultural patterns of functional activation of social processing illustrate the functional basis of culture in brain and behavior.

The interconnectivity of cortical brain regions demonstrates the intrinsic and extrinsic connectivity of large-scale cultural brain networks. The bidirectional processing of cultural and social information illustrates the information processing across cortical layers of representation. The feed-forward and feedback processing of cortical brain regions depicts the automatic detection and control or regulation of the processing of social information. The feedforward processing of core brain regions shows the automatic detection of social information of the cultural environment. The functional basis of core brain regions of cultural brain networks illustrates the automatic detection of social information that is culture-specific. The feedback processing of core brain regions shows the control or regulation of cultural and social processing. The feedback or inhibition of cortical brain regions illustrates the control or regulation of the cultural processing of social information. The bidirectional processing of functional mechanisms shows the intrinsic and extrinsic processing of cultural brain networks.

The intrinsic and extrinsic processing of cultural brain networks facilitates the information processing of cortical layers of representation. The representation and transformation of sensory input into motor output relies on the information processing of functional mechanisms. The processing of cultural and social information across cortical layers of representation facilitates the functional significance of intrinsic and extrinsic processing. The relaying of cultural and social information from the cortical input layer to the cortical hidden layer to the cortical output layer illustrates the transformation of the representational content of cultural and social processing. The cultural processing of the cortical input layer to the hidden or context layer and the cortical output layer facilitates the interpretation of cultural input that is of importance to social processing. The processing of cultural and social information of cortical brain regions shows the intrinsic and extrinsic processing that is essential to cultural brain networks.

The causal interaction of core brain regions that comprise cultural brain networks illustrate the causal-functional role of cortical brain activity on the functional processing of cultural capacities. The causal influence of culture on the functional brain activity of social processing illustrates the role of top-down and bottom-up influences on functional processes. The cultural influence on the functional brain activity of social processing illustrates the functional processing of cultural capacities in social context. The causal role of cultural brain activity on the cultural task activity of social processes depicts the functional significance of the cultural patterns of cortical brain mechanisms. The understanding of the causal interaction of core brain regions is essential to the functional basis of cultural brain networks.

The computational modeling of cultural brain networks is fundamental to the understanding of the cultural processing of social information. The computational modeling of the causal interaction of core brain regions that are responsible for cultural and social processing shows the dynamical aspects of cultural brain activity. The dynamical basis of cultural brain activity illustrates the causation of the functional activation of interconnected cortical brain regions and their functional significance for cultural and social processing. The dynamical properties of cultural brain activity depict the causation of the functional brain activation of core brain regions that predict the cultural task activity of mental function. The computational modeling of cultural brain networks is foundational to the characterization of the dynamical mechanisms of the cultural brain networks of social processing.

The understanding of the spatiotemporal dynamics of cortical brain regions that are of importance to cultural and social processing is fundamental to the functional basis of cultural brain dynamics. The spatiotemporal dynamics of cortical brain regions show how the fluctuations of electrophysiological signaling and mental activity comprise the functional basis of cultural processing in social context. The functional relation of electrophysiological signaling and the cultural task activity of social processing details the functional basis of cultural brain activity. The dynamics of spatiotemporal processing illustrate how cultural patterns of cortical brain activity are essential to the processing of cultural and social information.

The autonomous brain dynamics of large-scale cultural brain networks depict the spontaneous and intrinsic processing of cultural brain regions that are of importance to social processing. The endogenous brain activity of cultural brain networks illustrates the spontaneous and intrinsic activity of core brain regions that are involved in cultural and social processing. The cortical brain activity of cultural brain regions at rest describes the patterns of activation that are of importance to the endogenous brain activity of cortical brain regions. The spontaneous and intrinsic activity of cortical brain regions show how endogenous brain activity comprises the cortical brain activity of intrinsic cultural processing.

The intrinsic cultural processing of cortical brain regions serves as a functional basis of the autonomous brain dynamics that contribute to the entrainment of the cultural brain. The intrinsic properties of cultural brain activity detail the spontaneous fluctuations of endogenous brain activity that is stimulus-independent. The spontaneous fluctuations of cultural brain activity illustrate how the intrinsic processing of cortical brain regions facilitate the dynamical aspects of cultural processing. The endogenous brain activity of cultural brain function shows the autonomous dynamics that are based on the intrinsic properties of cortical brain regions. The intrinsic cultural processing of cultural brain networks is of importance to the entrainment of the cultural brain.

The study of the cultural brain dynamics of social processing is of importance to the understanding of the structural and dynamical basis of cultural brain activity. The autonomous brain dynamics of cultural and social processing show the functional basis of the intrinsic processing of cultural brain activity. The entrainment of the cultural brain facilitates the intrinsic and extrinsic processing of cultural brain activity. The characterization of the dynamical mechanisms of cultural brain networks facilitates the understanding of the structural and dynamical basis of cultural brain function.

Culture and Autonomous Brain Dynamics

The study of culture and autonomous brain dynamics describes the exploration of the dynamical principles that underlie the causal interaction of cultural brain activity and its effects. The dynamical mechanisms of cultural brain activity detail the intrinsic properties of the cortical brain that comprise the functional basis of cultural processing. The causal interaction of cultural brain regions illustrates the functional relation of the cultural processing of cortical brain mechanisms and its causal effects. The entrainment of the cultural brain demonstrates the intrinsic and extrinsic processing that arises based on the structural and functional properties of cortical brain function. The characterization of the dynamical principles of cultural brain dynamics illuminates on the functionality and causality of cultural brain activity as a functional basis of the endogenous activity of the cultural brain.

The dynamical principles of cultural brain dynamics postulate on the notion of the autonomous and self-organizing principles of cortical brain function. The fundamental principles of cultural brain dynamics describe the intrinsic processing of cultural brain activity that is spontaneous and is based on the endogenous activity of cortical brain function. The intrinsic properties of cultural brain activity entail the autonomous brain dynamics of cortical and cultural processing. The endogenous activity of the brain at rest depicts the intrinsic processing of the cultural brain that is independent of stimulus-driven activity. The dynamical aspects of cultural brain activity detail the spontaneous and intrinsic activity of cortical brain regions that comprise the autonomous brain dynamics of cultural brain function.

The entrainment of the cultural brain describes the inherent dynamics of the cultural brain that is based on endogenous brain activity. The spontaneous and intrinsic processing of the cultural brain facilitates the maintenance of active and itinerant representation of cultural processing that are of importance to the activation patterns of brain dynamics. The endogenous brain activity of the cultural brain details the patterns of cultural brain activity that comprise the representation of cultural processing that is stimulus-independent. The intrinsic properties of cultural brain activity arise in response to a multitude of causal roles that are of functional significance to the cultural processing of autonomous brain dynamics.

The structural properties of the neuroanatomical structures that comprise cultural brain networks serve as a causal impetus of the intrinsic connectivity of cortical brain regions. The neuroanatomical structures of the cultural brain detail the structural properties that are of importance to the intrinsic connectivity of the cultural brain. Structural properties of cultural brain networks comprise a multitude of features that are characteristic of the intrinsic connectivity of the cultural brain. The anatomical connectivity of cultural brain networks shows the functional importance of the structural properties of intrinsic connectivity. The neuroanatomical basis of cultural brain networks facilitates the intrinsic processing of cultural brain activity.

The structural properties of cultural brain networks describe the intrinsic processing of cultural brain activity that is based on neuronal activity and its response properties. The intrinsic processing of cultural brain activity arises from the cortical brain regions that are responsible for cultural processing. The neuroanatomical structures of cortical brain regions located within distinct brain areas demonstrate the intrinsic properties of neuronal activity that are characteristic of the population-level neuronal activity of cortical brain regions. The intrinsic properties of cultural brain activity reflect the distinct types of patterns of neuronal activity and its response properties that are distinct across cortical brain regions. The intrinsic connectivity of cultural brain networks shows the structural basis of intrinsic processing that is characteristic of autonomous brain dynamics.

The structural basis of intrinsic processing of cultural brain activity describes the distinct types of neuronal activity and response properties. The structural properties of distinct types of neurons reflect the patterns of neuronal response that comprise the intrinsic activity of cortical brain regions. The patterns of neuronal activity of cortical brain regions show how the responsivity of neuronal populations reflects spontaneous and intrinsic activity of cultural brain activity at rest. The intrinsic properties of neurons show the electrophysiological signaling of neuronal activity and its response properties that are stimulus-independent. The structural properties of distinct types of neurons detail the intrinsic processing of cultural brain activity.

The neuroanatomical structures of cultural brain activity facilitate the intrinsic processing of endogenous brain activity. The neuroanatomical location of cortical brain regions contributes to the intrinsic properties that are of importance to the cultural processing of autonomous brain dynamics. The structural properties of neuronal activity that comprise cortical brain regions details the patterns of neuronal activity that are of importance as intrinsic properties of cultural brain networks. The neuroanatomical basis of cultural brain activity shows causal-functional significance as a structural basis of the intrinsic connectivity of cultural brain activity.

The functional properties of autonomous brain dynamics further considerations of the activation patterns that are of functional significance to cultural brain activity and its dynamical mechanisms. The encoding of cultural processing describes the patterns of activation states that are of importance to the cultural brain activity that is based on the entrainment of cortical brain function. The encoding of cultural information that is based on the intrinsic processing of endogenous brain activity shows the activation patterns of cultural brain activity. The intrinsic processing of endogenous brain activity entails the encoding of cultural processing that facilitates the entrainment of cultural brain function. The activation patterns of cultural brain activity that are based on intrinsic processing show the functional significance of endogenous brain activity on the understanding of cultural brain dynamics.

The intrinsic processing of cultural brain activity facilitates the encoding of cultural patterns of activation that are based on functional properties of cultural brain networks. The cultural patterns of activation of cultural brain networks show how endogenous brain activity is a causal mechanism of the encoding of cultural information. Endogenous brain activity demonstrates the causal-functional significance of spontaneous fluctuations of cortical brain activity on inherent cultural mental function. The endogenous brain activity of cultural brain networks facilitates the causation of intrinsic processing as a dynamical mechanism of cultural information processing. The encoding of the cultural patterns of brain activity illuminate on the causal-functional significance of the endogenous brain activity of cultural brain networks.

The spontaneous and intrinsic activity of cultural brain networks serves as a dynamical mechanism of the functional processing of cultural brain networks. The spontaneous fluctuations of cultural brain activity that comprise intrinsic processing illustrate the dynamical aspects of cultural brain activity. The causal-functional role of intrinsic properties on the cultural processing of endogenous brain activity illustrates the dynamic causal role of cultural brain activity. The intrinsic properties of endogenous brain activity serve as a functional basis of the intrinsic cultural processing of higher-level processing. The functional relation of endogenous brain activity and intrinsic cultural processing shows the encoding of cultural information as patterns of activation that comprise cultural brain dynamics.

The functional properties of cultural brain networks show the causation of dynamical mechanisms that comprise the functional processing of cultural brain activity. The intrinsic processing of cultural brain activity facilitates the patterns of activation of that comprise the prediction of cultural processing. The dynamical aspects of cultural brain activity show how endogenous brain activity contributes to the generation of the activation states of cultural processing that are of importance to the intrinsic processing of cultural brain activity. The endogenous brain activity of cultural brain networks shows the generation of activation patterns that serve as a causal mechanism of states of activation of cultural processing. The dynamical aspects of cultural brain activity not only facilitate intrinsic cultural processing, but also the generation of activation patterns of cultural processing. The functional basis of cultural brain networks is of importance to the causal-functional significance of dynamical mechanisms of cultural brain activity.

The endogenous brain activity of cultural brain networks shows the generation of activation patterns that facilitate cultural processing and its causal effects. The functional properties of cultural brain activity show how endogenous brain activity has causal-functional significance as a dynamical mechanism of the cultural patterns of activation. The generation of activation patterns that comprise cultural processing show the functionality and causality of cultural brain networks. The causal-functional role of cultural processing illustrates the functional properties of endogenous brain activity that are of importance to cultural patterns of activation. The functional properties of cultural brain activity show the causation of activation patterns that is based on the cultural processing of endogenous brain activity.

The autonomous brain dynamics of cultural brain networks detail the structural and functional properties of cultural brain activity that comprise cultural processing. The endogenous brain activity of cultural brain networks facilitates the causation of the spontaneous fluctuations of cortical brain regions that comprise brain function at rest. The intrinsic processing of cultural brain activity shows functional relation to inherent cultural mental function. The endogenous brain activity of cultural brain networks shows the causal-functional role of intrinsic cultural processing as a functional basis of the cultural brain activity of cortical brain networks.

Discussion

The study of cultural brain dynamics encompasses a wide realm of considerations on the structural and dynamical principles of cortical brain organization. The dynamical mechanisms of the cultural brain entail the characterization of the causal interaction of interconnected brain regions that are responsible for cultural processing. The dynamical principles of cortical brain function show how interconnected brain regions facilitate the bidirectional processing of cultural information flow that is essential to cultural capacities. The structural and dynamical principles of cortical brain organization are fundamental to the understanding of the simple to complex dynamics of cultural brain activity.

Computational principles of cortical brain organization provide a foundation for the understanding of the structural and functional properties that comprise the dynamical basis of cultural brain function. The functional architecture of the cultural brain shows the hierarchical structure of cortical layers that facilitate the information processing of cultural adaptation.

The structural and functional connectivity of cortical brain regions demonstrate the information processing of cortical layers that facilitate the information flow and activation patterns of importance to cultural capacities. The attractor dynamics of cultural brain function show how cultural processing facilitates the patterns of activation that facilitate the functional processing of cultural task activity. The cultural patterns of functional brain activity of cortical brain networks detail the functional role of cultural processing on cortical brain function.

The study of cultural brain dynamics provides a foundation for the understanding of the dynamical principles of cultural brain function. The characterization of the causal interaction of cultural brain regions shows the information flow and activation patterns that comprise cultural processing. The dynamical aspects of cultural brain function detail the functionality and causality of cortical brain processing that is of importance to the functional processing of cultural capacities. The causal interaction of cultural brain regions demonstrates the functional relation of cultural processing and functional activation states of cultural brain function. The casual–functional role of cultural patterns of activation on cultural processing depicts the functional significance of activation patterns on the pattern completion of cultural processing.

The autonomous and self-organizing principles of cultural brain dynamics describe the spontaneous and intrinsic activity of cortical brain function that is based on the entrainment of the cultural brain. The neuronal activity of cortical brain areas shows the spontaneous and intrinsic activity of cortical brain function that is inherent of the cultural brain at rest. The functional properties of cultural brain dynamics demonstrate how the response properties of neuronal activity facilitate the spontaneous and intrinsic processing of cultural brain activity that is based on cultural entrainment. The understanding of the endogenous activity of cortical brain function at rest is of importance to the demonstration of the autonomous and self-organizing principles of the cortical brain.

Conclusion

The scholarly and scientific interest on scientific observation and empiricism shows the interest on the discernment of the lawlike principles of patterns and regularities of simple to complex systems of the natural world. The understanding of the causation of basic mechanisms is of importance to the characterization of the functional and emergent properties of complex systems. The patterns and regularities of causal mechanisms detail the parts and particulars that comprise the causal impetus of functional and emergent properties. The causal interaction of basic mechanisms shows the functional relations of the aggregation of parts of mechanisms and their causal effects. The lawlike principles of patterns and regularities of causal mechanisms facilitate the characterization of the mechanistic basis of the causal structure of the natural world.

The study of cultural brain dynamics encompasses the dynamical aspects of cultural brain function. The computational approaches to the study of cultural brain dynamics entails the computational principles that facilitate the prediction and explanation of the dynamical mechanisms of cultural brain networks. The characterization of the causal interaction of cortical brain regions contributes to the understanding of the functionality and causality of cultural brain function. The structural and dynamical principles of cultural brain networks show how the intrinsic and extrinsic processing of cortical brain activity is sufficient as a functional basis of the dynamical mechanisms of cultural brain function.

The autonomous and self-organizing principles of cultural brain dynamics reveal the causation of the intrinsic processing of cultural brain activity as a causal mechanism of cortical brain function. The causal role of intrinsic properties of cultural brain activity detail the spontaneous and intrinsic activity of cortical brain function that comprises the functional processing of the cultural brain. The autonomous brain dynamics of intrinsic processing show how the entrainment of the cultural brain is linked to patterns of dynamical mechanisms that have causal–functional significance. The cultural brain activity at rest illustrates the causation of intrinsic processes that are fundamental to the autonomous and self-organizing principles of cultural brain dynamics.

The study of cultural brain dynamics is of importance to the understanding of the structural and dynamical principles of cultural brain function. The cultural brain activity of cortical brain regions comprises fluctuations of physiological signaling and functional activity that are of importance to the dynamical basis of cultural brain function. The cultural brain dynamics detail the dynamical mechanisms of cultural brain activity that are fundamental to cultural processing. The dynamical principles of cultural brain activity show the structural and functional properties that are of importance to the cultural patterns of functional brain activity and its causal effects.

Implications

Research on cultural brain dynamics has translational impact on the research and development on culture and neuroscience in health, medicine and related fields. The systematic study of cultural brain dynamics generates novel knowledge of the structural and dynamical basis of cultural and cognitive processing. The research and development on cultural brain dynamics contributes to the scientific and technological innovation on computational cultural neuroscience. The building of scientific infrastructure and research capabilities of the study of cultural brain dynamics contributes to the development of computational tools and technologies for the study of the cultural brain.

The scholarly and scientific understanding of the origin and function of cultural brain dynamics provides scientific knowledge for the scientific and societal enrichment on computational cultural neuroscience and its practical applications on culture and health. The scientific and societal enrichment on cultural brain dynamics is beneficial to the broader understanding of the dynamical principles of the cultural brain. The building of humanistic and scientific understanding of the fundamental principles of the cultural brain has implications for scientific and societal enrichment. The broader understanding of computational cultural neuroscience is beneficial to the building of societal awareness of the importance of neuroscience in culture and health.

8
CULTURAL MACHINE LEARNING

Introduction

The philosophical and scientific tradition of empiricism of the 17th century builds on the notion of the importance of experience as a basis of truth confirmation of the world (Locke, 1689). The philosophical and scientific notion of empiricism places emphasis on observation and experience as a foundation of the determination of truth and explanation of the world. Empiricism highlights the importance of experience as a conceptual basis of the generation of metaphysics and epistemology of the world. The postulation on empiricism as a source of knowledge of the world encompasses broader notions of the role of observation and experience as a foundation of the theory confirmation and truth verification that informs understanding of the causal structure of the world. The philosophical and scientific interest in empiricism illuminates on the role of observation and experience as a foundation of the theory confirmation and hypothesis generation on the causation of the world.

The scientific work of the 19th century on the empiricism of the mind led to the advancement of the theoretical notion of the mind as an inference engine. The early scientific work on perception formulates scientific theories of empiricism that ascertain on the role of experience as a foundation of the acquired knowledge or learning (Dayan et al., 1995). The notion of mental function as based on learning and experience places emphasis on the role of experience as fundamental to the discernment of mental inference. The mental states of learning and experience are essential to the formulation of mental inference on the causal structure of the world.

DOI: 10.4324/9781003384236-9

The postulation on the learning and experience of the world informs the formulation of inference as observation and explanation of the causal structure of the world. The theoretical notion of the mind as an inference engine illuminates on the functionality and causality of mentality and thought as a basis of learning and experience.

Culture as computation entails inferences on the causal structure of the world. Culture as computation comprises the mentality and thought that ascertain on the causation of cultural and mental states and their functional purpose. Culture as mental states informs the representation and interpretation of the environment. The discernment of culture as mental function encompasses the understanding of culture as mental inference that is based on the learning and experience of the environment. Culture serves as a fundamental basis of the learning and experience on the causal structure of the world.

The cultural computation of minds and machines entails pontification on the notion of culture as mentality and thought on the causal structure of the world. The notion of culture as learning and experience is of importance to the generation and acquisition of knowledge of the world. Culture as mental inference builds on the notion of learning and experience as a functional basis of the formulation of the inherent causation of the world. The functional role of culture as learning and experience illustrates how the learner is a model of the environment. The internal architecture of the cultural mind comprises a functional basis for the learning and experience of the environment. The functional basis of culture of minds and machines illuminates on the learning and experience of the inherent causation of the world.

The goal of the chapter is to provide a review of the fundamentals of cultural machine learning. The study of cultural machine learning investigates learner models of the prediction and inference of the cultural environment. Cultural machine learning is of importance to the understanding of the lawlike principles of cultural patterns in the natural world. The notion of culture as minds and machines implies that cultural computation relies on the internal architecture of computing machines. The functional architecture of the cultural brain serves as a basis for the prediction and inference of the cultural environment. The cultural brain as an inference machine facilitates the prediction and inference on the inherent causation of culture in the world. The study of cultural machine learning is fundamental to the understanding of the computational foundations of the cultural brain.

Culture as Minds and Machines

The study of culture as minds and machines encompasses a wide realm of considerations of culture as processes and mechanisms of minds and machines with causal-functional significance. The computation of culture of minds and machines entails the understanding of the processing of cultural information and its functional basis. The mental computation of culture illustrates the cultural and mental states that are of importance to the cultural processing of information and its functional purpose. The machine computation of culture depicts the cultural and machine states that demonstrate a functional role in the processing of cultural information. The cultural computation of minds and machines describes the cultural states that comprise a functional basis of cultural performance and its functional relation.

The culture of minds and machines depicts the casual-functional role of cultural processing on mental and machine computation. The cultural processing of the mind comprises a functional basis of cultural capacities. The mental computation on the cultural patterns of the world contributes to the understanding of the lawlike principles of culture. The detection of cultural patterns of the environment facilitates the functional processes that comprise the cultural mind. The cultural mind demonstrates the functional capacities for the detection of the cultural patterning of the environment. The functional processes of the cultural mind are essential to the understanding of culture in the mind and in the world. The functional basis of cultural processing demonstrates the importance of mental function in cultural computation.

The mind as a physical realizer of culture entails the causal-functional relation of cultural and mental states. Cultural states as mental states imply the manifestation of the cultural mind as a physical realizer of culture. The functional relation of cultural states and mental states illustrates the importance of mental function as states of cultural processing. The functional role of mentality and thought as a fundamental basis of cultural processing implies the causation of cultural states as mental states. Cultural states serve as a causal mechanism of mental states. The mental causation of culture illuminates on the causal-functional role of the cultural mind as a physical realizer.

The mental causation of culture shows the functional significance of cultural states as mental states. The role of cultural states as a causal impetus of mental states depicts the functional significance of cultural processing. Cultural states are a causal mechanism of mental states. Culture serves as a causal mechanism of the functional processes of the mind. Cultural processing is a causal influence on the functional performance of the mind. The mental causation of culture further suggests the causal role of mental states as a precursor of cultural states. The functional processes of the mind enact the causal influence of cultural capacities. Mental states serve as a causal precursor to the functional role of cultural states. The functional role of mental states as a causal mechanism of cultural states illustrates the mental causation of culture.

The cultural mind encompasses the casual interaction of cultural and mental states and its functional purpose. The causal interaction of cultural and mental states illuminates on the causal-functional role of cultural and mental causation. The interaction of cultural and mental states comprises the internal properties of the cultural mind and its causal effects. The causal-functional significance of cultural and mental states demonstrates the role of functional properties in the functional processing of the cultural mind. The causal interaction of cultural and mental states illustrates the functional purpose of the cultural mind and its causal effects.

The cultural computation of machines illuminates on the notion of cultural states as machine states. The cultural processing of machine computation entails the functional performance of culture. Culture as a computing machine implies simple to complex machines with the functional capabilities for the mentality and thought of culture. The machine computation of culture relies on computing machines with an internal architecture that facilitates the functional capabilities for mentality and thought that is cultural. The cultural processing of machine computation details the functional role of internal components as a functional basis of the cultural computation of computing machines.

The internal architecture of computing machines shows the functional role of internal components that are fundamental to cultural performance. The internal components of computing machines demonstrate the importance of levels of capabilities on functional performance that is culture. The internal states of machine computation detail the initial and final states of functional performance that are of importance to the demonstration of cultural computation. The functional capabilities of machine computation for memory, internal processing and algorithmic function illustrate the internal components that are of importance to the functional performance of cultural machine computation.

The cultural computation of machines is essential to the discernment of the cultural patterns and regularities of the world. The functional performance of machine computation facilitates the detection of cultural information in the casual structure of the world. The cultural computation of computing machines contributes to the understanding of cultural patterns and regularities and their causal effects. The functional capabilities of machine computation further the functional performance on the causal interaction of cultural states and their functional significance. The computation of simple to complex devices illustrates the functional capabilities of machines for cultural computation.

The cultural computation of machines shows the functional performance of machine states that is deterministic. The functional performance of machine states illustrates the initial and final states of computation that are of importance to the automated production of cultural input–output.

The initial and final states of the computation of machine states detail the functional performance of machines that is deterministic. The cultural computation of machines describes the functional capabilities for the automated production of cultural input–output. The computation of machine states that show functional performance that is deterministic depicts the automated production of machine computation based on the presence of the machine states of cultural input–output.

Cultural machine computation also demonstrates how the functional performance of machines is conditional. The functional performance of machine states and machine state transitions relies on standard criteria or conditionals of machine computation. The manifestation of machine states and their transitions illustrates how the functional performance of machine computation is based on the presence or absence of machine states with conditionals. The functional performance of machine states and machine state transitions at a predetermined rate demonstrates the computation of machine states that is conditional rather than deterministic.

The cultural computation of the brain illustrates the causal role of functional mechanisms as machine computation. The cultural brain is a biological computing machine that is capable of cultural capacities. The internal processing of the cultural brain illustrates the functional architecture of the cortical brain and its casual-functional significance. The functional architecture of the cultural brain comprises processes and mechanisms that are of importance to cultural capacities. The functional components of cultural capacities detail the specific processes and mechanisms that facilitate cultural task performance. The functional role of cortical brain mechanisms as a mechanistic basis of cultural processing demonstrates the feasibility of the cultural brain as a biological computing machine.

The cultural processing of the brain demonstrates the functional basis of cultural capacities. The information processing of cultural brain function shows the structural and functional basis of cultural processing. The structural and functional properties of the cultural brain demonstrate how the internal components of cortical brain function facilitate cultural processing. The internal properties of the cultural brain detail the structural and functional basis of the processing of cultural information. The cultural computation of the brain illustrates the functional mechanisms that are essential for the performance of cultural tasks. The characterization of the functional mechanisms of the cultural brain contributes to the understanding of the brain basis of culture.

The functional processing of the cultural brain is fundamental to the representation of the cultural environment. The cultural brain as an internal model of the environment demonstrates the role of functional mechanisms in cultural processing. The functional basis of the cultural brain illustrates the representation and transformation of sensory data into motor output. The cultural patterns of cortical brain function depict the levels of processing that are of importance to cultural capacities. The functional correspondence of cortical brain function and cultural task performance illustrates a functional relation of culture as mental and functional states. The cultural processing of cortical brain function is essential to the understanding of the cultural patterns and regularities of the environment.

Culture as Computation

Culture as computation comprises the computation of minds and machines and its functional performance. The cultural computation of minds and machines illustrates how the function of minds and machines facilitates the functional performance of culture. The notion of culture as computation encompasses a multitude of notions on the physical realization of minds and machines as devices that perform a mental function that is cultural. The demonstration of mental and machine computation that is cultural shows how the mind and machine serve as a functional basis of cultural computation. The multiple realizability of minds and machines that perform mental function that is cultural illuminates on the plausibility of the cultural computation of minds and machines.

The physical instantiation of cultural computation as mental function shows the causal role of mental states as a functional basis of culture. The computation of the cultural mind comprises functional components that facilitate the performance of cultural tasks. The functional components of cultural mental computation detail the algorithmic function that serves as a functional basis of cultural task performance. The functional performance of cultural tasks illustrates the core capacities of mental function that are of importance to cultural mental function. The cultural computation of the mind demonstrates how the mind is a functional basis of cultural capacities.

The multiple realizability of cultural computation as machine function implies the causal-functional role of machine states as a mechanistic basis of culture. The causal role of machine states as a physical realizer of culture illustrates how machine states and their transitions are fundamental to functional performance that is cultural. Cultural states as machine states entail the internal states of computing machines that are of importance to the performance of cultural computation. The cultural machine computation of computing machines illustrates the machine states that are of importance to the functional performance of culture. Machine computation as a physical instantiation of cultural computation details the causal-functional role of simple to complex machines as a mechanistic basis of functional performance that is cultural.

Culture as computation entails the functional performance of minds and machines that is based on learning and experience. The cultural computation of minds and machines illustrates how learning and experience comprise a functional basis of mental function. Culture as learning and experience demonstrates the functional role of the generation and acquisition of knowledge of the world. The generation and acquisition of knowledge of the inherent causal structure of the world informs the formulation of mental inference. Culture as learning and experience facilitates the formulation of mental inference as prediction and explanation. The generation and acquisition of cultural knowledge informs the formulation of the prediction and explanation of mental inference. The learning and experience of culture and the environment is fundamental to the generation and acquisition of knowledge.

The Cultural Mind as a Statistical Engine

The study of the cultural mind as a statistical engine refers to the functional performance of mental function for the prediction and inference of the causal structure of the environment. The functional capacities of the cultural mind allow for the inference of the cultural patterns and regularities of the environment and its inherent causation. The prediction and inference of the cultural environment based on sensory data entails the understanding of the causal relations of cultural patterns that hold. The functional performance of the cultural mind as a statistical engine illustrates the functional capacities that contribute to the understanding of the inherent causation of the cultural patterns of the environment. The notion of the cultural mind as a statistical engine implies the functional role of the observer as a basis for the prediction and inference of the inherent causation of the world.

Culture as mental states details the causal role of mental function as a functional basis of statistical inference or learning. The cultural mind demonstrates the functional capacities for the prediction and inference of the cultural patterns of the environment and its causal effects. The statistical inference of the cultural patterns of the environment depicts the functional capacities of the cultural mind to infer the inherent causation of culture of the mind and of the world. The prediction and inference based on sensory data facilitates the understanding of the causal structure of the world. Culture as mental states illustrates the functional basis for the statistical inference or learning on the cultural patterns of the environment and its causal-functional role. The postulation of the cultural mind as a statistical engine connotates on the notion of mental function as machine computation.

The cultural mind as a statistical engine demonstrates the functional capacities of mental function to infer on the cultural patterns of the world. The statistical inference or learning of the cultural patterns of the environment facilitates the understanding of the lawlike principles of culture and its causal effects. The cultural patterns of the environment contribute to the understanding of the causal relation of cultural and mental states of the world that hold. The statistical inference or learning on the cultural environment entails the understanding of the causal arrangements of cultural patterns and their interaction. The functional capacities of cultural mental function encompass the prediction and inference on the regularities and probabilities of cultural patterns that describe actualities of the natural world. The cultural mind as a statistical engine illustrates the functional capacities of mental function to infer on the cultural patterns of the environment.

The cultural mind demonstrates the functional capacities of the prediction and inference on the cultural patterns of the environment based on sensation. The prediction and inference on cultural patterns entail the formulation of learning rules that describe the conditions in which patterns and regularities occur. The mental function of learning depicts the algorithmic function of the cultural mind to infer on the patterns and regularities of the world. The rule-based learning of the environment facilitates the inferences on cultural patterns and their causal effects that are conditional or deterministic based on learning algorithms. The algorithmic function of the cultural mind details the rule-based learning of the environment that is of importance to the formulation of mental inference.

The generation of knowledge of the cultural mind is based on the statistical inference or learning of the environment. The mental inference based on learning rules describes the prediction of patterns and regularities from learning algorithms. The formulation of mental inference of learning rules details the conditions of cultural patterns and regularities that comprise the rule-based learning of mental inference. The mental inference of learning that is based on algorithmic function illustrates the mental function that is essential for the learning of the cultural environment. The prediction and inference on the cultural environment as learning rules facilitates the generation of knowledge.

The statistical inference or learning on the cultural environment based on sensory data demonstrates the functional performance of the cultural mind and its functional role. The cultural patterns of the environment facilitate the prediction and inference on the causal arrangement of the culture of the natural world. The statistical inference or learning on the cultural patterns of the environment facilitates the generation of knowledge on the causal arrangements of cultural patterns that hold. The lawlike principles of culture enact cultural patterns on the environment that facilitate the prediction and inference of the environment. The functional capacities of the cultural mind to infer on the inherent causation of the cultural environment illustrate the functional importance of statistical inference or learning on the generation of knowledge.

The generation of knowledge on the inherent causation of the cultural environment comprises mental states that facilitate the prediction and inference of the environment. The generation of knowledge of cultural patterns informs the prediction and inference on the conditional or probabilistic states of the world under conditions of uncertainty. Prior states of knowledge inform on the conditional or probabilistic states that contribute to prediction and inference. Knowledge generation comprises prior states of patterns and regularities that inform conditional or probabilistic states that predict the actualities of the world. The learning of the cultural mind generates knowledge that informs the prediction and inference on the patterns and regularities of the cultural environment.

The generation of cultural knowledge facilitates mental inference of the inherent causation of the cultural environment. The generation and acquisition of cultural knowledge based on learning and experience contributes to an optimal model of the learner and the environment. Cultural knowledge is of importance to the generation and maintenance of expectations and predictions that reduce conditions of uncertainty. Culture as learning and experience facilitates the patterning of the cultural environment. Cultural patterns inform on the optimal states of the internal representation of the environment. The functional role of cultural knowledge as prior expectations that inform future predictions is of importance to the optimization of the learner model of the environment. The formulation of prediction and inference based on learning and experience is essential to the generation of knowledge of the cultural environment.

Cultural patterns serve as a functional basis of the internal representation of the environment and its causal-functional significance. The cultural patterning of the environment illustrates the cultural states of the world and its causal effects. The interaction of cultural states of the world depicts the inherent causation of cultural patterns. The inherent causation of cultural patterns details the causal interaction of cultural states with other states of the world. Cultural patterns of the environment facilitate the prediction of mental inference. The detection of cultural patterns and their causal-functional significance facilitates the prediction and inference of the cultural environment.

The cultural mind as a statistical engine broadens the notion of the cultural mind as machine computation. The cultural mind as a statistical engine informs the prediction and inference of the cultural environment based on sensation. The prediction and inference on patterns and regularities contribute to the generation of knowledge that informs the formulation of mental inference on the inherent causation of the world. The cultural patterns and regularities of the world contribute to the formulation of prediction and inference on the cultural environment. The cultural mind as machine computation demonstrates the functional basis of mental function as a learner model of the world.

The Cultural Brain as an Inference Machine

The study of the cultural brain as an inference machine connotes on the notion of the brain as a biological computing machine that is capable of inference and learning. The cultural brain implies the manifestation of the computing machine of biological organisms and their cultural capacities. The notion of the cultural brain as an inference machine details how the internal architecture of cortical brain function facilitates the prediction and inference on the cultural environment. The internal components of cortical brain function demonstrate the physical instantiation of functional mechanisms that facilitate the inference and learning of the environment. The cultural brain as an inference machine depicts how the brain as a biological computing machine is capable of the prediction and inference on the cultural environment.

The cultural brain comprises the functional architecture of the brain that is fundamental to cultural processing. The internal components of cortical brain function consist of neurons and networks of neurons that comprise a mechanistic basis of the functional processing of mental function. The cultural processing of cortical brain mechanisms details the computation that comprises a mechanistic basis of cultural mental function. The internal states of cortical brain function are comprised of the intrinsic and functional properties of the information processing of culture. The cultural processing of cortical brain function demonstrates the functional significance of the intrinsic and functional properties of the cultural brain.

The brain basis of culture describes cortical brain mechanisms that facilitate cultural processing. The representation and transformation of sensory data into motor output describes the functional processing of cortical brain function. The functional processing of cortical brain regions contributes to the manifestation of cultural input–output relations. The cortical brain function of cultural processing illustrates the states of activation that comprise the transformation of representational content from cultural input to cultural output. The cortical processing of cultural information illustrates cultural processing based on internal and external input. The functional processing of cortical brain regions is essential to the understanding of the brain basis of culture.

The internal states of cultural brain function facilitate the prediction and inference of the cultural environment. The internal states of cultural brain function comprise the cortical processing based on the active maintenance of representations. The cultural processing of cortical brain function shows how active maintenance of internal representations facilitates the interpretation of cultural input based on prior states of activation. The representation and interpretation of cultural input is fundamental to the functional role of prior states of activation to inform the prediction of activation states that produce cultural output. The internal states of cultural brain function facilitate the generation and maintenance of active representations that contribute to cultural processing.

The information processing of cultural brain function depicts the functional mechanisms that facilitate the cortical processing of cultural information. The states of activation of cultural brain function show the causal-functional role of internal and external input on functional output. The functional processing of cortical mechanisms illustrates the transformation of the representational content based on internal and external inputs. The internal representation and its transformation based on internal and external inputs facilitates the recurrence of activation states of cultural brain function. The recurrence of states of activation of cultural brain function shows the functional role of the maintenance of active and itinerant representations in the functional processing of the brain. The maintenance of active and itinerant representation serves as a functional basis of the cultural processing of cortical brain function.

The functional processing of the cultural brain shows the casual-functional significance of the intrinsic and functional properties of the cultural processing of cortical brain function. The functional basis of cultural processing depicts the patterns of cortical brain function that detail the inherent causation of activation states as prior prediction of mental inference. The patterns of cortical brain function depict the causal role of activation states as a prediction of mental inference. The cultural patterns of the brain show the causal-functional role of activation states of cortical brain function as cultural mental inference. Cultural brain patterns as predictors of mental inference comprise the activation states of cortical brain function that serve as a mechanistic basis of cultural processing.

The recurrence of activation states of cultural brain function demonstrates the causal-functional role of activation states as prediction and mental inference. The functional processing of cultural brain function based on internal and external inputs shows the recurrence of activation states that serves as a facilitation of functional processing toward optimal states of activation. The recurrence of activation states of cultural brain function depicts the facilitation of information processing based on the prior prediction of activation states as mental inference. The information processing of cultural brain function details the functional role of activation states as facilitation of the prior prediction as mental inference. The functional processing of the cultural brain illustrates how the recurrence of activation states serves as a causal mechanism of prediction and mental inference.

The patterns of activation states of cultural brain function illustrate the causal-functional significance of activation states as optimal states of brain function. The activation states of the cultural brain comprise optimal states of activation that demonstrate the cultural processing of functional mechanisms. The cultural patterns as optimal states of brain function comprise the activation states that predict the representation and transformation of cultural input into cultural output. The activation states of the cultural brain that are optimal states of activation facilitate the prediction and inference of the cultural environment. The casual-functional role of activation serves as a mechanistic basis of the cultural processing of cortical brain function.

The internal architecture of cultural brain function illuminates on the functional processes based on learning and experience. The cultural patterns of cortical brain function show the functional processing that is based on learning algorithms. The cultural influences on cortical brain function demonstrate the patterns of activation states that are in response to learning rules. The rule-based learning of cortical brain mechanisms details the cultural processing that is based on learning algorithms. The cultural processing of cortical brain function shows the patterns of activation that are responsive to the functional tasks of ruled-based learning. The cultural patterns of cortical brain function illustrate the functional processing that facilitates learning and experience.

The cultural patterns of cortical brain function demonstrate functional processing that is based on learning algorithms. The cultural processing of cortical brain mechanisms shows patterns of activation and response properties that are responsive to learning rules. The functional processing of cortical brain function contributes to the detection of cultural patterns and regularities of the environment. The processing of cultural information that facilitates rule-based learning illustrates the functional processing of cortical brain function based on learning algorithms. The cultural processing of the brain shows the functional role of algorithmic function on the information processing mechanisms of cortical brain function. The cultural patterns of activation of cortical mechanisms illustrate the functional basis of learning and experience.

Cultural brain function as internal states of the inference machine comprises a mechanistic basis of the functional states of the cultural brain. The internal states of cultural brain function facilitate the representation and transformation of cultural information of the environment. The transformation of the representational content of the environment details the information processing toward optimal states of activation. The processing of cultural information to the optimal states of cultural brain activation shows the functional processing of the cultural brain. Cultural patterns of functional mechanisms depict the optimal states of activation of cultural brain function.

Cultural patterns of activation states of cultural brain function are optimal as a functional basis of the internal representations of the environment.

Functional Architecture of the Cultural Brain

The functional architecture of the cortical brain describes the structural and functional properties of cortical brain organization. The structural and functional principles of the cortical brain detail the functional basis of the information processing of the cortical brain. The structural and functional properties of the cortical brain illustrate the structural and functional basis of cortical brain function. The structural properties comprise the interconnection of cortical brain regions and their causal effects. The functional properties detail the patterns of neurons and networks of neurons and their response properties. The understanding of the structural and functional basis of information processing is essential for the characterization of the multilevel processing of cortical brain function.

The neuroanatomical structures located within the main cortical lobules comprise the cortical brain areas. The main cortical lobules such as the occipital, temporal, parietal and frontal cortices demonstrate the neuroanatomical basis of cortical brain function. The hierarchical structure of cortical brain organization illustrates the columnar organization of cortical layers that comprise the functional pathways of information processing. The cortical layers of representation of columnar organization detail the distinct types of neurons that comprise the structural and functional basis of information processing. The functional pathways of cortical brain function show the specialized information processing of cortical brain regions.

The neuroanatomical basis of cultural brain function depicts the main cortical lobules that serve as a structural and functional basis of cultural processing. The cortical brain regions located within the occipitotemporal, limbic, parietal and frontal lobes serve as a neuroanatomical basis of large-scale cultural brain networks. The cortical brain function of neurons and networks of neurons located within cultural brain networks describes the activation patterns and brain dynamics of the core brain regions and their interconnection. The structural and functional properties of cortical brain regions located within cultural brain networks are of importance to the information processing of cultural brain function. The structural and functional properties of cortical brain regions show the anatomical and functional basis of the interconnection of cortical brain regions and their causal effects.

The functional specialization and functional integration of cortical brain organization describes the functional processing and mechanisms of cortical brain regions of behavior. The functional basis of core brain regions shows the specialized information-processing mechanisms of core capacities. The specialized processing of cortical brain regions illustrates the processing of neuronal mechanisms and their functional properties. The functional specialization of core brain regions illustrates the dedicated and specialized mechanisms for information processing. The functional processing of core brain regions is of importance to the understanding of the functional basis of core capacities.

The functional integration of interconnected brain regions details the structural and functional basis of the interconnection of core brain regions. The interconnectivity of cortical brain regions depicts the interconnection of cortical brain regions and their structural and functional properties. The structural and functional connectivity of interconnected brain regions comprises cortical brain networks that demonstrate the relaying of information processing across neural networks. The core brain regions of cortical brain networks describe the functional processing that is of importance to core capacities. The interconnectivity of cortical brain regions serves as a functional basis of cortical brain networks.

The structural and functional properties of cortical brain organization are of importance to the understanding of the higher-level processes of brain function. The structural and functional basis of cortical brain organization describes how neurons and networks of neurons serve as a functional basis of the information processing of cortical brain function. The structural and functional properties of cortical brain function detail the functional role of functional mechanisms and interconnectivity in the functional processing of the brain. The structural and functional principles of cortical brain organization serve as a foundation of the computational basis of the cortical brain function.

Cultural Machine Learning

The study of cultural machine learning explores the prediction and inference on the causal structure of the world. The cultural brain serves as an internal model of the prediction and inference on the causation of the world. The brain as an inference machine posits on the causal structure of the world based on sensory data. The functional architecture of the cultural brain facilitates prediction and inference of the environment based on sensation. The representation and transformation of sensory data into motor output relies on the information-processing mechanisms of the cultural brain. The information processing of cultural neural networks facilitates the functional relations of cultural input–output. The study of cultural machine learning investigates how the brain is an internal model of the prediction and inference on the cultural environment.

The cultural brain serves as an internal model of prediction and inference based on the sensory data of the cultural environment. The representation and transformation of sensory input into motor output relies on the functional processing of cortical brain mechanisms. The cortical processing of cultural brain function facilitates the relaying of information of cortical mechanisms. The transformation of representational content to motor output illustrates the functional processing of cortical layers of representation. The cultural brain serves as a mechanistic basis of the cultural patterns that are essential to the prediction and inference of the environment.

The functional processing of the cultural brain illustrates the representation and transformation of sensory data of the cultural environment. The functional basis of cultural brain function entails the representation and transformation of cultural input to cultural output. The cortical processing of the cultural brain demonstrates the transformation of the representational content of cultural input into cultural output with feedback. The functional processing of cultural information of cortical brain mechanisms that consist of internal and external inputs demonstrates a mechanistic basis of cultural processing. The functional basis of cultural processing shows the functional role of cortical processing that is based on the integration of internal and external inputs. The functional role of cortical processing with feedback shows the causal-functional significance of feedback processing on the production of cultural output. The functional processes of cortical brain regions detail the patterns of activation states that serve as a mechanistic basis of rule-based learning.

The information-processing mechanisms of cultural brain networks detail the causal-functional role of bidirectional processing in the functional relation of cultural input–cultural output. The bidirectional processing of neural networks shows the feedforward and feedback processing of cultural information that facilitates the production of cultural input–cultural output. The feedforward processing of cultural information demonstrates the unidirectional processing of cultural input into cultural output. Feedback processing comprises the inhibitory processing of cortical layers of representation that facilitate the transformation of cultural input into cultural output. The feedback or inhibitory processing of cortical layers of representation shows the functional role of feedback or inhibition as the facilitation of the production of cultural output. The information-processing mechanisms of cultural brain function illustrate the importance of bidirectional processing of cultural brain networks as a functional basis of cultural processing.

The bidirectional processing of cultural brain networks illustrates the functional basis of information processing mechanisms. The interconnection of networks of neurons depicts the neural transmission that comprises the functional processing of neural networks. The excitatory and inhibitory processing of neural networks shows the functional basis of the bidirectional processing of cortical brain mechanisms. The information processing of neural networks based on excitatory and inhibitory processing facilitates the transformation of representation from input to output. The functional processing of cultural brain networks shows the functional significance of the bidirectional processing of cortical brain mechanisms.

Feedforward processing of cultural brain function describes the functional basis of the specialized processing of the cultural brain. The unidirectional processing of cultural input to cultural output shows the functional basis of the dedicated and specialized processing of cortical brain function. The feedforward processing of cultural information is based on the excitatory processing of cortical brain mechanisms. The amplification of the features of cultural input facilitates the encoding of cultural information of cortical brain mechanisms. The encoding of cultural information consists of the neuronal firing patterns of cortical mechanisms and their response properties. The transformation of representational content depicts the functional role of amplification as a causal mechanism of the functional processing of cultural input to cultural output.

Feedback processing of the cultural brain demonstrates the functional role of feedback or inhibition on the cortical processing of cultural information. The feedback processing of the cultural brain details the inhibitory processing of cortical brain networks that contribute to the control or inhibition of the feedforward processing of cultural input. The inhibitory processing of cortical brain networks shows the control of feedforward processing across cortical layers of representation and its causal-functional significance. The relaying of information across cortical layers of representation with internal and external input shows the causal role of control or inhibition on feedback processing. The inhibitory processing of cultural information of cortical brain mechanisms facilitates the control or inhibition of cultural processing with feedback.

The feedback processing of cultural brain networks demonstrates the functional role of inhibition in the functional processing of cortical brain mechanisms. The functional processing of cultural information of cortical brain function depicts the inhibitory processing of cortical brain networks with feedback. The inhibition of cultural processing based on feedback shows the causal-functional role of the internal states of activation on the functional processing of cultural information. The internal states of activation serve as a functional basis of inhibition of the cultural processing of cortical brain mechanisms. The functional role of activation states as inhibition provides feedback to the information processing of cortical mechanisms that has functional purpose. The activation states of inhibition contribute to the feedback processing of cortical mechanisms.

Inhibitory processing comprises a multitude of activation states of cortical brain mechanisms. The feedback processing of cortical brain function facilitates the control or regulation of activation states of cortical brain mechanisms. The inhibition of functional processing acts as a selection filter on activation states that contribute to the feedforward processing of input. The inhibitory processing of activation states contributes to the selection of activation states that anticipates the feedforward processing of input. Feedback processing also exerts control or inhibition on the cortical layer of information processing. The control or inhibition of feedback processing enacts regulation of the information processing that interacts with the cortical hidden layer. The feedback processing of activation states ensures the control of information processing across cortical layers of representation.

The cultural brain as an internal model is fundamental to the understanding of the causal structure of the environment based on sensory data. The information processing mechanisms of cultural brain function serve as a mechanistic basis of cultural processing. The functional processing of the cultural brain facilitates the transformation of the representational content of the cultural environment. The detection of cultural patterns as information processing illustrates a functional basis of prediction and inference. The functional mechanisms of the cultural brain serve as a mechanistic basis of the prediction and inference on the cultural environment. The characterization of the functional processing of cultural brain networks is essential for the prediction and mental inference on the environment.

Machine Learning Algorithms

The functional basis of large-scale cultural brain networks describes the information processing of cortical brain regions that comprise learning and experience. The functional basis of cultural processing depicts cortical mechanisms and functional processes that are of importance to learning and experience. The functional patterns of cortical brain function show the activation states that are consistent with the learning and experience of culture. The cultural processing of cortical brain function details the cultural patterns of activation that comprise learning and experience. The cultural patterns of cortical brain function depict the processes and mechanisms that serve as a functional basis of large-scale cultural brain networks.

The hierarchical structure of cortical brain organization demonstrates the structural and functional properties that are of importance to cortical brain function. The structural and functional properties of the cortical brain organization show the internal states of functional processing that comprise cortical brain function. The structural and functional basis of cortical brain function details the internal properties that facilitate the functional processing of the brain. The structure and function of neuroanatomical structures and their interconnection comprise the functional architecture of the cortical brain. The hierarchical structure of cortical brain organization is fundamental to the characterization of the structural and functional properties of cortical brain function.

The cortical layers of representation detail the information processing mechanisms that facilitate the functional processing of algorithmic function. The relaying of information across cortical layers of representation illustrates the bidirectional processing that contributes to the activation states of cultural input–output relations. The bidirectional processing of cultural information of cortical brain regions depicts the states of activation that comprise the functional processing based on algorithmic function. The identification of the core brain regions that are of importance to the functional processing of learning is fundamental to the characterization of the multilevel processing of algorithmic function. The characterization of the multilevel processing of cortical brain regions is essential to the functional processes of learning.

The functional processes of cortical brain regions detail the patterns of activation states that serve as a mechanistic basis of rule-based learning. The patterns of activation states of cortical brain function depict the states of activation that predict mental inference. The functional patterns of the activation of cortical brain function show how prior states of activation facilitate the prediction of mental inference. The causal relation of prior activation states to other states of activation details the higher-level processing that is linked to the prediction of mental inference. The patterns of coactivation states depict states of higher-level processing that facilitate the cooccurrence or correlation of activation states that represent features of the environment. The higher-level processing of internal representation demonstrates the coactivation of states of cortical brain networks that contribute to the prediction and inference of the environment. The functional processes of higher-level processing of learning rules facilitate the cooccurrence or correlation of activation states that are based on learning.

The functional processing of large-scale cultural brain networks shows the cultural patterns of activation that comprise the internal processing of algorithmic function. The functional patterns of cortical brain function demonstrate the patterns of activation that comprise functional processing based on learning rules. The cultural processing of cortical brain function depicts the activation states that show the functional basis of the algorithmic function of rule-based learning. The patterns of the activation of cortical brain function detail the functional role of core brain regions that are of importance to information processing based on learning rules. The cultural patterns of activation of cortical brain function show the functional significance of learning algorithms for the processing of cultural information.

Recurrent Neural Networks

The distributed representation of large-scale cultural brain networks illustrates the relaying of information across cortical layers of representation. The information processing of cortical layers of representation demonstrates the information flow and activation patterns that comprise cultural processing. Large-scale cultural brain networks show the structural and functional basis of information-processing mechanisms that are of importance to the prediction and inference of the cultural environment. The prediction and inference on the environment based on sensory data relies on functional mechanisms that show the types of information processing of recurrent neural networks. Recurrent neural networks describe neural networks based on bidirectional information processing and their functional role in the control or inhibition of activation patterns.

The structural properties of recurrent neural networks show the types of information- processing mechanisms that facilitate the control or inhibition of activation patterns. Inhibitory or feedback processing demonstrates functional processes that show the anticipation and control of the feedforward processing of activation patterns. The inhibitory processing of feedforward processing based on cultural input illustrates the anticipation and inhibition of activation states that comprise the information processing of the cortical input layer. The inhibition of the activation patterns of the cortical input layer demonstrates the functional role of feedback as a selection filter on the feedforward processing of cultural input. Feedback processing as a selection filter facilitates the control or inhibition of activation patterns based on cultural input. The functional role of feedback processing contributes to the relative weighing of the featural characteristics of cultural input. The selection filter of activation patterns based on cultural input details the functional role of inhibition or feedback processing on the functional processing of cortical layers of representation.

The cultural processing of cortical brain function facilitates the prediction and inference on the cultural environment. The cultural patterns of cortical brain activation show the activation states that comprise the bidirectional processing of functional mechanisms based on cultural input. The internal representation of the cultural environment encodes the featural characteristics of sensory representation as cultural input. The feedforward processing of cortical brain mechanisms demonstrates the activation patterns that comprise the amplification of featural characteristics that facilitate cultural output. The feedback processing on the information processing of cultural input further shows the inhibition or control of the amplification of activation patterns based on cultural input. The selection and control of patterns of activation based on featural characteristics illustrates the functional role of inhibition as a selection filter on the information processing of cultural input. The inhibition of the feedforward processing of cultural input details the activation patterns that facilitate the relative weighing of the activation states of cultural input. The bidirectional processing of cultural input demonstrates the information flow and activation patterns that are of importance to the production of cultural output.

Inhibition or feedback processing also serves as a functional mechanism of the control or regulation of the patterns of activation of context representations. Context representations comprise the internal representations located within the hidden or context layer of representation. The activation patterns of the hidden or context layer comprise the prior activation states that predict mental inference. The prior activation states of functional patterns describe the internal representations of the hidden or context layer that serves as internal input. The prior prediction of the activation states of cortical brain function illustrates the functional role of internal or context representations as internal input to recurrent neural networks. The internal representations of prior activation states serve as a causal mechanism of the recurrence of activation states that predict mental inference. The inhibitory processing of the hidden or context layer facilitates the control or inhibition of the activation patterns that comprise the activation states of internal input. The inhibition or feedback processing of context representations shows the control of the activation patterns from internal states of cortical representation. The feedback and inhibition on the hidden or context layer of representation facilitates the control of activation patterns based on internal input.

The representation and transformation of sensory data of the cultural environment relies on the interpretation or context representation of cortical processing. The context representation of cortical processing facilitates the transformation of representational content into cultural output. The cortical processing of context representations demonstrates the integration of prior activation states as states of prior prediction of mental inference. The recurrence of prior activation states within the hidden or context layer of representation facilitates the interpretation of representational content based on cultural input and its transformation into cultural output. The recurrence of prior activation states serves as internal input to the information processing of cortical layers. The internal input to the context layer of representation demonstrates the inhibition or control of information processing that facilitates the completion of patterns of activation of cultural input–output relations.

The inhibitory or feedback processing of cortical brain function further shows the prediction and inference of recurrent neural networks based on internal and external inputs. The activation patterns of cortical brain function depict the activation states of information processing based on internal and external inputs. The activation patterns of internal and external inputs describe the role of feedback processing in the internal representation of cultural input–output relations. The activation states of feedback processing demonstrate the control or inhibition of the information processing of internal and external inputs. The recurrence of activation states of the hidden or context layer facilitates the pattern completion of initial activation states toward the final states of activation of cultural output. The prior prediction of activation states as context representations facilitates the transformation and interpretation of the internal representation of cultural input to cultural output. The inhibition or control of feedback processing illustrates the facilitation of recurrence on the pattern completion of neural networks.

The functional processing of recurrent neural networks details the inhibition or feedback processing on cultural input based on internal and external inputs. The cortical processing of the hidden or context layer facilitates the inhibition of feedback processing on cultural input. The context representations of the cortical hidden layer serve as the prior activation states of internal input. The recurrence of prior activation states facilitates the interpretation of representational content that is based on internal and external inputs. The patterns of prior activation states contribute to the recurrence of states of activation that facilitate pattern completion of information processing. The recurrence of prior activation states comprises activation patterns that ensure the prior prediction of mental inference. The functional role of prior activation states as control or regulation in the interpretation of representation content demonstrates the causal-functional significance of the inhibition of information processing mechanisms.

The cultural processing of functional mechanisms depicts the functional role of inhibition or feedback in cultural mental inference. The cultural patterns of cortical brain activation show the information processing that is of importance to the prediction of mental inference that is cultural. The activation patterns of cultural processing demonstrate the functional role of inhibition or feedback in the representation and interpretation of cultural input. The cultural patterns of function activation of neural networks show that feedback processing is a causal mechanism of the inhibition or control of cultural processing. The inhibition or feedback of cortical brain activation serves as a functional basis of cultural mental inference.

The causal-functional significance of inhibition or feedback processing illustrates a multitude of functional states that arise based on the inhibition or feedback of activation patterns of recurrent neural networks. The prior activation states of cultural information are a causal basis of the processing of cultural information that facilitates the representation and interpretation of cultural input–output. The inhibition of activation patterns facilitates the sequential processing of cultural information based on the presence of internal and external input. The temporal sequences of information processing based on internal and external inputs illustrate the causal-functional role of activation states in the information processing of cultural information. The sequential processing of cultural information facilitates the inhibition or control of cultural information processing. The multitude of functional states that comprise the cortical processing of cultural information with feedback depicts the causal-functional significance of inhibition.

The cortical processing of cultural information based on internal inputs demonstrates the initiation of the self-organization of cortical brain function. The functional role of inhibition on information processing based on internal feedback demonstrates a causal mechanism for the recurrence of activation patterns. The information processing of cortical brain mechanisms based on internal feedback illustrates the computation principle of self-organization. The inhibitory processing of cortical brain function with internal feedback shows the initiation of prior activation states that facilitate the self-organizing principle of cortical brain function. The initiation of prior activation states as internal feedback shows a functional basis of cortical brain function that demonstrates the recurrence of activation states. The reliance on the inhibitory processing of cortical brain function with internal feedback illustrates the functional role of prior activation patterns as internal input. The recurrence of activation patterns as internal input comprises a causal mechanism of the self-organization of cortical brain function.

The functional processing of large-scale cultural brain networks illustrates the causal-functional role of the bidirectional processing of cortical brain function. The hierarchical structure of cortical brain organization entails the cortical layers of representation that contribute to the bidirectional processing of cortical brain function. The functional architecture of cultural brain networks comprises a functional basis of the relaying of information flow and activation patterns of cultural processing. The multitude of functional states of cortical brain function demonstrates the causal-functional role of information-processing mechanisms on the prediction of mental inference. The characterization of the information-processing mechanisms of the cultural brain contributes to the understanding of the mechanistic basis of cultural computation.

Conceptual Foundations

Conceptual foundations of cultural machine learning abound of the postulation on the cultural brain as an inference machine. Cultural machine learning entails the functional capacity of the cultural brain to learn based on input and feedback. The notion of the cultural brain as an inference machine implies that the brain as a biological computing machine shows the capacity to learn based on internal input and feedback. The conceptual notion of the cultural brain as an inference machine entails the characterization of the functional mechanisms of cultural processing and its causal-functional role in the prediction and inference. The functional capacity of the brain to detect the cultural patterns and regularities of the environment is of importance to the understanding of the brain basis of culture.

The study of cultural machine learning entails the conceptual development of scientific concepts and language that contribute to theory building and hypothesis testing on the notion of the cultural brain as an inference machine. The early scientific work on machine computation postulates on the notion of the brain as an inference machine ("Helmholtz machine") (Helmholtz, 1867). The brain as a biological computing machine shows the capacity to predict and infer on the casual structure of the world based on acquired knowledge or learning. The cultural brain as an inference machine demonstrates the capacity to learn based on input and feedback. The cultural brain as an internal model of the environment illustrates how the brain infers and predicts on the causal structure of the world.

The scholarly and scientific interest in the cultural brain as an inference machine entails the conceptual development of scientific theories and hypotheses testing of the capacity of the brain to learn based on input and feedback. The wide realm of considerations on cultural machine learning arises from the theory building and hypothesis testing on the ontogenetic and phylogenetic basis of the cultural brain and the role of learning as a functional basis of cultural brain function. From cultural brain evolution and neurodevelopment to theoretical neuroscience, scientific theories on the cultural brain aim to characterize the origin and nature of the cultural brain and its structure and function. The postulation on the scientific theories of evolution, neurodevelopment and computation serves as a conceptual foundation for theory building on cultural machine learning. Theory building on cultural machine learning entails interdisciplinary approaches to the understanding of the cultural brain and its role in learning and experience.

The development of scientific concepts and language on cultural machine learning places emphasis on the function and causality of cultural brain function and its role as a learner model of the cultural environment. The cultural brain as a conduit of learning and experience comprises the functional mechanisms that facilitate the prediction and inference on the environment. The characterization of the functional mechanisms of the cultural brain demonstrates how the brain detects the cultural patterns and regularities of the environment. The functional capacity of the cultural brain to infer on the patterns and regularities of the cultural environment illustrates the role of functional mechanisms in cultural processing.

The functional capacity of the cultural brain to learn based on internal input and feedback implies the presence of information-processing mechanisms that perform cultural processing.

The systematic study of cultural machine learning builds on the notion of the brain basis of culture based on learning and experience. One aim of the study of cultural machine learning is the identification of the functional mechanisms that contribute to the capacity of the brain to predict and infer on the cultural environment. The characterization of functional mechanisms that facilitate the prediction and inference of cultural processing is essential to the understanding of the mechanistic basis of cultural brain function. A second aim is the characterization of the multilevel mechanisms of cultural and cortical processing that facilitate learning with feedback. The characterization of the multilevel mechanisms of the cultural brain contributes to the understanding of cultural processing based on learning and its causal effects. The systematic investigation of the brain basis of culture facilitates the understanding of the functional architecture of the brain that is fundamental to prediction and inference.

Conceptual foundations to cultural machine learning illustrate the broad notion of the cultural brain as a biological computing machine capable of learning and experience. The interdisciplinary approaches to theory building on cultural machine learning place emphasis on the identification of causal mechanisms that are of importance to the prediction and inference on the cultural environment. The scientific theories and hypothesis generation on the causal mechanisms of the cultural brain that learn with feedback facilitate the understanding of computational approaches to the study of the brain basis of culture. The broader interest in cultural machine learning contributes to the humanistic and scientific understanding of the cultural brain.

Methodological Foundations

The methodological foundations of cultural machine learning describe the computational techniques that are of importance to the study of the causal interaction of core brain regions of large-scale cultural brain networks. The systematic investigation of the causal interaction of core brain regions that comprise cultural brain networks contributes to the computational modeling of the cultural brain and its functional purpose. The identification of the causal relations of core brain regions facilitates hypothesis testing on the cortical brain mechanisms that are responsible for the facilitation and inhibition of cultural processing. Computational techniques are of interest to the systematic study of the causality and directionality of the functional patterns of large-scale cultural brain networks.

The development of mathematical modeling based on nonlinear dynamics and information theory to describe how the brain performs prediction and inference of the environment is of importance to the understanding of the computational foundations of the cultural brain (Rabinovich et al., 2012). The mathematical modeling of the brain dynamics of neurons and networks of neurons is of importance to the understanding of the patterns of cortical brain function and its response properties that are of importance to prediction and mental inference. The development of mathematical models of brain dynamics provides an idealized model of the causal interaction of functional mechanisms that predict the patterns of cortical brain function and mental inference. The use of mathematical models of brain dynamics in the study of the patterns and regularities of cortical brain function and its casual effects is essential to the data fitting and extraction of idealized models and large-scale datasets of theoretical and computational neuroscience.

Neuroscience methods serve as a methodological foundation of the observation and measurement of patterns of cortical brain function and its spatiotemporal properties across multiple time scales. Neuroscience techniques such as single-cell recording, electrophysiology and functional brain imaging among others provide a methodological foundation for the development of large-scale bioinformatic datasets in neuroscience. From single-cell recording to functional brain imaging, the development of large-scale bioinformatic datasets in neuroscience presents a computational feat that requires the use of mathematical models and computational tools for data extraction and modeling. The development of large-scale bioinformatic datasets entails the use of computational techniques for the data fitting and data extraction of patterns and regularities of cortical brain function and its response properties and causal effects. The use of mathematical modeling with neuroscience techniques is fundamental to the computational modelling of large-scale datasets in neuroscience.

The study of large-scale cultural brain networks with neuroscience techniques is of importance to the identification and characterization of the functional mechanisms of cultural processing and its causal effects. The identification and characterization of the core brain regions that comprise cultural brain networks is fundamental to the study of cultural patterns of brain activity. The computational modeling of the causal interaction of core brain regions of large-scale cultural brain networks is essential to the characterization of the multilevel mechanisms of cultural processing. The use of neuroscience techniques is beneficial to the systematic study of large-scale cultural brain networks with mathematical modeling.

Computational tools and techniques facilitate the modeling of large-scale cultural brain networks based on bioinformatic datasets. The development of computational approaches to the modeling of the causal interaction of cultural brain networks is essential to the understanding of the patterns and regularities of cultural brain function and its causal effects. Computational approaches are also important to the modeling of the dynamical basis of cultural brain function. The study of cultural brain dynamics relies on the use of computational tools and techniques for the data fitting and data extraction of the dynamical basis of patterns and regularities of cultural brain function. Computational tools and techniques are essential to the modeling and data fitting of large-scale datasets that describe how the brain performs prediction and inference of the environment and its causal relation to higher-level processes.

Computational Modeling of Cultural Neural Networks

The study of the computational modeling of cultural neural networks illustrates computational techniques that are of importance to the understanding of the causal interaction of cortical brain regions and their functional significance. The computational modeling of cultural neural networks facilitates the identification of cultural patterns of cortical brain function and its causal effects. The characterization of the causal interaction of cultural patterns of cortical brain function is of importance to prediction and mental inference. Computational techniques are beneficial to the study of the cultural patterns of cortical brain function that serve as predictors of mental inference. The computational modeling of cultural neural networks details a multitude of multilevel approaches to the understanding of the causal interaction of cultural neural networks and their functional significance.

Computational approaches to the study of cultural neural networks are fundamental to the identification of core brain regions that comprise the structural and functional basis of cultural brain function. The core brain regions of cultural neural networks demonstrate the structural and functional properties of core brain regions and their interconnection on cultural processing. The identification of core brain regions contributes to the characterization of neurons and networks of neurons that comprise the fundamental elements of cultural brain function. The characterization of neurons and networks of neurons and their functional properties is of importance to the understanding of the structural and functional basis of the brain basis of culture.

The cultural patterns of cortical brain function illustrate the functional role of core brain regions in cultural processing. Computational modeling of cultural neural networks ascertains the functional relation of cortical brain function and cultural processing. The functional relation of cortical brain function and cultural processing shows the causal interaction of core brain regions that facilitate prediction and mental inference. Computational models of cultural brain function consider the directionality and causality of the interaction of core brain regions and cultural processing. The modeling of the causal interaction of cultural brain function and its functional significance is essential to the characterization of the information processing of core brain regions of cultural brain networks.

Computational models of cultural neural networks consider systematic approaches to the characterization of the causal interaction of core brain regions and their causal effects. The study of the causal relation between cortical brain function and cultural processing entails a wide range of conceptual models. First, computational models seek to identify the causal interaction of core brain regions and the functional relation of activation patterns and cultural task activity. Second, computational models aim to characterize the directionality of the causal interaction of core brain regions and its functional relation to the performance of cultural tasks. Third, computational models seek to determine the causal interaction of core brain regions and their relation to levels of cultural processing. Fourth, computational models aim to identify the cultural patterns of cortical brain function based on resting function. Finally, computational models aim to determine the cultural patterns of cortical brain function that predict mental inference.

Computational techniques are fundamental to the study of the functional processing of cultural brain networks and their role in prediction and mental inference. The use of computational techniques is of importance to the modeling of the cultural patterns of cortical brain function as a predictor of mental inference ("multivoxel pattern analysis"). The multivariate coding of representation across cortical brain regions facilitates the distributed representation of information. Computational techniques are important to determine the functional role of the distribution of representation of cultural information as cultural patterns of cortical brain function. The characterization of the distributed representation of cultural information across cortical brain regions of the cultural neural network is of importance to the understanding of the multivariate coding of cortical representation.

The study of computational approaches to cultural neural networks is fundamental to the characterization of the causal interaction of core brain regions of cultural neural networks. The computational modeling of cultural neural networks is fundamental to ascertain on the multitude of functional relations of core brain regions and their functional significance. The systematic study of cultural neural networks with computational models contributes to the understanding of the mechanistic basis of the brain basis of culture. The characterization of the cultural patterns of cortical brain function is fundamental to the study of cultural neural networks.

Computational approaches to the study of cultural neural networks are fundamental to the broader understanding of the computational principles of the cultural brain.

Computational Modeling of Cultural Brain Dynamics

The study of computational modeling of cultural brain dynamics is of importance to the characterization of the cultural brain dynamics of cortical brain function. The dynamical basis of cortical brain function entails the understanding of the causal interaction of core brain regions and their intrinsic and functional connectivity. The study of cultural brain dynamics is fundamental to the characterization of the brain dynamics of cultural brain function that is based on intrinsic and extrinsic processing. The characterization of the dynamical mechanisms of cultural brain function contributes to the understanding of the structural and dynamical principles of cortical brain organization.

The global and local dynamics of cultural brain networks describe the global and local representation of the cultural processing of neurons and networks of neurons. The global representation of cortical brain dynamics depicts the functional role of interconnectivity in intrinsic and extrinsic processing. The local representation of cortical brain dynamics entails the understanding of cortical brain mechanisms and their response properties as a functional basis of specialized information processing. The global and local dynamics of cortical brain regions details the patterns of cortical brain function that facilitate the higher-level function of cultural processing.

Computational modeling of cultural brain dynamics is fundamental to the understanding of the global and local dynamics of cultural brain networks. The modeling of the global dynamics of cultural brain networks considers the interconnectivity of core brain regions and their functional purpose. The computational modeling of the global dynamics of cultural brain networks considers the overall interconnectivity of core brain regions and their causal effects. The causal interaction of core brain regions of cultural brain networks demonstrates the interconnectivity of cultural brain function and its function. The study of the global dynamics of cultural brain networks contributes to the understanding of cultural brain dynamics.

The computational modeling of the local dynamics of cultural brain networks ascertains the dynamical properties of cortical brain function. The structural and dynamical properties of cortical brain mechanisms illustrate the intrinsic and extrinsic processing of neurons and their response. The dynamical basis of neuronal mechanisms illustrates the aggregation of population-level neuronal activity that comprises the local dynamics of cortical brain regions. The dynamical properties of cortical brain regions illustrate the functional role of the cortical brain activity of neuronal populations on local brain dynamics. The computational modeling of the local dynamics of cultural brain networks facilitates the understanding of the structural and dynamical properties of cortical brain function.

The study of cultural brain dynamics is fundamental to the understanding of the cortical brain dynamics of cultural processing. The structural and functional properties of core brain regions that comprise cultural brain networks facilitate the characterization of the global and local dynamics of cultural brain function. The dynamical mechanisms of cultural brain function show how the functional processing of neurons and networks of neurons are sufficient as a mechanistic basis of cultural processing. The characterization of cultural brain dynamics is essential to the understanding of the structural and dynamical basis of the cortical brain function of cultural processing.

Practical Applications

Computational tools and technologies are fundamental to the scientific discovery of the cultural patterns of cortical brain function and their functional significance. The development of computational tools and technologies contributes to the methodological approaches to theory building and hypothesis testing on the prediction and inference of the cultural patterns of cortical brain function. Computational tools and technologies provide methodological approaches to the data extraction of cultural patterns of cortical brain function of large-scale datasets. The computational tools and technologies for the modeling of the cultural patterns of cortical brain function illustrate the computational techniques for the data fitting of models and the data extraction of datasets. The innovation of computational techniques is beneficial to the computational discovery of the cultural patterns of cortical brain function.

The development of computational tools and technologies contributes to the methodological approaches of computational cultural neuroscience. Computational tools and technologies contribute to the techniques of computational discovery that are fundamental to the study of the cultural patterns and regularities of cortical brain function and its causal-functional role in the prediction of mental inference. The use of computational techniques for the scientific discovery of patterns and regularities based on large-scale bioinformatic datasets contributes to the innovation of computational tools and technologies in health and medicine. The innovation of computational techniques is beneficial to the data extraction of large-scale bioinformatic datasets in health, medicine and related fields.

Computational techniques contribute to the quality of data and evidence that promote evidence-based approaches to culture and health. Computational tools and technologies demonstrate the advancement of observational techniques for the study of the root causes and mechanisms of health and disease. The development of computational techniques is fundamental to the theory building and hypothesis testing of the field of study. The scientific and technological innovation of evidence-based research on computational cultural neuroscience is beneficial to the building of scientific and societal resources that inform on culture and health.

Discussion

The study of cultural machine learning investigates the prediction and inference of the causal structure of the world. The cultural mind as a statistical engine connotates on the notion of mental function as a functional basis of the predictive inference of the cultural environment. The mental function of cultural capacities illustrates the detection of the cultural patterns and regularities in the world and their causal effects. The detection of cultural patterns informs the prediction and inference of the cultural environment. The interaction of the learner model with the cultural environment illustrates the importance of learning and experience of the generation and acquisition of knowledge of the world. Cultural thought as mental inference postulates on the inherent causation of culture in the natural world.

The pontification on the cultural brain as an inference machine broadens the notion of the cultural computation of minds and machines. The notion of the cultural brain as an inference machine entails the understanding of cultural brain function as a mechanistic basis of learning and experience. Cultural learning and experience comprise functional mechanisms that are of importance to the acquisition of knowledge. The functional mechanisms of the cultural brain illustrate a mechanistic basis of learning and experience. The generation and maintenance of cultural knowledge facilitates the prediction and inference of the cultural environment. The cultural brain demonstrates how the cultural computation of minds and machines performs based on learning and experience.

The cultural brain shows how learning and experience are fundamental to cultural thought. The functional basis of cultural brain function demonstrates the role of information processing mechanisms on mental inference based on learning and experience. The functional processing of cultural brain function illustrates the inherent causation of the cultural environment. The functional role of the cultural brain as an inference machine underlies the importance of the functional architecture of the brain as a computational basis of prediction and mental inference. The functional basis of the cultural brain details the causal-functional significance of information processing mechanisms as a mechanistic basis of prediction and inference on the environment.

Conclusion

The study of cultural machine learning explores the foundations of the cultural brain as a model of the prediction and inference on the environment. The characterization of the mechanistic basis of the machine learning of the cultural brain illustrates the functional role of neural networks in cultural processing. The functional basis of recurrent neural networks serves as a causal mechanism of cultural processing. The demonstration of the information flow and cultural patterns of recurrent neural networks shows the information processing of cultural information based on feedback. The feedback or inhibition of cultural processing contributes to the states of activation that contribute to the pattern completion of information processing mechanisms. The information processing of recurrent neural networks demonstrates the plausibility of the cultural brain as an inference machine.

The scientific study of cultural machine learning generates knowledge that informs the scientific and educational enrichment of culture as minds and machines. The demonstration of cultural machine learning shows how the cultural brain is a learner model of the cultural environment. The understanding of the cultural brain and its causal-functional role in learning and mental inference contributes to the scientific and educational enrichment on the cultural brain as an inference machine. The multiple realizability of culture as a physical realizer details the functional significance of the cultural brain as a machine capable of prediction and mental inference based on input from the cultural environment. Cultural machine learning demonstrates the higher-level processing of the cultural brain and its functional role in the formulation of the prediction and inference on the inherent causation of the world.

Implications

The study of cultural machine learning has practical applications that are beneficial to the scientific and educational enrichment on culture and health. The scientific and technological innovation of computational tools and technologies that promote the computational discovery of the cultural brain and its function is of importance to the research and development on culture and health in health, medicine and related fields. Research and development on computational tools and technologies advance the research capabilities and scientific infrastructure that is essential to the generation of scientific knowledge. The research and development on cultural machine learning broadens understanding of the computational techniques that are beneficial to the advancement of computational discovery of the root causes and mechanisms of health and disease. The scientific and technological innovation of computational tools and technologies shows the practical application and societal impact of computational science and bioinformatics on the advancement of research and development for the study of the root causes and mechanisms of health and disease.

The use of computational discovery to inform the observation and empiricism of scientific work is beneficial to the scientific and societal enrichment on culture and health. Computational discovery serves as an evidence-based approach to the building of scientific resources that inform on the root causes and mechanisms of health and disease. Computational tools and technologies provide quality of evidence and data to inform scientific research as well as to broaden societal and educational enrichment on culture and health. The scientific and technological innovation of computational tools and technologies is beneficial to the promotion of scientific research that informs on areas of culture and health and its scientific and societal impact.

Reference

Dayan, P., Hinton, G. E., Nel, R. M., & Zemel, R. S. (1995). The Helmholtz machine. *Neural Computation, 7*, 889–904.

Helmholtz, H. (1867). *Handbuch der physiologischen Optik*. Leipzig.

Locke, J. (1689). *An essay concerning human understanding*. Oxford: Claredon Press.

Rabinovich, M. I., Friston, K. J., & Varona, P. (Eds.). (2012). *Principles of brain dynamics: Global state interactions*. Cambridge: MIT Press.

9

CULTURAL CONNECTOME

Introduction

The history of neuroscience illustrates the scientific importance of the discovery of the neuron and its broader impact. The scientific discovery of the nineteenth century of the neuron as a cellular and molecular mechanism led to the advancement of the understanding of the structure and function of the nervous system. The neuron as a fundamental unit of the electrophysiological signaling of neural transmission shows the functional importance of the cellular and molecular mechanisms of the brain. The systematic study of the cellular and molecular basis of the nervous system contributes to a broader understanding of the structural and functional principles of cortical brain organization. The contemporary developments in neuroscience of the higher-level function of the brain are an advancement of the understanding of the fundamentals of the nervous system.

The innovation of computational tools and technologies for the study of the brain has further led to the development of research capabilities for scientific discovery in neuroscience. Computational tools and technologies provide a multilevel approach to the observation and measurement of the brain function and its structural and dynamical basis. The computational discovery of the structural and dynamical basis of neurons and networks of neurons facilitates the understanding of the computational foundation of cortical brain function. The use of computational tools and technologies to investigate the higher-level processes of brain function is fundamental to the understanding of the brain basis of culture. The development of computational approaches to the study of the brain basis of culture is of importance to the innovation of computational foundations in neuroscience.

DOI: 10.4324/9781003384236-10

The goal of the chapter is to provide a review on the computational foundations of the cultural connectome. The study of the structural and dynamical basis of the cultural brain entails the structural and functional properties of cultural brain function. The structural and dynamical principles of the cultural brain contribute to the understanding of the global and local dynamics of large-scale cultural brain networks. The characterization of the dynamical mechanisms of cultural brain function illustrates the structural and functional basis of the interconnectivity of the cultural brain as a connectome. The multilevel approaches to the study of the cultural connectome contribute to the computational foundations of cultural brain dynamics. Implications of research on the cultural connectome for health, medicine and related fields are discussed.

Computational Principles of Cortical Brain Organization

The computational principles of cortical brain organization serve as a fundamental basis for the understanding of the structural and functional properties of the cortical brain. The functional specialization and functional integration of cortical brain function details the structural and functional properties of cortical brain function. From neurons to networks of neurons, the information processing of cortical brain regions serves as a functional basis of the cortical processing of brain function. The functional specialization of cortical brain regions shows the specialized and dedicated processing of cortical brain mechanisms. The functional integration of cortical brain regions illustrates the integration of the functional processing of higher-level processes. An understanding of the computational foundations of cortical brain organization is essential to the characterization of the multilevel processing of cortical brain function.

The hierarchical structures of cortical brain organization illustrate the columnar organization of cortical brain organization. The columnar organization of the cortical brain organization comprises cortical layers of representation that constitute the structural and functional basis of cortical brain organization. The columnar organization of cortical layers of representation describes the distribution of types of representation across distinct areas of cortical organization. The distribution of representation across cortical brain areas illustrates the distributed processing streams of large-scale brain organization. The hierarchical structure of cortical layers of representation is of importance to the information-processing streams of functional pathways.

Large-Scale Brain Organization

The large-scale brain organization of cortical brain function describes the distributed representation of information across interconnected brain regions. The distribution of representation of cortical brain regions illustrates the hierarchical structure of cortical brain organization as functional pathways. The distribution of information-processing streams across distinct cortical areas demonstrates the multiple functional pathways that comprise large-scale brain organization. The distributed processing of information across multiple functional pathways facilitates the hierarchical structure of cortical brain organization. The distribution of representation of functional pathways across distinct cortical areas illustrates the hierarchical structure of cortical brain function.

The large-scale brain organization of brain function entails the functional processing of information streams that is based on the interconnection of cortical brain regions. The functional processing streams of interconnected cortical brain regions located within cortical brain areas located within occipitotemporal cortices demonstrate the processing of information that is content-specific. The information- processing streams of functional pathways demonstrate the interconnection of cortical brain regions that comprise a functional basis of content-specific information processing. The interconnection of cortical brain regions demonstrates the functional basis of the information- processing streams of cortical brain areas. The interconnection of cortical brain regions that comprise distinct types of information processing facilitates the specialized mechanisms of cortical brain areas.

The interconnection of cortical brain regions furthers the functional pathways of information processing of distinct types of processing. The hierarchical structure of large-scale cortical brain organization illustrates the functional processing of information that is process-specific. The information processing of functional pathways of cortical brain function of cortical brain areas illustrates functional processing that comprises a process-specific information processing. The functional processing of interconnected brain regions located within the parietal lobule facilitates information processing streams that are process-specific. The distributed processing of functional pathways illustrates the distinct types of processing that comprise the information process streams of interconnected brain regions.

The large-scale brain organization of cortical brain function demonstrates the functional role of hierarchical structure as a computational principle of the distributed representation of information processing. The interconnection of cortical brain regions facilitates the distributed processing of representation across multiple functional pathways. The distribution of representation across cortical brain areas facilitates the information processing of functional pathways across interconnected brain regions. The functional pathways of information processing demonstrate the importance of distributed processing on the types of representation of cortical brain function.

Interconnectivity

The interconnectivity of cortical brain regions is fundamental to the understanding of the structural and functional principles of cortical brain organization. The interconnection of cortical brain regions illustrates the structural and functional basis of the structural and functional connectivity of cortical brain function. The interconnectivity of cortical brain regions demonstrates a functional basis of the relaying of information of cortical brain regions. The flow of information across cortical brain regions demonstrates the functional significance of the interconnection of cortical brain areas. The structural and functional connectivity of cortical brain function shows the functional purpose of the interconnectivity of cortical brain organization.

The structural and functional connectivity of cortical brain function illustrates the functional processing of neurons and networks of neurons based on the interconnection of cortical brain areas. The functional processing of neuronal networks demonstrates the causal-functional significance of the cortical processing of interconnected brain regions. The structural connectivity of cortical brain function illustrates the structural basis of the interconnection of cortical brain regions. The structural basis of cortical brain regions facilitates the functional processing of cortical brain function. The structural properties of cortical brain regions serve as a causal impetus for the functional processing that is based on the structural basis of cortical brain areas.

The functional connectivity of cortical brain function demonstrates the functional processing of neuronal networks based on the interconnection of cortical brain areas. The functional processing of neuronal networks comprises a functional basis of the information-processing pathways of the brain. The functional basis of information-processing streams facilitates the distributed representation of information across multiple functional pathways. The functional properties of cortical brain regions depict the distribution of representation that comprises the information processing of interconnected brain regions. The functional connectivity of cortical brain function illustrates the functional significance of interconnection on the information-processing mechanisms of functional pathways.

The structural and functional connectivity of cortical brain function comprises a functional basis for the patterns of cortical brain activity of brain function. The patterns of cortical brain activity of functional pathways illustrate the activation states that are based on the structural properties of cortical brain areas. The functional patterns of cortical brain activity describe the states of functional activation that show the causal role of structural properties on functional processing. The patterns of cortical brain activity depict the activation states that comprise optimal states of functional activation based on structural properties.

The functional patterns of cortical brain activity also depict the flow of information and patterns of activation that are of importance to the functional properties of cortical brain areas. The patterns of cortical brain activity show the types of functional processing that correspond with the patterns of activation of cortical brain regions. The functional processing of cortical brain regions demonstrates the levels of activation of cortical brain function that coincide with the levels of processing of performance of functional tasks. The patterns of activation show the functional properties of cortical brain areas that are of importance to the functional processing of interconnected brain regions. The patterns of activation of cortical brain activity demonstrate the functional correspondence of cortical brain regions and functional processing. The functional connectivity of cortical brain function comprises the patterns of functional brain activity that are of importance to the relaying of information of cortical brain regions. The functional basis of the interconnection of cortical brain regions shows the functional processing of information processing mechanisms.

The interconnectivity of cortical brain regions shows the functional significance of information flow and activation patterns on the functional processing of core capacities. The structural and functional connectivity of cortical brain function illustrates the functional significance of interconnection as a functional basis of the cortical processing of information. The interconnection of cortical brain regions shows the functional processing of cortical brain area that comprises the causal-functional significance of the information flow of cortical brain regions as activation patterns of cortical brain function. The flow of information and patterns of activation of interconnected brain regions illustrate the interconnectivity of cortical brain function.

The interconnectivity of cortical brain function comprises a functional basis for the structural and dynamical aspects of cortical brain networks and their causal effects. The interconnection of cortical brain regions serves as a structural and dynamical basis of the cortical mechanisms that show the patterns of activation that comprise information processing mechanisms. The structural and dynamical basis of cortical brain networks depicts the information flow and activation patterns that are essential for functional processing. The interconnectivity of cortical brain networks illustrates the functional role of the causal interaction of core brain regions in the functional processing of cortical brain regions.

Multiple Constraint Satisfaction

The computational principle of multiple constraint satisfaction refers to the functional processing of cortical brain function that illustrates the satisfaction of multiple constraints. The functional processing of cortical brain function shows the states of activation that comprise the optimal states of the information processing of functional brain activation. The patterns of activation of cortical brain function illustrate the activation states that show optimal levels of activation. The optimal levels of activation of cortical brain function are based on the satisfaction of the constraints based on internal and external input to information processing. The patterns of functional brain activation of cortical brain function demonstrate the optimal states that are important for functional processing.

The computational foundations of cortical brain organization are fundamental to the understanding of the structural and functional principles of cortical brain function and its relation to higher-level function. The computational principles of cortical brain organization facilitate the understanding of cortical brain function and the higher-level function of cultural processing. The computational basis of structural and functional principles comprises a structural and functional basis of cortical and cultural processing. The computational approaches to the understanding of the structural and functional basis of cortical brain function and its relation to cultural processing show the functional significance of the structural and functional principles on the cortical brain function of cultural processing. The computational principles of cortical brain organization are foundational to the characterization of the cortical brain function of cultural processing.

The Brain Basis of Culture

The study of the brain basis of culture investigates the functional mechanisms of cultural capacities. The notion of the brain basis of culture postulates on the cultural brain as a biological computing machine. The functional processes and mechanisms of the brain constitute a functional basis of cultural capacities. The study of the functional mechanisms of cultural capacities contributes to the characterization of the multilevel mechanisms of the cortical brain and cultural processing. The characterization of the multilevel mechanisms of cultural brain function is fundamental to the understanding of the brain basis of culture.

The computational principles of cortical brain organization are fundamental to the understanding of the brain basis of culture. The brain basis of culture comprises the cortical brain function of functional mechanisms that are important to cultural processing as a higher-level function. The systematic study of the brain basis of culture explores the functional mechanisms of cortical and cultural processing. From neurons to networks of neurons, the study of cortical brain organization illustrates the multilevel processing of the nervous system and its structural and functional principles. The structural and functional properties of cortical brain organization serve as a functional basis of higher-level processes of behavior. The multilevel mechanisms of cortical and cultural processing illustrate a functional basis of the processing of cultural information. The characterization of the multilevel mechanisms of cortical brain function facilitates the understanding of the mechanistic basis of cultural capacities.

The functional specialization of cortical brain areas for the specialized and dedicated processing of information illustrates the specialized processing of functional mechanisms. The functional processing of cortical mechanisms located within cortical brain areas shows the functional specificity of neuronal response that is specific to culture. The patterns of neuronal response located within cortical brain areas demonstrate the specialized processing of cortical mechanisms. The neuronal response to the cultural environment illustrates the core capacities of cortical brain regions for the specialized and dedicated processing of cultural information.

The functional integration of cortical brain networks demonstrates the functional purpose of the interconnection of cortical brain regions for the integration of functional processing. Networks of neurons comprise the neuronal activity of populations of neurons located within cortical areas. The interconnection of networks of neurons of cortical brain regions describes the structural and functional connectivity of cortical brain networks. The interconnection of cortical brain regions illustrates the levels of activity of neuronal populations and their functional properties. The population-level activity of networks of neurons details the cortical brain dynamics of interconnected brain regions. The interconnectivity of interconnected brain regions demonstrates a functional basis for the relaying of information across cortical layers of representation.

Neuronal networks serve as a functional basis of the cultural processing of the brain. Cultural brain networks consist of core brain regions and their interconnection. The functional processing of cultural brain networks shows the relaying of information across interconnected brain regions that comprise cultural information processing. The interconnectivity of neuronal networks describes the role of intrinsic and extrinsic processing on the structural and dynamical basis of cortical and cultural processing. The interconnection of neuronal networks facilitates the structural and functional basis of cultural brain networks. The interconnectivity of neuronal networks is fundamental to the cultural processing of the brain.

The structural properties of cultural brain function describe the structural connectivity of the core brain regions that comprise cultural brain networks. The interconnection of core brain regions responsible for cultural processing serves as a functional basis of cultural brain networks. The structural connectivity of the cultural brain demonstrates the structural basis of the interconnection of core brain regions. The structural properties of core brain regions contribute to the functional processing of interconnected brain regions and their functional purpose. The structural properties of cultural brain function depict the intrinsic activity of core brain regions based on the characteristics of neuroanatomical structures.

The functional properties of cultural brain function detail the functional connectivity of cultural brain networks. The functional connectivity of cultural brain networks comprises the functional properties of the cortical brain function of cultural processing. The functional and emergent properties of cultural brain function illustrate the causal-functional significance of cultural processing. The interconnection of core brain regions facilitates the functional processing of cultural brain networks. The functional connectivity of core brain regions illustrates the extrinsic processing of cultural brain networks. The functional properties of cultural brain function are of importance to the understanding of the functional basis of cultural processing.

The interconnectivity of the cultural brain demonstrates the intrinsic and extrinsic connectivity of cultural brain function. The intrinsic connectivity of cultural brain function details the intrinsic processing of core brain regions based on the interconnection of neuroanatomical structures. The intrinsic processing of core brain regions describes the intrinsic activity of cultural brain networks based on the characteristics of neuroanatomical structures. The intrinsic activity of cultural brain networks shows the functional processing of cultural information that is based on the internal states of cultural processing. The internal generation of cultural processing illustrates the inherent cultural function of cultural brain networks. The intrinsic connectivity of cultural brain networks shows the causal role of interconnection on the functional processing of cortical brain function.

The extrinsic connectivity of cultural brain networks is comprised of the extrinsic processing of interconnected brain regions. The extrinsic connectivity of core brain regions facilitates the functional processing of cultural brain networks. The functional processing of cultural brain networks shows the extrinsic processing of cultural information. The causal role of extrinsic processing as a functional basis of cultural brain function illustrates the functional significance of interconnection on the processing of cultural information. The extrinsic connectivity of core brain regions comprises the functional basis of cortical and cultural processing.

The extrinsic processing of cultural brain function details the information flow and activation patterns that comprise the processing of cultural information. The functional processing of cultural brain activity shows the patterns of cortical brain activation that coincide with the information flow of cultural processing. The patterns of functional brain activation demonstrate functional correspondence with the processing of cultural information. The information flow and activation patterns of cultural brain function show how extrinsic processing facilitates cultural processing. The cultural patterns of functional brain activation contribute to the extrinsic processing of the cultural brain.

The cultural patterns of cortical brain function demonstrate the activation patterns of cultural brain function that are important to cultural processing. The activation patterns of cultural brain function illustrate activation states as optimal states of cultural processing. The cultural patterns of functional brain activation depict the states of activation that comprise the pattern completion of cultural processing from cultural input-output. The pattern completion of cultural processing of cortical brain function shows the processing of cultural information to the optimal states of activation. The processing of cultural information to optimal states of activation comprises the functional patterns of activation cultural processing.

The activation patterns of cultural brain function show the functional processing of cultural information as states of activation of cultural processing. The cultural patterns of cortical brain function illustrate the activation states that show multiple constraint satisfaction. The patterns of cultural brain function demonstrate states of activation as optimal states that show the satisfaction of multiple constraints. The cultural patterns of functional activation at optimal levels of activation show constraints of the levels of functional activation of cortical brain regions. The cultural patterns of cortical brain function demonstrate the activation patterns that are essential to cultural processing.

The structural and dynamical basis of cultural brain networks illustrates the causal interaction of cortical brain dynamics and its functional significance. The structural and dynamical properties of cultural brain networks depict the understanding of the cortical brain dynamics of cultural processing. The causal interaction of core brain regions details the causality of information flow and activation patterns of interconnected brain regions based on the mental function of cultural task activity. The patterns of functional activation of core brain regions comprise the core brain regions that are responsible for the processing of cultural task activity. Cultural brain networks are fundamental to the understanding of the structural and dynamical basis of cultural brain function.

Scholarly and scientific interest on the brain basis of culture pontificates on the processes and mechanisms of cultural brain function. The characterization of the multilevel mechanisms of cortical brain function and cultural processing describes a functional basis of cultural capacities. The study of the structural and dynamical basis of cortical brain function and cultural processing is essential to the fundamental understanding of cultural brain networks. The multidisciplinary approaches to the study of the cultural brain illustrate the innovation of computational approaches to the understanding of the brain basis of culture.

Cultural Brain Dynamics

ultural brain dynamics describes the structural and dynamical basis of the cortical brain function of cultural brain networks. The cultural brain dynamics of large-scale cortical brain networks describes the functional processing of dynamical mechanisms of cortical brain activity and their inherent cultural function. The functional processing of dynamical mechanisms of cortical brain activity shows how interconnection comprises a structural and functional basis for the functional processing of cultural brain networks. The dynamical basis of the functional processing of cultural brain networks illustrates the information flow and activation patterns that comprise cultural brain activity. The cultural brain dynamics of large-scale cortical brain networks show the dynamical mechanisms that are fundamental to the intrinsic and extrinsic processing of cortical brain function.

The study of cultural brain dynamics is fundamental to the characterization of the structural and functional properties of cultural brain function. The structural and functional properties of cultural brain dynamics comprise the structural and dynamical basis of the causal interaction of core brain regions that are responsible for cultural processing. The causal interaction of core brain regions and their interconnection facilitates cultural processing. The functional properties of cultural brain function show the structural and dynamical properties of cultural processing. The characterization of the causal interaction of core brain regions of cultural brain networks illustrates the functionality and causality of interconnection.

Cultural brain dynamics are essential to the understanding of the intrinsic and extrinsic processing that comprises cortical brain dynamics and its inherent cultural function. The intrinsic and extrinsic processing of interconnected brain regions of cultural brain networks shows the information flow and activation patterns of cultural processing. The intrinsic processing of cultural brain dynamics demonstrates the spontaneous and intrinsic activity of cortical brain areas responsible for cultural processing. The extrinsic processing of cultural brain dynamics illustrates the extrinsic connectivity of cortical brain networks and the functional role of interconnection on cultural processing. The intrinsic and extrinsic processing of cultural brain networks formulates a functional basis of the types of processing that are essential to cultural brain function.

Large-Scale Cultural Brain Networks

Large-scale cultural brain networks comprise the core brain regions and their interconnection that are responsible for cultural processing. The causal interaction of core brain regions of cultural brain networks shows the directionality and causality of information flow and activation patterns that comprise cultural processing. The characterization of the functionality and causality of cultural brain networks is fundamental to the understanding of the functional basis of the cultural processing of the brain. The functionality and causality of cultural brain networks show the information flow and activation patterns that have causal-functional significance. The interconnection of the core brain regions of cultural brain networks is fundamental to cortical and cultural processing.

Large-scale cultural brain networks consist of core brain regions that show the patterns of activation that are fundamental to the processing of cultural information. The information flow and patterns of activation of cultural brain networks are fundamental to the functional processing of the cultural brain. The patterns of activation of cultural brain networks illustrate the activation states of functional processes that are optimal states of cultural processing. The cultural patterns of activation comprise of the optimal states of activation that facilitate cultural input–output. Large-scale cultural brain networks demonstrate the optimal states of activation patterns that are essential to the processing of cultural information.

The global states of local attractor networks of core brain regions illustrate the global dynamics of the cultural brain. The global dynamics of cultural brain function depict the extrinsic processing based on the interconnectivity of core brain regions. The global states of local attractor networks demonstrate the activation states that are optimal at a global minimum. The core brain regions of local attractor networks comprise the global states that facilitate extrinsic processing that is cultural. The global brain dynamics of core brain regions are fundamental to the understanding of the global states of local attractor networks.

Interconnectivity of the Cultural Brain

The interconnectivity of the cultural brain serves as a structural and functional basis of the global and local brain dynamics of cultural processing. The interconnectivity of core brain regions illustrates the intrinsic and extrinsic processing of large-scale cultural brain networks and its causal effects. The global and local brain dynamics of cortical brain function detail the intrinsic and extrinsic processing of core brain regions. The structural and functional properties of interconnectivity serve as a mechanistic basis of the intrinsic and extrinsic processing of cultural brain networks. The global and local brain dynamics of cultural processing are fundamental to the characterization of the functional basis of cultural brain networks.

The causal interaction of the core brain regions of cultural brain networks illustrates the functional role of interconnectivity as a functional basis of intrinsic and extrinsic processing. The interconnection of core brain regions demonstrates the interconnection of neurons and networks of neurons based on white matter tracts. The interconnection of core brain regions illuminates the structural and functional properties of intrinsic and extrinsic processing. The causal interaction of core brain regions facilitates the functional basis of cultural processing and their intrinsic and extrinsic properties.

Interconnectivity is essential to the characterization of the functional processing of cortical brain networks and their inherent cultural function. The interconnection of core brain regions facilitates the global states of local attractor networks that comprise cultural brain networks. The global and local states of local attractor networks describe the intrinsic and extrinsic connectivity of core brain regions and their functional role in cultural processing. The global brain dynamics of core brain regions and their interconnection demonstrate the functional significance of cultural brain networks.

Intrinsic and Extrinsic Processing

The intrinsic and extrinsic processing of core brain regions of cultural brain networks is essential to the understanding of the functional processing of cultural information. The intrinsic and extrinsic connectivity of cultural brain networks illustrates the interconnection of core brain regions based on intrinsic and extrinsic activity. The intrinsic processing of core brain regions describes the spontaneous and intrinsic activity of cortical brain function. The extrinsic processing of core brain regions depicts the extrinsic activity of cultural brain networks. The intrinsic and extrinsic processing of cultural brain networks illustrates a functional basis of cortical and cultural processing.

The intrinsic connectivity of core brain regions demonstrates the functional basis of the spontaneous and intrinsic activity of cultural brain networks. The intrinsic connectivity of cultural brain networks shows the structural basis of the intrinsic processing of core brain regions. The autonomous dynamics of endogenous brain activity depict the functional role of intrinsic processing on the entrainment of cortical brain function. The endogenous brain activity of cultural brain networks illustrates the intrinsic processing of core brain regions that is based on the autonomous and self-organization of cortical brain activity.

The extrinsic connectivity of core brain regions describes the functional role of extrinsic processing on cultural brain networks. The extrinsic connectivity of core brain regions illustrates the global states of local attractor networks and their interconnection. The extrinsic activity of interconnected brain regions shows the overall aggregation of population-level neuronal activity that is stimulus-driven and based on the extrinsic connectivity across cortical brain areas. The characterization of the causal interaction of core brain regions encompasses the intrinsic and extrinsic processing of cultural brain networks.

Computational Foundations of Cultural Brain Dynamics

The computational foundations of cultural brain dynamics encompass the structural and dynamical basis of large-scale cultural brain networks. The study of the computational principles of cultural brain dynamics entails the characterization of the brain dynamics of cultural processing. The computational approaches to the characterization of cultural brain dynamics investigate the functional processing of large-scale cultural brain networks and their causal interaction. The causal interaction of core brain regions illustrates the structural and dynamical basis of large-scale cultural brain networks. The computational foundations of cultural brain dynamics facilitate consideration on the functional processing of large-scale cultural brain networks and their causal-functional significance.

The structural and functional properties of large-scale cultural brain networks illustrate the structural and functional basis of cortical and cultural processing. The structural and functional basis of cortical and cultural processing describes the functional processing of large-scale cultural brain networks and their causal effects. The structural and functional properties of cultural brain regions encompass the dynamical basis of the functional processing of cultural information. The dynamical mechanisms of cortical and cultural processing are based on the structural and functional properties of large-scale cultural brain networks.

The functional specialization and functional integration of cortical brain function demonstrates the distinct types of processing. The functional specialization of core brain regions depicts the specialized and dedicated processing of cortical mechanisms. The specialized and dedicated processing of cortical mechanisms illustrates the functional processing that is content-specific. The functional integration of core brain regions shows the functional pathways of information processing. The functional integration of cortical pathways details the functional processing of cortical mechanisms that is process-specific. The functional specialization and functional integration of cortical brain mechanisms shows the distinct types of processing that comprise cortical pathways.

The causal interaction of cultural brain regions describes the causal-functional role of the cultural processing of interconnected brain regions. The causal interaction of cortical brain regions that are responsible for the processing of cultural information depicts the brain dynamics of cortical and cultural processing. The interconnection of cultural brain regions facilitates the causal interaction of information flow and activation patterns of cultural brain networks. The causal interaction of interconnected brain regions shows the relaying of information that coincides with the cultural patterns of functional brain activation. The interconnection of cultural brain regions serves as a functional basis of the cultural processing of the brain.

The structural and functional connectivity of cultural brain networks demonstrates the functional role of interconnection on the cortical brain dynamics of cultural processing. The structural connectivity of interconnected brain regions based on white matter tracts shows the interconnection of cortical brain regions that facilitates the processing of cultural information. The structural connectivity of cultural brain regions comprises a structural basis of the cortical brain function of cultural processing. The functional connectivity of cultural brain regions shows the functional processing of interconnected brain regions. The functional connectivity of cultural brain regions serves as a functional basis of cortical and cultural processing. The structural and functional connectivity of cultural brain networks illustrates the importance of interconnection of the dynamical basis of the cultural brain function.

The interconnectivity of cultural brain networks illuminates on the functional processing of cultural brain regions. The interconnection of cultural brain regions depicts the relaying of cultural information across interconnected brain regions. The interconnectivity of cultural brain regions demonstrates the information flow and activation patterns of cultural processing. The strength of the interconnection of cultural brain regions facilitates the information processing and patterns of activation that comprise cultural brain networks. The functional processing of cultural brain regions shows the functional significance of the interconnection of cortical brain regions.

The interconnectivity of cultural brain networks illustrates a structural and functional basis for the cortical brain dynamics of cultural processing. The structural and functional connectivity of cortical brain regions shows how the causal interaction of core brain regions facilitates the dynamical mechanisms of cortical and cultural processing. The causal interaction of core brain regions illustrates the directionality and causality of information flow and activation patterns that comprise cultural processing. The characterization of the causal interaction of core brain regions based on interconnectivity contributes to the understanding of the structural and dynamical basis of cultural brain dynamics.

Global and Local Brain Dynamics

The global and local brain dynamics of cortical brain regions demonstrate the dynamical properties of neurons and networks of neurons and their functional processing. The global dynamics of cortical brain regions comprise the global representation of the aggregation of the population-level activity of networks of neurons. The global dynamics of cortical brain activity describe the causal interaction of core brain regions and their functional purpose. The cortical brain activity of interconnected brain regions shows how the information flow and activation patterns of functional processing are sufficient to comprise the global dynamics of cortical brain function. The information flow and activation patterns of interconnected cortical brain regions contribute to the global dynamics of cortical brain networks.

The local brain dynamics of cortical brain regions describes the structural and functional properties of local cortical brain areas. The cultural brain activity of cortical brain areas depicts the local representation of neuronal dynamics that comprise the functional processing of core brain regions. The neuronal dynamics of core brain regions describe the patterns of activation states that comprise functional processing. The dynamical basis of neuronal activity and its response properties illustrates the functional role of neuronal mechanisms in the local brain dynamics of cortical brain regions.

The global and local dynamics of cortical brain activity detail the dynamical basis of the functional processing of neurons and networks of neurons. The global and local dynamics of cultural brain function describe the information-processing mechanisms that illustrate the dynamical mechanisms of cortical brain activity and multilevel processing. From synaptic dynamics to large-scale cortical brain networks, the dynamical mechanisms of global and local representations illustrate the patterns of cortical mechanisms that comprise a functional basis of the intrinsic and extrinsic processing of the brain.

Attractor Dynamics

The attractor dynamics of cortical brain function describe how the brain facilitates the functional processing of cortical brain activity. The cultural patterns of cortical brain activity of interconnected brain regions describe the patterns of activation states that are linked to cultural processing. The relaying of information across cortical layers of representation illustrates the functional processing of cultural information. The information flow and activation patterns of cultural brain networks detail the states of activation that facilitate cultural input-output. The states of activation of cultural patterns illustrate the attractor states of functional brain activity that is based on its inherent cultural function.

Attractor states of patterns of functional activation show the specific activation states that are of importance to the pattern completion of functional processing. The attractor states of functional brain activation illustrate the global or local minimum as optimal states of activation that facilitate the pattern completion of functional processing. The optimal states of functional brain activation consist of the patterns of activation states that facilitate pattern completion. The patterns of cortical brain activity at optimal states of activation show the functional processing that satisfies multiple constraints in the production of cultural input-output.

The local networks of attractor states comprise the patterns of activation states that describe the local brain dynamics of cortical brain areas. The local network of attractor states details the activation states of the functional processing of local cortical brain areas. Local networks of attractor state consist of the activation states of neuronal activity that are specific to distinct types of neurons and their response properties. The functional processing of activation states describes the dynamical mechanisms of cortical brain activation. The synaptic dynamics of neurons and their interconnection constitute the dynamical mechanism of cortical brain activation. The density of the synaptic connection of neurons facilitates the dynamical basis of activation states of local attractor networks.

The global model of local attractor networks describes the pattern of activation that is based on the interconnection of cortical brain areas. The global states of local attractor networks entail the global brain dynamics of the functional processing of cortical brain areas. The functional processing of cortical brain areas shows how interconnection is a functional basis of the dynamical mechanisms of cultural brain networks. The interconnection of cortical brain areas facilitates the global states of local attractor networks. The global model of local attractor networks illustrates the extrinsic processing of global brain dynamics.

Autonomous Brain Dynamics

Autonomous brain dynamics consist of the brain dynamics and intrinsic processing of cortical brain function. The autonomous brain dynamics of cultural processing encompass the spontaneous and intrinsic activity of cultural brain regions. The study of autonomous brain dynamics is fundamental to the understanding of the functional role of endogenous brain activity as a functional basis of intrinsic cultural processing. The endogenous brain activity of cortical brain function describes intrinsic processing based on resting-state function. The cortical brain function of intrinsic processing details spontaneous and intrinsic activity that is stimulus-independent. The autonomous brain dynamics of cortical brain function illustrate the intrinsic processing of cortical brain regions that is cultural.

The autonomous brain dynamics of cultural processing comprise the intrinsic processing that demonstrates the functional role of endogenous brain activity on inherent cultural function. The intrinsic processing of endogenous brain activity describes the functional basis of the spontaneous fluctuations of cultural brain regions. The functional basis of the spontaneous and intrinsic activity of cultural brain regions shows the intrinsic processing of cortical brain function that arises based on the maintenance of active and itinerant representations. The cultural brain activity of cortical brain regions shows the intrinsic activity that facilitates the maintenance of active representations of cortical brain areas.

The autonomous brain dynamics of intrinsic cultural processing describes the endogenous brain activity that is based on the entrainment of cortical brain function. The endogenous brain activity of intrinsic processing describes the spontaneous fluctuations of cultural brain activity as internal states of brain activity. The spontaneous and intrinsic activity of cultural brain regions shows the autonomous brain dynamics of cultural brain networks. The cultural brain activity of cortical brain regions that comprise intrinsic processing illustrates the functional role of the structural and functional properties of cortical brain networks. The intrinsic processing of cultural brain networks demonstrates the functional basis of spontaneous and autonomous activity based on endogenous brain activity.

The computational foundations of cultural brain dynamics are fundamental to the understanding of the structural and dynamical basis of cultural brain networks. The computational principles of cortical brain function demonstrate the lawlike principles that predict the causal interaction of cultural brain regions and their causal effects. The causal interaction of cultural brain regions contributes to the functional processing of cultural brain dynamics. The cultural patterns and regularities of cortical brain activity are essential to the characterization of the structural and dynamical basis of cultural brain networks.

Cultural Connectome

The study of the cultural connectome consists of the understanding of the structural and dynamical basis of large-scale cortical brain networks and their inherent cultural function. The notion of the cultural connectome encompasses the consideration on the structural and functional properties of large-scale cortical brain networks and their inherent cultural function. Large-scale cortical brain networks serve as a functional basis of the cultural processing of the brain. The interconnection of cortical brain areas facilitates the information flow and activation patterns of cultural brain networks that are linked to cultural processing. The strength of the interconnection of cortical brain areas contributes to the direct and indirect effects of anatomical connectivity on cultural and cortical processing.

The cultural connectome entails consideration of the anatomical connectivity of large-scale cortical brain networks and their functional purpose. The anatomical connectivity of large-scale cortical brain networks encompasses the structural and functional properties of that are of importance to cortical brain dynamics. The anatomical connectivity of the brain consists of the interconnection of neurons and networks of neurons that facilitate the functional processing of cortical brain function. The anatomical connectivity of large-scale cortical brain networks has direct and indirect effects on cortical processing. The understanding of the anatomical connectivity of the brain is fundamental to the consideration of the higher-level processes of cortical brain function.

The anatomical connectivity of large-scale cortical brain networks contributes to cultural processing and its functional significance. The interconnection of neurons and networks of neurons facilitate the processing of cultural information. The anatomical connectivity of the cultural brain refers to the interconnection of cortical brain regions that are of importance to cultural brain function. The strength of the interconnection of cortical brain regions contributes to the direct and indirect effects of cortical and cultural processing. The anatomical connectivity of the cultural brain facilitates the causal interaction of cortical brain regions and their dynamical basis. The characterization of large-scale cortical brain networks is fundamental to cultural processing.

The intrinsic and extrinsic processing of cortical brain networks shows the facilitation of cultural brain activity based on anatomical connectivity. The intrinsic processing of cortical brain regions depicts the spontaneous and intrinsic activity of cultural brain activity. The cultural brain activity demonstrates the intrinsic processing of cortical brain networks that is spontaneous and intrinsic based on resting-state function. The connection strength of interconnected brain regions facilitates the intrinsic processing of cortical neural networks. The functional processing of cortical brain networks illustrates how interconnection facilitates the intrinsic processing of cultural brain activity.

The extrinsic processing of cortical brain networks further illustrates how anatomical connectivity facilitates the functional processing of cultural brain activity. The functional processing of cultural brain activity is comprised of the functional brain activity that is task-based and stimulus-driven. The cultural brain activity that is task-based and effortful depicts the functional patterns of cortical brain function that is linked to cultural processing. The strength of interconnection of cortical brain regions facilitates the functional processing of cultural brain activity. The connection strength of interconnected brain regions contributes to the information flow and activation patterns that comprise the functional processing of cortical brain networks. The anatomical connectivity of the cultural brain serves as a functional basis of the extrinsic processing of cultural brain activity.

The anatomical connectivity of large-scale cortical brain networks is linked to cultural processing. The strength of the interconnection of cortical brain regions facilitates cortical and cultural processing. The interconnection of cultural brain regions illustrates the information flow and activation patterns that are linked to cultural processing. The connection strength of cultural brain regions shows the functional significance of interconnection on the relaying of information across interconnected brain regions. The interconnection of cortical brain regions serves as a structural basis of the information flow and activation patterns of cultural processing. The anatomical connectivity of large-scale cortical brain networks facilitates the cortical brain activity of cultural processing.

The functional processing of the cultural brain contributes to the brain dynamics of cortical brain networks. The interconnection of cortical brain regions facilitates the dynamical basis of cultural brain function. The global representation of cortical brain activity describes the aggregation of the neuronal activity of populations of neurons. The global level of the interconnection of cortical brain regions shows how the interconnection of cortical brain regions facilitates the information flow and activation patterns of cortical brain dynamics. The global dynamics of cortical brain networks demonstrate the functional role of anatomical connectivity on the dynamical basis of cortical brain regions and their link to cultural processing.

The anatomical connectivity of cultural brain function also serves as a structural basis of the endogenous brain activity of autonomous brain dynamics. The strength of the interconnection of cortical brain regions has causal effects on the endogenous brain activity of cultural brain function. The spontaneous and intrinsic activity of cultural brain function shows the functional role of structural properties on endogenous brain activity. The anatomical connectivity of cultural brain regions shows the functional significance of interconnection on the spontaneous and intrinsic activity of cultural brain function. The structural and dynamical properties of the cultural brain have causal-functional significance on the cortical and cultural processing of the brain.

The notion of the cultural connectome is of importance to the understanding of the structural and dynamical basis of large-scale cortical brain networks and their inherent cultural function. The interconnection of cultural brain regions facilitates the intrinsic and extrinsic processing of cultural information. The structural and dynamical properties of cultural brain function serve as a mechanistic basis of cultural processing. The strength of the interconnection of cultural brain regions is linked to the cultural processing of brain function. The characterization of the anatomical connectivity of large-scale cortical brain networks and its causal effect is fundamental to the broader considerations of the brain basis of culture.

Conceptual Foundations

Conceptual foundations on the cultural connectome posit on the structural and dynamical basis of cortical brain networks and their inherent cultural function. The functional processing of cortical brain networks illustrates the structural and dynamical basis of the intrinsic and extrinsic processing of cortical brain regions. The interconnection of neurons and networks of neurons comprise a mechanistic basis for the structural and functional connectivity of cortical brain networks and their causal effects. The structural and functional properties of neurons and networks of neurons affect the strength of interconnection of cortical brain areas. The characterization of the structural and functional connectivity of cortical brain networks is fundamental to the understanding of cultural brain dynamics.

The conceptual development of scientific concepts and language on the cultural connectome provides a foundation for theory building and hypothesis testing on the cultural brain and its causal effects. The scientific concepts and language of the cultural connectome broaden the understanding of the interconnection as a mechanistic basis of cultural brain dynamics and its causal-functional significance. The understanding of the structural and functional basis of cultural brain dynamics is fundamental to the characterization of cortical brain networks and their structural and functional properties. The conceptual approaches to the study of the cultural connectome depict the formulation of scientific theories and hypothesis testing on the structural and dynamical basis of the cultural brain.

The computational principles of cortical brain organization serve as a conceptual foundation for the understanding of the structural and functional connectivity of cortical brain networks and their functional purpose. The structural and functional principles of cortical brain organization illustrate a conceptual foundation for the understanding of the cortical brain and its interconnection. The structural and functional properties of neurons and networks of neurons describe the computational principles that are of importance to the characterization of cortical brain networks and their causal effects. The structural and dynamical basis of cortical brain networks details the functional role of internal properties on cortical brain dynamics. The computational principles of cortical brain organization are fundamental to the understanding of cortical brain networks and their causal functional significance.

Conceptual approaches to the cultural connectome consider the causal role of interconnection on the functional processing of cortical brain networks and its inherent cultural function. The interconnection of neurons and networks of neurons comprises a mechanistic basis of the functional processing of cultural brain regions. The strength of the interconnection of cortical brain areas contributes to the direct and indirect effects on the functional processing of cultural brain networks. The consideration of the structural and functional basis of cultural brain networks is fundamental to the multilevel approaches to the cultural connectome and its functional purpose.

The cultural connectome describes how the interconnection of cortical brain regions affects the intrinsic and extrinsic processing of large-scale cultural brain networks. The anatomical connectivity of cortical brain networks serves as a structural basis of the intrinsic and extrinsic processing of cultural information. The structural connectivity of cortical brain networks demonstrates a structural basis of intrinsic cultural processing. The interconnection of cortical brain regions further illustrates the functional role of cortical brain networks in the extrinsic processing of cultural information. The anatomical connectivity of cortical brain networks is fundamental to the structural and dynamical basis of cortical brain networks.

The functional processing of cortical brain networks describes the intrinsic and extrinsic processing of cortical brain function and its functional purpose. The intrinsic processing of cortical brain networks depicts the spontaneous and autonomous activity of cortical brain regions that is based on resting-state function. The intrinsic processing of cortical brain networks details the spontaneous fluctuations of cortical brain activity that are based on the maintenance of active and itinerant representations that facilitate the entrainment of cultural brain function. The spontaneous fluctuations of cortical brain activity and its intrinsic properties depict the patterns of endogenous brain activity that is stimulus-independent. The intrinsic processing of cultural brain networks illustrates the causal role of cortical brain function as endogenous brain activity. The spontaneous and autonomous activity of cortical brain regions shows the intrinsic properties of cortical brain activity based on the structural properties of the cortical brain network.

The structural properties of cortical brain networks affect the intrinsic processing of cultural brain activity. The interconnection of cortical brain regions serves as a structural and functional basis of the cortical brain network. The strength of the interconnection of cortical brain regions contributes to the intrinsic cultural processing of cultural brain regions. The connection strength of cortical brain areas is of importance to the causal interaction of cultural brain function and its intrinsic activity. The functional role of connection strength as a structural basis of the causal interaction of interconnected brain regions illustrates the functional relation of anatomical connectivity and the intrinsic cultural processing of cultural brain networks. The anatomical connectivity of cortical brain networks shows the functional role of interconnection as a structural basis of intrinsic cultural processing.

The anatomical connectivity of cortical brain networks serves as a structural basis of the functional processing of cortical brain activity. The functional processing of cortical brain networks illustrates the relaying of information across interconnected brain regions. The information flow and patterns of the activation of cortical brain networks detail the causal-functional role of connection strength in the functional processing of interconnected brain regions. Cultural brain networks demonstrate the functional role of connection strength on the extrinsic cultural processing of cortical brain regions. The information flow and patterns of activation show the functional significance of connection strength on the extrinsic cultural processing of interconnected brain regions.

Cortical brain networks show the facilitation and inhibition of intrinsic and extrinsic processing based on the interconnectivity of interconnected brain regions. The functional processing of cultural brain networks shows how the facilitation of intrinsic and extrinsic cultural processing is based on the connection strength of interconnected brain regions. The facilitation of intrinsic and extrinsic cultural processing of cultural brain networks shows the functional significance of the strength of interconnection in cortical processing. The inhibition of functional processing of cortical brain networks further demonstrates the functional role of interconnection as a functional basis of the control or feedback of cultural brain activity. The facilitation and inhibition of functional processing based on interconnectivity depicts the functional significance of the interconnection of cortical brain networks.

The anatomical connectivity of cultural brain networks contributes to the intrinsic and extrinsic processing of cultural information. The strength of interconnection of cultural brain networks illustrates a structural basis for the intrinsic and extrinsic processing of the cultural brain. The structural and functional connectivity of cultural brain function shows the direct and indirect effects of interconnection on cultural processing. The anatomical connectivity of cortical brain areas affects the information flow and activation patterns of cultural brain networks. The connection strength of cortical brain networks is linked to the cultural processing of cortical brain networks. The anatomical connectivity of the brain is of importance to the characterization of cultural and cortical processing of brain function.

The interconnection of cortical brain areas serves as a structural basis of cultural brain dynamics and its causal effects The connection strength of neurons depicts the density of synaptic connections between neurons of cortical brain areas that are of importance to local brain dynamics. The local brain dynamics of cortical brain areas describes the local representation of cortical brain mechanisms and their response properties. The local dynamics of neuronal activity located within cortical brain areas describes the patterns of neuronal activity of cortical brain mechanisms and its response properties. The interconnection of neurons comprises a structural basis of the local brain dynamics of cortical mechanisms.

The anatomical connectivity of large-scale cortical brain networks describes a structural basis for global brain dynamics. The length of extrinsic connections between cortical areas details the functional basis of the global brain dynamics of cortical brain networks. The global representation of brain dynamics depicts the overall aggregation of population-level neuronal activity of cortical brain areas and their interconnection. The connection strength between cortical areas facilitates the information flow and activation patterns of interconnected brain regions of cortical brain networks. The global dynamics of cortical brain networks comprises the overall levels of activity that are based on the interconnection of cortical brain areas and their extrinsic activity. The extrinsic activity of cortical brain networks describes the cortical brain activity that is stimulus-driven. The anatomical connectivity of cortical brain networks facilitates the global representation of cortical brain dynamics.

The autonomous brain dynamics of cortical brain networks describes the spontaneous and intrinsic properties of endogenous brain activity. The anatomical connectivity of cortical brain networks contributes to the structural basis of autonomous brain dynamics. The interconnection of cortical brain areas facilitates the intrinsic activity of cortical brain regions. The strength of connections of cortical brain areas is of importance to the intrinsic cultural processing of cortical brain regions. The anatomical connectivity of cortical brain networks illustrates the structural properties of interconnected brain regions that facilitate autonomous brain dynamics. The spontaneous and intrinsic activity of endogenous brain activity is fundamental to autonomous brain dynamics.

The conceptual foundations of the cultural connectome broaden the understanding of the structural and functional basis of the brain dynamics of cortical brain networks and their inherent cultural function. The conceptual development of scientific concepts and language on the cultural connectome contributes to the theory building and hypothesis testing on culture and cortical brain dynamics and its functional purpose. The conceptual formulation of scientific theories on the cultural brain contributes to the multilevel approaches to the understanding of the structural and functional basis of cultural brain function. Conceptual approaches to the study of the cultural connectome serve as a foundation for the broader understanding of the cultural brain and its functional purpose.

Methodological Foundations

Methodological foundations on the cultural connectome describe neuroscience techniques for the study of the structural and functional connectivity of cortical brain networks and their inherent cultural function. Neuroscience techniques such as diffusion tensor imaging and functional brain imaging facilitate the observation and measurement of the structural and functional properties of the cortical brain. Diffusion tensor imaging is a neuroscience technique that allows for the indirect observation and measurement of the magnitude of interconnection of cortical brain regions of interest based on white matter tracts. The structural connectivity of cortical brain regions is based on the synaptic density of white matter tracts that comprise the interconnection of cortical brain areas as regions of interest. The neuroscience technique of diffusion tensor imaging is fundamental to the study of the structural connectivity of the cortical brain and its functional purpose.

Functional brain imaging is a neuroscience technique that allows for the indirect observation and measurement of the functional brain activity of the cortical brain. Functional brain imaging is of importance to the characterization of the functional connectivity of cortical brain regions. The causal interaction of cortical brain regions based on functional tasks comprises patterns of functional brain activity. The causality and directionality of the coactivation of cortical brain regions based on functional task activity demonstrates the functional connectivity of the cortical brain. The computational modeling of the causal interaction of cortical brain regions is fundamental to the characterization of the functional integration of cortical brain function.

Methodological foundations of research on the cultural connectome describe neuroscience techniques for the study of the neuroanatomical structure and function of the brain and its interconnection. The development of computational techniques in neuroscience facilitates the use of data fitting and data extraction of the patterns and regularities of cortical brain structure and function and its causal effects. The computational discovery of the interconnection of the cortical brain comprises systematic approaches to the study of the structure and function of cortical brain areas. The use of mathematical modeling with neuroscience techniques facilitates the computational modeling of large-scale datasets in neuroscience.

Computational techniques are of importance to the modeling of the anatomical connectivity of the cortical brain. The development of computational techniques for the modeling of anatomical connectivity considers parameters such as the magnitude of interconnection or connection strength between cortical brain areas and the strength of connection between target and source areas across cortical areas (Rabinovich et al., 2012). The use of computational techniques is fundamental to the characterization of the interconnection of cortical brain areas and its causal effects. Computational approaches are fundamental to the study of the anatomical connectivity of the cortical brain.

Computational tools and technologies are fundamental to the study of the anatomical connectivity of cortical brain areas and their functional purpose. Computational modeling comprises the design of idealized models and parameter estimation based on the data fitting and data extraction of large-scale datasets. The model estimation of the connection strength of target and source areas across regions of interest of the brain demonstrates the data fitting and data extraction of idealized models to large-scale datasets. The use of computational techniques for the study of anatomical connectivity of cortical brain areas is fundamental to the characterization of neuroanatomical structure and function.

Methodological foundations are of importance to the study of the cultural connectome. The development of mathematical modeling and neuroscience techniques is essential to the systematic study of structural and functional connectivity and its role in the functional processing of cultural brain networks. Computational techniques are also important to the data fitting and data extraction of large-scale bioinformatic datasets. The development of computational tools and technologies is beneficial to the innovation of methods. The multimethod approach to the study of the cultural connectome illustrates the complementarity of methodological approaches on computational cultural neuroscience.

Computational Approaches to the Cultural Connectome

Computational approaches to the cultural connectome encompass multilevel approaches to the understanding of the structural and dynamical basis of the cultural brain and their causal effects. The study of the cultural connectome entails the characterization of the anatomical connectivity of cortical brain networks and its causal effects. The use of mathematical modeling and neuroscience techniques illustrates the role of anatomical connectivity of cortical brain networks as a structural basis of the global brain dynamics of cortical brain networks. The interconnection of cortical brain areas further serves as a functional basis of autonomous brain dynamics and endogenous brain activity. The structural basis of cortical brain networks depicts the role of anatomical connectivity in intrinsic and extrinsic processing. The understanding of the anatomical connectivity of cortical brain networks is fundamental to the characterization of the structural and dynamical basis of the cortical brain and its inherent cultural function.

The study of the structural and dynamical basis of the cultural brain entails the use of mathematical modeling and neuroscience techniques to characterize the dynamics of the cultural brain and its structural and functional properties. Mathematical modeling provides idealized models of the global and local brain dynamics of neurons and networks of neurons that comprise large-scale cortical brain networks. Neuroscience techniques facilitate the observation and measurement of the structural and functional properties of the cortical brain and its causal effects. Computational tools and technologies further contribute to the use of idealized models for the data fitting and data extraction of the patterns and regularities of cultural brain dynamics and its causal interaction based on large-scale datasets. Multilevel approaches to the study of cortical brain dynamics are essential to the characterization of the structural and functional properties of cultural brain function.

Computational approaches to the cultural connectome demonstrate the use of computational tools and technologies for the observation and measurement of the anatomical connectivity of cortical brain networks and their causal effects. Computational tools and technologies are essential to the identification of the patterns and regularities of the structure and function of cortical brain networks. The modeling of anatomical connectivity as connection strength of cortical brain areas has functional significance for the cortical brain dynamics and tis inherent cultural function. The functional operationalization of the connection strength of cortical brain networks as the length of interconnection between cortical brain areas and the density of synaptic connection between neurons is fundamental to the characterization of global and local brain dynamics. The structural properties of cortical brain networks describe the connection strength of cortical brain networks and its causal effects.

The study of the cultural connectome contributes to the computational foundations of computational cultural neuroscience. Research on cultural connectome is fundamental to the understanding of the structural and dynamical basis of cortical brain networks and their inherent cultural function. The characterization of the functional role of anatomical connectivity as a mechanistic basis of intrinsic and extrinsic processing is essential to the understanding of the functional basis of cortical brain networks. The interconnection of cortical brain networks describes the structural basis of the intrinsic and extrinsic processing of interconnected brain regions. The anatomical connectivity of interconnected brain regions serves a causal-functional role in the intrinsic and extrinsic processing of cortical brain networks. The study of the cultural connectome facilitates the understanding of the functional role of anatomical connectivity of the intrinsic and extrinsic processing of cultural information.

Practical Applications

Computational tools and technologies are fundamental to the computational discovery of the cultural brain. The development of computational tools and technologies for the study of the structural and functional properties of the brain is of importance to the understanding of cultural brain function. Computational techniques are fundamental to the study of large-scale bioinformatic datasets. The development of computational techniques is of importance to the building of bioinformatic databases that contribute to the knowledge-based resources on culture in health and medicine. The innovation of knowledge-based resources in neuroscience on culture and health serves as a foundation for research on computational cultural neuroscience.

The development of bioinformatic databases is fundamental to the research and development of computational cultural neuroscience. The building of knowledge-based resources such as bioinformatic databases contributes to the innovation of computational tools and technologies in neuroscience. The innovation of computational tools and technologies in neuroscience advances the research capabilities of health and medicine. The development of computational tools and technologies that are culturally- and ethnically appropriate is fundamental to the building of knowledge-based resources on culture and health. The development of bioinformatic databases is of importance to the building of research capabilities and scientific infrastructure that advances the quality of information of knowledge-based resources.

The scientific and technological innovation of computational cultural neuroscience is of importance to the building of knowledge-based resources on culture and health. The development of large-scale bioinformatic datasets contributes to the quality of data and evidence in neuroscience on culture and health. The building of bioinformatic datasets contributes to the research and development of computational tools and technologies that advance the quality of knowledge-based resources on computational cultural neuroscience. The scientific and technological innovation of bioinformatic databases that inform on the development of computational techniques that are culturally and ethnically appropriate is essential to research and development on culture and health.

Practical applications of research and development on computational cultural neuroscience are essential to the evidence-based research of culture and health. The development of computational tools and technologies in computational cultural neuroscience advances the research capabilities of health and medicine. The innovation of research capabilities on culture and health in health and medicine is beneficial for the practical applications of scientific and technological innovation of the field of study. The innovation of neuroscience techniques benefits the practical impact of research and development on computational cultural neuroscience.

Discussion

The study of the cultural connectome is fundamental to the understanding of the structural and dynamical basis of the cultural brain. The structural and functional properties of the cortical brain facilitate the understanding of the dynamical mechanisms that are fundamental to cortical brain function. The brain as a cultural connectome comprises the structural and functional properties of the cortical brain and its inherent cultural function. The detailed characterization of the intrinsic and extrinsic connectivity of the cortical brain based on anatomical connectivity comprises a functional basis of the modeling of the global and local dynamics of the cultural brain. The characterization of the structural and functional basis of cortical brain function and its causal effects is essential to the modeling of cultural brain dynamics.

The systematic study of the cultural connectome serves as a foundation for the understanding of the computational foundations of cultural brain dynamics. The anatomical connectivity of cortical brain networks is fundamental to the intrinsic and extrinsic properties of cultural brain function. The intrinsic and extrinsic processing of cultural brain function based on neuroanatomical structure and function illustrates the dynamical basis of the cultural brain. The global and local dynamics of cultural brain function demonstrate the structural and functional properties of its inherent cultural function. The characterization of cultural brain dynamics contributes to the understanding of the autonomous and self-organization principles of the cultural brain.

The study of cultural brain dynamics is fundamental to the characterization of the cortical brain dynamics of cultural function. The dynamical mechanisms of cultural processing serve as a mechanistic basis of the neural computation of culture. Theoretical foundations to the study of the brain dynamics of culture posit on the functional role of structural and dynamical properties of the cortical brain as a functional basis of cultural processing. The structural and dynamical principles of cultural brain dynamics are fundamental to the understanding of the neural computation of cultural brain function. From global and local dynamics to autonomous brain dynamics, the study of the dynamical basis of the cultural brain is essential to the characterization of the computational foundations of culture and cortical brain dynamics.

Conclusion

The scholarly and scientific interest in the cultural connectome illuminates on the importance of the structural and functional basis of the cortical brain and its inherent cultural function. The study of the cultural connectome broadens the understanding of the computational foundations of the cultural brain. The role of anatomical connectivity as a structural basis of the dynamical mechanisms of cultural brain function shows the functional significance of structural properties on cortical brain function. The characterization of the structural basis of cortical brain function and its relation to inherent cultural function is fundamental to the understanding of the role of the cortical brain as a mechanistic basis of cultural processing. The systematic study of the cultural connectome contributes to the computational foundations of the cultural brain and its structural and dynamical basis.

The systematic study of the neuroanatomical structures of the cortical brain and its inherent cultural function is fundamental to the understanding of the brain basis of culture. The characterization of the anatomical connectivity of the cortical brain and its causal effects entails the functional significance of the structural and functional properties of cortical brain function. The causal role of anatomical connectivity as a structural basis of intrinsic cultural processing illustrates the functional processing of the cultural brain. The neuroanatomical structures of cortical brain function show the importance of structural and functional properties in the modeling of cultural brain dynamics.

Scholarly and scientific interest in the cultural connectome contributes to the understanding of the computational principles of the cultural brain. Computational approaches to the study of the cultural connectome demonstrate multilevel approaches to the study of the structural and dynamical basis of the cultural brain. The theory building and hypothesis testing on the cultural connectome broadens the conceptual development of scientific concepts and language on the computational foundations of the cultural brain. The dynamical basis of cultural brain function generates scientific knowledge of the brain dynamics of culture. The systematic study of the cultural connectome is fundamental to the breadth of perspectives on the cortical brain and its inherent cultural function.

Implications

Research on the cultural connectome and its practical applications informs the practice and policy of culture and health. The study of the cultural connectome generates novel knowledge on the structural and dynamical basis of the cultural brain. The generation of scientific knowledge on the cultural connectome demonstrates the research and development on computational cultural neuroscience. The scientific and technological innovation on computational cultural neuroscience contributes to the building of evidence-based resources on culture and health. The building of scientific infrastructure and research capabilities in health and medicine is beneficial to the scientific and societal enrichment on culture and health.

Evidence-based research on the cultural connectome informs the building of evidence-based resources on computational cultural neuroscience. Evidence-based approaches to the study of computational cultural neuroscience inform the development of evidence-based resources on culture and health. The generation of scientific knowledge on computational cultural neuroscience contributes to the building of scientific infrastructure and research capabilities in health and medicine. The development of evidence-based resources on computational cultural neuroscience is beneficial to the scientific and societal enrichment on culture and health.

The study of computational cultural neuroscience informs the translation of scientific research into practice and policy. The research and development on computational cultural neuroscience is of importance to the practical applications on culture and health. The scientific and technological innovation of computational tools and technologies is fundamental to the advancement of the research capabilities of health and medicine. The development of evidence-based resources on culture and health benefits from the scientific and technological innovation of health and medicine at the intersection of culture, technology and society. The broadening of humanistic and scientific understanding on culture, technology and society and its societal impact has implications for health, medicine and related fields.

Reference

Rabinovich, M. K., Friston, K. J., & Varona, P. (2012). *Principles of brain dynamics: Global state interactions.* Cambridge: MIT Press.

10

CULTURE AND TECHNOLOGY

Introduction

The scholarly and scientific interest in culture and machine computation illuminates on the functional performance of machines that show the capabilities of performance that is cultural. The notion of culture as a computing machine highlights the role of computation as a fundamental basis of culture. From computational biology to artificial life, the scientific discovery of the computational foundations of culture in the world demonstrates the functional capabilities of modern computing and its practical applications. The understanding of the computational foundations of culture formulates a conceptual basis for the simulation and construction of culture in the world. The advancement of the scholarly and scientific interest in culture and machine computation is fundamental to the study of culture and modern computation and its societal impact.

The advancement of the modern computer from the 19th century illustrates the scientific and societal interest in the scientific and technological innovation of culture and computation. The development of modern computation illustrates the innovation of the capabilities of modern devices for the advancement of cultural and computational performance of technology. From simple to complex devices, the innovation of modern computation and its capabilities broadens the breath of understanding of culture and technology and its societal impact. The ubiquity of the capabilities of modern computing for scientific and technological innovation is fundamental to the practical applications of modern computation and the breadth of its societal impact.

DOI: 10.4324/9781003384236-11

The goal of the chapter is to provide a review on culture and technology and its societal impact. The study of computational cultural neuroscience builds evidence-based approaches to the study of the fundamental principles of cortical brain organization. The systematic study of computational cultural neuroscience contributes to evidence-based research on the foundations of the brain basis of culture. Computational tools and technologies advance the innovation of research and delivery science capabilities in health and medicine. Evidence-based approaches to computational cultural neuroscience build research capabilities and scientific infrastructure on culture, technology and health. Policy-based approaches to culture and technology facilitate the innovation of scientific and research capabilities for the advancement of discovery and delivery science in health and medicine. The building of research capacity advances the evidence-based and knowledge-based approaches for discovery and delivery science on culture and health.

Translational Research on Computational Cultural Neuroscience for Practice and Policy

Translational research on computational cultural neuroscience informs the development of practice and policy. The integrative research on computational cultural neuroscience builds evidence-based resources on culture and health. The use of computational tools and technologies for data mining and data extraction of large-scale datasets on culture and brain function build evidence-based approaches on culture and health. Computational tools and technologies contribute to the research and development of large-scale databases that advance the bioinformatics of culture and health in health and medicine. The advancement of research capabilities and scientific infrastructure on culture and health informs the development and implementation of practice and policy.

Research on computational cultural neuroscience investigates the computational foundations of cultural neuroscience. The computational principles of cortical brain organization illustrate the structural and functional principles that are of importance to cortical brain function. The study of the computational foundations of cultural neuroscience demonstrates the systematic investigation of the structural and functional principles of cultural brain function. The use of computational tools and technologies for the study of the cultural patterns of brain function and its functional purpose contributes to the evidence-based approaches of computational cultural neuroscience. The evidence-based research on computational cultural neuroscience is fundamental to the understanding of the computational foundations of the cultural brain.

Evidence-based approaches to computational cultural neuroscience demonstrate the empirical study of the computational approaches to the study of cultural brain function. Computational approaches to the study of cultural brain function consider a wide realm of multilevel approaches to the characterization of the cultural patterns of brain function. The multilevel approaches to the study of cultural brain function show the systematic investigation of neurons and networks of neurons and their functional significance to cultural brain function. The characterization of the cultural patterns of brain function details the causal relation of processes and mechanisms that are fundamental to culture and adaptation.

Multilevel approaches to the study of computational cultural neuroscience consist of systematic programs of research on cultural brain function with a range of approaches. The programs of research on computational cultural neuroscience incorporate systematic investigation of cultural brain function with mathematical modeling, computational tools and neuroscience techniques. The development of programs of integrative research seeks to investigate the cultural patterns of brain function and its causal effects. The use of multilevel approaches to the study of cultural brain function advances the modeling of the structural and functional basis of cultural brain function.

The use of computational tools and technologies to investigate cultural brain function relies on the development of bioinformatic databases. The development of computational tools and technologies contributes to the building of large-scale datasets on culture and brain function. Computational tools and technologies contribute to the building of large-scale datasets that facilitate the computational discovery of the cultural patterns of brain function. The development of large-scale datasets to investigate culture and brain function facilitates the use of computational tools and technologies. The computational discovery of culture and brain function based on large-scale datasets illustrates the research capabilities of computational tools and technologies for the data extraction and knowledge generation of bioinformatic databases.

The design of bioinformatic databases further provides an evidence-based resource for the data extraction of the patterns and regularities of cultural brain function. Systematic investigation of computational cultural neuroscience benefits from the development of bioinformatic databases on the structural and functional basis of cultural brain function. The study of the structural and functional basis of cultural brain function integrates multilevel approaches to the discovery of the fundamental mechanisms of the functional architecture of culture and brain function. The computational discovery of the structural and functional basis of cultural brain function advances the data extraction and knowledge generation of the fundamental mechanisms of culture and brain function.

The bioinformatic databases on culture and brain function facilitate the use of computational tools and technologies for the data extraction and knowledge generation of the cultural patterns of brain function and its relation to health and disease. The development of bioinformatic databases provides evidence-based resources for the understanding of root causes, prevention and intervention of complex disease in health and medicine. The development of systematic programs of research shows integrative approaches to the study of the culture and brain function and its relation to complex disease. The study of culture as a strategy of prevention and intervention in health and medicine is fundamental to the prevention and treatment of complex disease. The computational discovery of culture as an evidence-based intervention is beneficial to the prevention and intervention of complex disease. The use of computational tools and technologies advances the precision of discovery and delivery science in health and medicine.

Bioinformatic databases are fundamental as an evidence-based resource for the knowledge generation on culture and health. The development of bioinformatic databases provides evidence-based resources on the structural and functional basis of culture and brain function. Evidence-based resources facilitate the research and development on the fundamental principles of culture and brain function. Computational tools and technologies are essential to the scientific discovery of the computational principles of culture and brain function. The development of bioinformatic databases promotes the use and availability of computational tools and technologies for the advancement of discovery and delivery science on culture and health in health and medicine.

The development of evidence-based resources on computational cultural neuroscience advances the use of computational tools and technologies for the understanding of culture and brain function. Computational tools and technologies build the research capabilities in health and medicine for the development of large-scale bioinformatic databases to study the cultural patterns of brain function. Large-scale datasets and computational tools contribute to the computational discovery of the cultural patterns of brain function. The development of computational tools and technologies for the data mining of large-scale datasets contributes to the evidence-based research on culture and brain function. The building of evidence-based resources on culture and brain function contributes to the scientific enrichment of culture and health promotion.

Policy-level approaches to the study of computational cultural neuroscience contribute to the development of knowledge-based resources on culture and technology in health and medicine. The development of policy-level approaches advances the building of knowledge-based resources that provide data and information on culture and technology that is beneficial to culture and health promotion. Knowledge-based resources on culture and health are fundamental to the understanding of the impact of technology on culture and health. The development of technology as prevention and intervention of culture and health contributes to the efficacy of strategies of culture and health promotion. The strategies of prevention and intervention benefit from the innovation of technology for the promotion of culture and health.

Policy-level approaches on culture and technology are fundamental to the prevention and intervention of culture and health promotion. Policy-level approaches to culture and health promotion provide programs and policies that advance the building of knowledge-based resources on culture and technology in health. The development of knowledge-based resources facilitates the understanding of the impact of technology on prevention and intervention in culture and health promotion. The strategies of health prevention and intervention affect the efficacy and responsiveness to culture and health promotion. The health promotion of technology as a strategy of prevention and intervention is essential to the advancement of the prevention and intervention in health and medicine. Culture and health promotion promotes the scientific infrastructure and research capabilities of culture, technology and health that advance scientific and societal enrichment and its public impact.

Culture and Technology

The scholarly and scientific interest in culture and technology encompasses a wide range of considerations on the innovation of culture and technology. The considerations of culture and technology center on the understanding of the fundamentals of culture and technology and its broad impact on society. The study of culture and technology encompasses a wide realm of considerations of the role of culture and technology on society, the impact of culture and technology and its societal impact, the innovation of culture and technology and its societal benefits. The scholarly and scientific interest in culture and technology entails multilevel approaches to the understanding of the impact of culture and technology on society. Multilevel approaches to culture and technology provide a foundation for systematic approaches to the study of culture and technology and its practical applications.

The advancement of the innovation of culture and technology broadens the research and development of computational tools and technologies to inform scientific and computational discovery. The science of computational discovery advances with the innovation of computational tools and technologies that broaden the use and availability of data and information to advance scientific approaches for the improvement of societal outcomes. Computational tools and technologies are essential to the scientific understanding of computational discovery and its wide implications. The innovation of computational tools and technologies broadens the impact of technology on culture and society. The implementation of scientific and technological innovation that arises from computational discovery leads to the improvement of societal outcomes.

The development of research on culture and technology illustrates the functionality and adaptability of computational tools and technologies for scientific and technological innovation. The research and development of computational tools and technologies informs the advancement of technological resources that are beneficial to the understanding of the innovation of culture and technology and its societal impact. Computational tools and technologies are fundamental to the advancement of computational discovery to inform scientific approaches to societal concerns. The scientific and technological innovation of computational tools facilitates the understanding of the functionality and adaptability of complex systems that show responsiveness to system-wide approaches for the benefit of societal outcomes.

Research on culture and technology informs on the development of practical applications in health, medicine and related fields. The innovation of culture and technology illustrates the research and development of computational tools and technologies such as devices and computers that show the functional capabilities of culture and computation. The development of cultural machines that show functional performance for cultural computation illustrates the innovation of computational tools and technologies that have practical applications for scientific and computational discovery. The innovation of scientific and computational discovery with computational tools and technologies facilitates the development of scientific infrastructure and research capabilities in health and medicine that are of importance to the research and development of culture and technology.

Computational tools and technologies advance the research capabilities of discovery and delivery science in health and medicine as well as related fields. Research capabilities such as computational tools and technologies comprise a fundamental aspect of the discovery and delivery science of complex systems. Computational tools and technologies facilitate the development of functional capabilities that advance the precision of computational discovery. The advancement of computational discovery as a source of information is beneficial to the improvement of system-wide performance. The development of functional capabilities contributes to the reliability and accuracy of computational tools and technologies for scientific observation and computational discovery. The development of computational tools and technologies is fundamental to the advancement of the discovery and delivery science in health and medicine of diverse cultures and populations. The building of research capabilities of computational tools and technologies facilitates the advancement of computational discovery that informs the discovery and delivery science of complex systems.

The building of scientific infrastructure with computational tools and technologies contributes to the advancement of the precision of computational discovery. Computational discovery is essential to the generation of knowledge based on data extraction and data-mining techniques. The generation of knowledge based on computational discovery is essential to the implementation of data techniques with large-scale databases. The functional capabilities of computational tools and technologies advance the conceptual approaches to computational discovery that facilitate the study of the cultural patterns and regularities of large-scale databases. Conceptual approaches to computational discovery are fundamental to the scientific and technological innovation of the knowledge-based research of computational discovery.

The generation of scientific knowledge based on computational techniques builds scientific and technological resources that are fundamental to computational discovery. Computational techniques such as data extraction and data mining are essential to the scientific discovery of the cultural patterns and regularities of large-scale databases. The use of computational techniques with large-scale databases provides quality of data and information that is fundamental to the research and development of evidence-based resources that inform diverse cultures and populations. The development of computational techniques illustrates the functionality and adaptability of computational tools and technologies for the generation of novel scientific knowledge.

The development of research capabilities with computational tools and technologies is of importance to the building of bioinformatic databases that provide quality of data and information on culture and health. Bioinformatic databases in health and medicine are essential to the quality of information and data that inform diverse cultures and populations. The use of bioinformatic databases facilitates the computational discovery of cultural patterns and regularities that are of importance to the development of evidence-based resources. Evidence-based resources that comprise bioinformatic datasets facilitate the generation of novel scientific knowledge based on computational techniques of diverse cultures and populations. The development of computational techniques for use with bioinformatic datasets is fundamental to the advancement of discovery and delivery science in health and medicine.

The integral use of computational techniques with large-scale databases is beneficial to the building of knowledge-based resources that inform diverse cultures and populations. The development of large-scale databases contributes to the quality of information and data of knowledge-based resources. The building of research capabilities such as large-scale databases is fundamental to the equitable access to information and data of diverse cultures and populations. Knowledge-based resources provide information and data that are essential to the research and development of computational tools and technologies of diverse cultures and populations. Knowledge-based resources facilitate the use and availability of information and data that inform the scientific and educational resources of diverse cultures and populations.

The development of knowledge-based resources is fundamental to the building of scientific infrastructure and research capabilities that impact diverse cultures and populations. The generation of novel scientific knowledge that is based on diverse cultures and populations is fundamental to the equitable use and availability of societal and scientific resources. The building of scientific infrastructure demonstrates the research capabilities that are necessary for the research and development of scientific knowledge. The development of the knowledge-based resources is fundamental to the building of scientific infrastructure and research capabilities that are fundamental to diverse cultures and populations.

The research and development on culture and technology impacts the equitable access to scientific and societal resources. Scientific resources such as evidence-based resources demonstrate the use and availability of computational tools and technologies for the computational discovery of cultural patterns and regularities. The development of evidence-based resources is beneficial to the quality of data and information that inform on scientific and societal issues. Societal resources such as knowledge-based resources are fundamental to the information and data that facilitate the building of awareness on scientific and societal issues. The integral use of cultural data and information on scientific and societal issues is beneficial to inform diverse cultures and populations. The integration of cultural data and information with evidence-based resources is fundamental to equitable access to scientific and societal resources. The development of evidence-based resources is essential to the research and development on culture and technology.

The advancement of culture and technology is foundational to the promotion of machines that perform cultural computation. The scientific and technological innovation of cultural devices and computers that perform computation that is cultural illustrates the importance of culture on the development of technology. The promotion of machines that show the functional performance of computation that is cultural demonstrates the technological innovation of cultural devices and computers and their internal architecture. The functional performance of cultural computation illustrates the promotion of the functional operations that facilitate the internal programming of cultural devices and computers. The innovation of the functional capabilities of cultural devices and computers is essential to the demonstration of the practical applications of research and development on culture and technology.

Conceptual Framework on Culture and Technology

Conceptual foundations of culture and technology center on the fundamentals of culture and technology and its societal impact. The functional capabilities of culture and technology have broad application to the functional performance of complex systems and their societal outcomes. The innovation of tools and technologies that show responsiveness to cultural contexts contributes to the efficacy of the functional performance of complex systems. The functionality and adaptability of tools and technologies to the cultural context demonstrates the role of computational approaches to the improvement of the reliability and responsiveness of culture and technology. The scientific and technological innovation of tools and technologies is beneficial to the quality of information and data that informs the model performance of system-wide approaches. The innovation of culture and technology is fundamental to the performance of complex systems and its societal impact.

The conceptual development of the framework on culture and technology encompasses the fundamentals of culture and technology and the considerations of its societal impact. The fundamentals of culture and technology consider the interests that are central to the development and implementation of culture and technology and the understanding of its societal and public impact (Table 10.1). The development and implementation of culture and technology consider the global policy environment that is necessary for the building of capacity in the areas of interest. The garnering of the interests on culture and technology with a wide range of stakeholders is beneficial to the setting of priorities on culture and technology. The development of an agenda on culture and technology and its goals is fundamental to the development and implementation of programs and policies on culture and technology.

The framework on culture and technology aims to build capacity on the innovation of culture and technology. The consideration of the scientific infrastructure that is essential to the building of capacity on culture

TABLE 10.1 Framework on Culture and Technology

Fundamentals	Goals and Outcomes
Building capacity	Policy environment
Setting priorities	Policy environment
Scientific infrastructure	Reliability
Research and development	Responsiveness
Data and information	Effectiveness
Products and technologies	Efficacy
Delivery science	Policy environment

and technology is beneficial to scientific and technological innovation. The building of scientific infrastructure on culture and technology advances the equitable access to scientific and societal resources that are essential to scientific and technological innovation. The consideration of the issues that arise in the building of scientific infrastructure is beneficial to the capacity development of culture and technology. The innovation on culture and technology benefits from the building capacity of scientific infrastructure.

The building of capacity and scientific infrastructure on culture and technology is fundamental to the generation of novel scientific knowledge. The generation of novel knowledge on culture and technology is essential to the building of capacity on areas of interest. Research and development on culture and technology provides quality of data and information on the conceptual foundations and practical applications of culture and technology for the improvement of societal outcomes. The considerations of research and development on culture and technology advance the quality of data and information to inform the building of capacity on areas of interest. The generation of novel knowledge is fundamental to the framework on culture and technology and the understanding of its broad impact.

The research and development on culture and technology is fundamental to the scientific and technological innovation of practical applications. The scientific and technological innovation of culture and technology is of importance to the improvement of societal outcomes. The design of products and technologies is essential to the effectiveness of performance goals and outcomes. The innovation of practical applications facilitates the design of products and technologies that show responsiveness to cultural contexts and demonstrate the functionality and adaptability of tools and technologies to advance system performance. The innovation of culture and technology benefits from the development of practical applications that show responsiveness to cultural considerations.

The generation of novel knowledge on culture and technology contributes to the quality of data and information that informs the practical applications of system performance. The discovery and delivery science of complex systems consists of the practical applications that are of importance to system-wide performance and its goals and outcomes. The generation of novel knowledge that is based on computational discovery contributes to the quality of data and information that informs the system-wide performance. The scientific approaches of computational discovery demonstrate the quality of data and information that is reliable. The building of evidence-based resources that consist of quality of data and information facilitates the functional capabilities that contribute to the system performance of goals and outcomes.

The building of capacity on culture and technology benefits from the strategies of the global policy environment. The generation of novel knowledge on culture and technology informs the development of areas of interests that inform the development of programs and policies. The building of scientific infrastructure contributes to the feasibility of building capacity that is essential to the development of policy areas and interests. The innovation of practical applications on culture and technology is fundamental to the demonstration of its societal benefits. The development of policy areas and interests furthers the programs and policies on culture and technology and the building of societal and public support. The building of capacity on culture and technology is fundamental to the development and implementation of the societal and policy outcomes of the global policy environment.

The conceptual framework on culture and technology provides a broad overview of the fundamentals of culture and technology and its societal impact. The framework relies on an evidence-based approach to the building of capacity development on culture and technology. The building of evidence-based resources on fundamental to the development and implementation of the programs and policies on culture and technology. The innovation of the practical applications of culture and technology is essential to the improvement of societal outcomes and the understanding of its public impact. The building capacity development on culture and technology is fundamental to the development and implementation of the discovery and delivery science of complex systems.

Research and Development on Culture and Technology

Research and development on culture and technology is fundamental to the generation of novel knowledge that contributes to evidence-based resources. Evidence-based resources are of importance to the quality of data and information that inform societal outcomes. The innovation of culture and technology encompasses a wide realm of considerations regarding the importance of computational discovery and its societal impact. Computational discovery provides evidence-based resources such as tools and technologies that are beneficial to the gathering of data and information that is beneficial to scientific and societal interests. The research and development of culture and technology generates scientific knowledge that contributes to scientific and technological innovation.

The research and development on culture and technology advances with the scientific and technological innovation of modern computing. The advancement of the functional capabilities of computing machines illuminates on the high-level performance and real-world advantages of modern computation. Modern computation is essential to the evidence-based approaches of scientific and medical discovery. The functionality and adaptability of modern computation demonstrates the high-level processing of data and information that facilitates scientific discovery with computational techniques. The high-level performance of computing machines shows the functional capabilities of computation that are fundamental to the precision of computational discovery. The use of computing machines for the purpose of computational discovery of large-scale datasets illustrates a real-world advantage of modern computation.

The advancement of modern computation for computational discovery facilitates the understanding of the cultural patterns of the natural world and its simulation. The research and development on culture and modern computation advances the computational tools and technologies for the scientific discovery of cultural life patterns in the world. The innovation of computational tools and technologies contributes to the computational discovery of cultural patterns and regularities that inform on the design of practical applications. The simulation of cultural patterns with digital environments and computational techniques facilitates the development of digital tools that are of importance to the design of computational models. The simulation and construction of cultural patterns with computational models illustrates the functional use of computational technologies and digital tools for the purpose of scientific and computational discovery. The understanding of culture and modern computation is fundamental to the research and development of tools and technologies that have practical benefits.

The innovation of modern computation is beneficial to the research and development on culture and technology. The understanding of the fundamentals of culture and modern computation contributes to the scientific and technological innovation on culture and technology. Research and development on culture and modern computation provides an understanding of the functional performance and real-world advantages of computing machines that show the functional capabilities of computational performance that is responsive to culture. The innovation of computing machines such as devices and computers demonstrates the capabilities of functional performance that show functionality and adaptability in response to cultural contexts. The research and development on culture and modern computation is fundamental to the innovation of culture and technology and its practical applications.

Cultural machines demonstrate the scientific and technological innovation of culture and technology that shows functional compatibility within cultural contexts. The design of cultural machines that show the high-level performance and real-world advantages of modern computation is of importance to the development of scientific and technological innovation. The understanding of the fundamentals of culture and modern computation contributes to the research and development of cultural machines and devices that show functional performance that is compatible with cultural life patterns. The functional compatibility of cultural machines and devices that are synchronous with cultural life patterns demonstrates the function-ality and adaptability of culture and modern computation. The scientific and technological innovation of culture and technology is fundamental to the understanding of the practical applications of tools and technologies that improve the functional performance of computing machines that are responsive to culture.

The research and development of artificial or synthetic devices contrib-utes to the functional capabilities of system-wide performance in health and medicine. The demonstration of the functional compatibility of biologi-cal machines and artificial devices facilitates the system-wide performance of discovery and delivery science in health and medicine. The functional capabilities of artificial or synthetic devices are fundamental to the demon-stration of the feasibility of the simulation and construction of biological computation as artificial devices. The advancement of the functional capa-bilities of artificial devices contributes to the medical capabilities of discov-ery and delivery science that is beneficial to the advancement of the research and development of computational tools and technologies of health and medicine.

In a broad sense, the simulation and construction of biological com-putation as artificial life patterns entails the design and modeling of the machine computation of naturalistic and artificial life. The development of artificial technologies that advance the design and modeling of machine computation contribute to the simulation and construction of the machine computation of artificial life patterns. The simulation and construction of the machine computation of artificial life patterns demonstrates the func-tional capabilities of artificial or synthetic devices for functional perfor-mance. The design and modeling of machine computation of naturalistic and artificial life patterns is fundamental to the advancement of the design of artificial technologies that promote the simulation and construction of natural and artificial life.

The simulation and construction of artificial or synthetic devices that demonstrate functional capabilities that are similar to biological and machine computation is of importance to the practical benefits of the com-putational modeling of biological computation. The simulation and natu-ralistic machines such as biological computation depict the computational modeling of fundamental components for the purpose of the demonstration of the compatibility of functional performance of biological machines and artificial devices. The construction of synthetic devices to show functional performance that is compatible with biological computation is essential to the advancement of the computational performance of synthetic devices and their functional capabilities.

The cultural computation of natural and artificial life illustrates the fea-sibility of the cultural computation of biological machines and artificial devices. The research and development of cultural machine computation as naturalistic and artificial life illuminates on the simulation and construction of machines that perform cultural computation. The design and modeling of the cultural computation of biological machines facilitates the demonstration of the cultural life patterns that are emblematic of naturalism. The development of the computational tools and technologies that show the functional performance of biological machines as cultural life patterns illustrates the functional capabilities of machine computation for cultural performance. The illustration of the cultural life patterns of biological machines is beneficial to the advancement of the cultural performance of naturalistic and artificial life.

The simulation and construction of artificial life that performs cultural life patterns shows the feasibility and practicality of artificial devices as cultural machines. The simulation and construction of artificial devices as machines that perform cultural computation demonstrates the feasibility of the functional performance of artificial life that is cultural. The demonstration of artificial devices that show the functional capabilities of cultural performance is beneficial to the advancement of artificial life patterns. The functional capabilities of artificial devices for cultural performance show the practicality of the artificial life patterns of cultural machines. The functional capabilities of cultural machines that show artificial life patterns are beneficial to the advancement of the simulation and construction of artificial life.

The promotion of the development of practical applications in health and medicine illustrates the feasibility of the functional capabilities of tools and technologies to provide simulation and construction of the functional performance of biological computation. The demonstration of the feasibility of the functional capabilities of tools and technologies for the simulation and construction of functional performance is essential to the advancement of the capabilities of system-wide performance in health and medicine. The usage and availability of bioengineering tools and technologies that illustrate the functional capabilities of artificial and synthetic devices are beneficial to the functional capabilities of machine computation and its simulation and construction of biological computation as artificial life patterns.

Scientific and Technological Innovation on Culture and Technology

The scientific and technological innovation on culture and technology advances the considerations of the functional capabilities and performance of culture and machine computation. From simple to complex devices, scientific and technological innovation advances the functional capabilities and performance of the computational foundations of machines that are responsive to culture. The development of scientific and technological innovation of a wide realm of tools and technologies benefits from the advancement of research and development on culture and technology. The innovation of culture and machine computation is fundamental to the design of practical applications that have societal benefits.

The innovation of culture and technology contributes to the design and construction of practical applications that have a real-world purpose. The design and construction of products and technologies that show the functional compatibility of system-wide performance is of importance to the effectiveness of practical applications. The scientific and computational discovery of culture and machine computation leads to the innovation of products and technologies that show the practical benefits of system-level performance. From vaccines to knowledge-based platforms, the scientific and computational discovery of culture and machine computation advances the research and development of products and technologies that show the interests and outcomes that are responsive to societal and public concerns. The research and development of products and technologies is fundamental to the dissemination of cultural information that leads to the improvement of outcomes that are in the public interest.

The design of digital tools and technologies is of importance to the usage and practicality of data and information that inform practical applications and impact societal outcomes. Digital tools and technologies are beneficial to the integral use of data and information that has practical importance in context. The design of digital tools and technologies such as software applications and hardware devices facilitate the interaction of system users and knowledge platforms for the sharing and monitoring of data and information. The design and construction of digital software and hardware generates knowledge platforms that contribute to the functional performance of complex systems.

The design and construction of digital tools and technologies that demonstrate functional capabilities that are cultural illustrate the practical applications of culture and technology. The design of digital tools and technologies that build and share on the cultural content demonstrates the feasibility of the construction of practical applications that are responsive to cultural contexts. The integral use of the cultural content of digital tools and technologies informed with cultural data and information demonstrates the functional compatibility of software applications and hardware devices with cultural systems. The functional capabilities of software applications and hardware devices to show functional performance that is compatible with cultural data and information are essential to the effectiveness of the cultural performance of digital tools and technologies. The functional capabilities of digital tools and technologies that are cultural show the functionality and adaptability of the practical applications of culture and technology.

The functionality and adaptability of the knowledge-based resource of digital tools and technologies is fundamental to the effectiveness of the practical applications of culture and technology. Knowledge-based resources comprise the cultural data and information that informs the content-based applications and devices and their functional performance. The design and construction of knowledge-based platforms that store and maintain cultural data and information to inform content-based applications and devices ("C-app", "C-ware") demonstrates the functionality of digital tools. The design of content-based applications and devices provides support for the sharing and monitoring of cultural data and information that inform digital tools and technologies. Knowledge-based platforms are essential to the storage and maintenance of cultural data and information of knowledge-based resources that have practical application.

The design and construction of content-based applications and devices that are compatible with cultural contexts show the functionality and adaptability of digital tools and technologies to cultural systems. Content-based applications provide interfaces that allow for the sharing of cultural content that is compatible with the cultural data and information of knowledge-based platforms. Content-based applications also allow for the adaptability of cultural content based on interaction with knowledge-based platforms. Content-based devices provide hardware for the sharing of cultural data and information with knowledge-based platforms. The design and construction of content-based devices allows for the storage and maintenance of cultural content that is interactive with knowledge-based platforms. The functionality and adaptability of content-based applications and devices demonstrates the effectiveness of digital tools and technologies as practical applications of culture and technology.

Hardware devices are fundamental as a knowledge base of cultural data and information that informs the functional performance of complex systems. The design and construction of hardware devices that include the functional specifications of cultural systems demonstrate the functional compatibility of digital tools and technologies with cultural contexts. The design of hardware devices that are compatible with the functional specifications of cultural systems facilitates the functional performance of content-based software. The demonstration of the functional compatibility of the design and construction of hardware devices with the functional specifications of cultural systems is of importance to the overall effectiveness of digital tools and technologies.

The functional compatibility of hardware devices as a knowledge base of culture ("C-base") demonstrates the functionality and adaptability of complex systems. The integral use of hardware devices as functional components of cultural systems demonstrates the functional performance of digital tools and technologies of cultural contexts. The usage and application of hardware as a knowledge base that is central to the storage and maintenance of the data and information of cultural systems shows the practical importance of digital tools and technologies. The functional performance of hardware devices as a knowledge base that supports the sharing of cultural data and information illustrates the adaptability of digital tools and technologies to the integral usage of complex systems.

The implementation of the knowledge base of cultural systems facilitates the coordination and sharing of knowledge-based resources as a complex system. The usage of knowledge-based resources as cultural modules ("C-module") of content-based applications and devices contributes to the effectiveness of digital tools and technologies. Knowledge-based resources that provide informatic content as cultural modules of content-based applications and devices facilitate the functionality and adaptability of digital tools and technologies. Cultural modules provide cultural informatics that are compatible with the usage and preferences of the cultural data and information of knowledge-based platforms. The integral use of cultural modules with content-based applications and devices contributes to the overall effectiveness of digital tools and technologies across settings.

The knowledge-based resources of cultural modules provide informatic content to content-based application and devices that are essential to the overall effectiveness of digital tools and technologies. The informatic content of cultural modules consists of the functional components of knowledge-based resources that are part of the system-level performance of knowledge-based platforms. The implementation of cultural modules disseminates the functional components of knowledge-based resources that are essential to the overall performance of knowledge-based platforms. The integral usage of cultural modules constructs system-wide options that are fundamental to the system-level performance of knowledge-based platforms.

The informatic content of cultural modules comprises functional components of knowledge-based resources. Knowledge-based resources consist of data informatics that consolidate the usage and preference of information of the knowledge-based platforms. The implementation of data analytic techniques on the usage and preference information from knowledge-based platforms facilitates the production of data informatics that inform knowledge-based resources. The building of data informatics serves as a knowledge base for the design and construction of digital tools and technologies that advance the implementation of products and technologies of knowledge-based platforms.

The scientific and technological innovation of culture and technology illustrates the wide realm of considerations that are of importance to the practical applications of culture and technology. The innovation of products and technologies that are cultural illustrates the implementation of practical applications that have societal benefits. The scientific and computational discovery of products and technologies that improve functional outcomes contributes to the advancement of the effectiveness of practical applications. Digital tools and technologies facilitate the integral use of scientific and societal resources that lead to the implementation of goals and outcomes. The building of knowledge-based resources is fundamental to the innovation of products and technologies as practical applications of culture and technology.

Culture and Technology in Health and Medicine

Culture and technology are fundamental to the practical applications of health and medicine. The use of computational tools and technologies for scientific and medical discovery on cures, preventions and interventions to complex disease illustrates the broad range of practical applications in health and medicine. Computational tools and technologies provide quality of data and information that informs the design and implementation of knowledge-based resources that have practical importance in health settings. The development of digital tools and technologies for health prevention and intervention contributes to the practical applications that inform the training and rehabilitation of complex disease. Knowledge-based resources provide data and information that is beneficial to the scientific and societal enrichment of the programs of culture and health promotion. The development of the practical applications of culture and technology demonstrates the societal benefits of computational tools and technologies in health and medicine.

Evidence-based research in health and medicine benefits from the use of computational methods in biomedical and biobehavioral research. Computational tools and technologies are fundamental to the precision of computational discovery in scientific and medical research. The development of computational tools and technologies for the study of diverse cultures and populations is essential to the precision of computational discovery. The study of diverse cultures and populations with computational tools and technologies contributes to the building of evidence-based resources that inform the scientific and medical discovery of cures, preventions and interventions to complex disease. The development of evidence-based resources with computational methods is beneficial to the advancement of biomedical and biobehavioral research.

Computational tools and technologies provide methods for the scientific and medical discovery of evidence-based research. The research and development of health and medicine with computational methods facilitates the data-mining techniques of large-scale datasets. The use of computational tools and technologies to study the cultural patterns of large-scale datasets is fundamental to the scientific and medical discovery of cures, preventions and interventions to complex disease. The systematic study of the cultural patterns of large-scale datasets contributes to the development of cures, preventions and interventions that are appropriate for diverse cultures and populations. The scientific and medical discovery with data mining techniques and large-scale datasets demonstrates the efficacy of computational methods as a multilevel approach to the strategies of prevention and intervention in health and medicine.

The research and development of the programs of research of diverse cultures and populations with computational tools illustrate the practical role of computational discovery in health and medicine. The research and development of diverse cultures and populations with computational tools contributes to the understanding of the cultural patterns of large-scale datasets. Computational tools advance the precision of scientific and medical discovery that is based on data-mining techniques. The development of computational tools that are culturally- and ethnically-appropriate facilitates the precision of the computational discovery of the cultural patterns of large-scale datasets. The understanding of the cultural patterns of large-scale datasets demonstrates the precision of computational discovery with diverse cultures and populations in health and medicine.

The development of computational tools and technologies that show the precision of computational discovery demonstrates the practical benefits of culture and technology. The development of computational methods that are culturally and ethnically appropriate illustrates the use and availability of tools and technologies for the amelioration of complex disease. Computational methods provide tools and technologies that facilitate the strategies of health prevention and intervention. Computational methods that demonstrate the precision of computational discovery that is culturally and ethnically appropriate show the overall effectiveness of tools and technologies for use with diverse cultures and populations. The use and availability of computational methods demonstrates the precision of computational discovery that improves the quality of evidence-based research and the overall efficacy of the strategies of health prevention and intervention.

Knowledge-based research contributes to the data and information that inform the strategies of health prevention and intervention. The building of knowledge-based resources advances the data and information that serve as scientific and societal resources to inform on societal interests. The development of knowledge-based resources facilitates the equitable access to scientific and societal resources that enrich the societal and public awareness of health and medicine. Knowledge-based resources contribute to scientific and societal enrichment that are beneficial to the building of public awareness of health prevention and promotion.

The development of knowledge-based resources is fundamental to the scientific and societal resources that advance the strategies of health prevention and intervention. The building of knowledge-based resources of diverse cultures and populations facilitates the data and information that contribute to the building of scientific and societal awareness of the strategies of health prevention and intervention. Knowledge-based resources that consist of the data and information of diverse cultures and populations show the functional capabilities of system performance to provide scientific and societal resources that have a wide breadth of public impact. The advancement of knowledge-based resources for the purpose of scientific and societal enrichment is beneficial to the procurement of public health interests.

Computational tools and technologies contribute to the overall effectiveness of health prevention and promotion. The use and availability of computational methods contributes to the equitable access to tools and technologies that are essential for the improvement of societal outcomes. Computational tools and technologies advance the precision of computational approaches to evidence-based research. Computational approaches build the information and data that are available as knowledge-based resources for health prevention and promotion. The integral use of computational tools and technologies with diverse cultures and populations illustrates the broad spectrum of practical benefits of culture and technology. The use and availability of computational tools and technologies for health prevention and promotion shows the importance of culture and technology in health and medicine.

Computational tools and technologies advance the medical system capabilities in health and medicine. The use and availability of computational tools and technologies facilitate the system capabilities of discovery and delivery science in health and medicine. The research and development of computational tools and technologies contribute to the discovery and delivery science of cures, prevention and intervention to complex disease. The development of computational tools and technologies with diverse cultures and populations advances the medical system capabilities for discovery and delivery science in health and medicine. The use and availability of computational tools and technologies is beneficial to the advancement of the quality of care and treatment in health and medicine.

Culture and Technology as Health Promotion

Research and development on culture and technology contributes to the strategies of health prevention and promotion. Evidence-based research provides information and data that enriches scientific and societal resources on culture and health. The research and development of computational tools and technologies facilitates the building of evidence-based resources that inform scientific and societal resources for the benefit of culture and health promotion. Evidence-based resources consist of information and data that are beneficial to the understanding of the study of the computational approaches of culture and health. The evidence-based resources of culture and health promotion show how scientific and computational discovery contributes to the evidence-based research that informs culture and health promotion.

Culture and technology as innovation for culture and health promotion facilitates the evidence-based practice and policy of culture and health. The building of the scientific base on culture and technology contributes to the development of the practical applications of information and communication technology. The practical applications of information and communication technology illustrate how digital tools and technologies facilitate the sharing of information and data that contributes to culture and health promotion. Digital tools and technologies are fundamental to the sharing of information and data that is beneficial to the promotion of culture and health. The innovation of the practical applications of culture and technology contributes to the usage and availability of digital tools and technologies that are fundamental to culture and health promotion.

Information and communication technology is beneficial to the sharing and maintenance of information and data of diverse cultures and populations. Digital tools and technologies enable the use and availability of system capabilities with the information and data of diverse cultures and populations in health and medicine. The functional performance of system capabilities contributes to the use and availability of the information and data of scientific and societal resources that are of importance to the promotion of culture and health. The innovation of digital tools and technologies facilitates the functional capabilities of system-level performance to inform on the societal interests of culture and health promotion. The building of the functional capabilities of system-level performance is essential to the promotion of the societal and public interests on culture and health.

Information and communication technology such as social media contributes to the building of societal and public health awareness on culture and health promotion. The building of scientific and societal resources on culture and health facilitates the enrichment of societal and public health awareness on culture and health. The scientific and societal resources on culture and health are fundamental to the knowledge-based resources that inform the building of public health awareness campaigns with social media. The development of cultural content for social media contributes to the information and data of information and communication technology. The dissemination of cultural content with social media promotes the societal and scientific enrichment that is beneficial to culture and public health awareness.

The development of cultural content for social media illustrates the societal and public health benefits of knowledge-based resources on culture and health. The dissemination of cultural content with social media shows the broad range of public health interests that are of interest to the programs of culture and health promotion. Cultural content in social media contributes to the breadth of public health information that is available with information and communication technology. The integral use of cultural content with social media facilitates the sharing and maintenance of public health information that promotes the building of public awareness of the importance of cultural enrichment as a societal interest.

Social media demonstrates how information and communication technology is beneficial to the promotion of cultural enrichment and its public health impact. The cultural content of social media provides a societal forum for the health communication on the societal benefits of cultural enrichment. The development of social media content on the societal benefits of cultural enrichment is fundamental to the promotion of culture and health. The cultural content of health communication provides data and information on the societal benefits of cultural enrichment to health prevention and promotion. The building of cultural content of social media enhances the health communication of public health campaigns on the importance of cultural enrichment and its societal benefits.

The development of programs on culture and health promotion benefits from the use and availability of big data that is based on information and communication technology. The use and availability of big data from information and communication technology shows the importance of information and data on the promotion of culture and health. The cultural content of big data provides knowledge-based resources that inform on health communication. The promotion of culture and health with the cultural content of big data facilitates the sharing and maintenance of health communication that is beneficial to the building of public health awareness.

Knowledge-based resources facilitate the use and availability of big data for health communication on culture. The understanding of that usage and preferences of tools and technologies contributes to the design of strategies of health communication that are efficacious in the promotion of culture and health. The integral use of big data to inform the strategies of health communication is beneficial to the development of tools and technologies that promote the cultural content of health. The knowledge-based resources of health information are fundamental to the tools and technologies that contribute to the building of public awareness on culture and health.

Practical Applications

The capabilities of modern computing show how scientific and technological innovation of modern computation has led to the advancement of the study of cultural and biological computation. The development of computational tools and technologies has led to the innovation of the understanding of the computational foundations of culture and the brain. The development of the research capabilities of modern computing for the scientific study of the computational foundations of culture and the brain is fundamental to culture and health promotion. The advancement of modern computing as computational tools and technologies contributes to the societal and public health impact of culture and health promotion.

The innovation of science and technology in health and medicine has practical benefits in health and medicine. From vaccines to knowledge-based platforms, the scientific and technological innovation of computational tools and technologies has led to the advancement of cures, prevention and intervention of culture and health promotion. The innovation of computational tools and technologies in health and medicine contributes to the development of scientific and societal resources that inform culture and health promotion. Computational tools and technologies contribute to the capabilities of research for the discovery and delivery science of prevention and intervention of culturally and ethnically diverse populations in health and medicine.

The practical applications of culture and technology illustrate the broad range of scientific and technological innovation of importance to the research and development of computational tools and technologies. The development of practical applications of culture and technology in health and medicine illustrates the computational tools and technologies that are beneficial to the prevention and intervention of health and disease. From digital tools to knowledge-based platforms, computational tools and technologies are beneficial to the functionality and adaptability of the system performance of discovery and delivery science in health and medicine. Digital tools facilitate the interface of medical and technological system capabilities that ensure the discovery and delivery science of health and medicine.

The research and development of digital tools and technologies is fundamental to the use of digital technologies that contribute to the functional performance of medical system capabilities. The development of digital tools such as software applications and hardware devices contributes to the use and availability of digital technologies that are beneficial to the reliability and efficiency of functional performance in health and medicine ("eHealth") (WHO, 2005). Digital tools such as software applications provide computational interfaces that advance to the quality of data and information of system capabilities. Hardware devices enable the system input and maintenance of quality of data and information that is beneficial to the functional performance of medical system capabilities. Medical system capabilities that interface with digital technologies contribute to how digital technologies transform the quality of information and data to inform the prevention and intervention of health and disease.

The design of digital tools and technologies contributes to the usage and availability of digital resources that provide quality of data and information of system capabilities. Digital tools and technologies such as eHealth facilitate the equitable access to digital resources that provide societal databases to inform on the functional performance of system capabilities in health and medicine. Digital tools and technologies provide the societal resources that facilitate the integral use of information and communication technologies. The integration of digital tools with information and communication technologies is beneficial to the development of system capabilities that are foundational to the discovery and delivery science of health and medicine.

The advancement of digital tools and technologies illustrates the practicality of the digital tools of discovery and delivery science for culture and health promotion. The equitable access to digital tools facilitates the usage and availability of information and communication technologies that broaden culture and health promotion. The usage and availability of digital tools contributes to the quality of information and data that are available based on information and communication technologies. The equitable access to digital tools advances the societal awareness of the benefits of culture and health promotion. Digital tools facilitate the communication of health information of diverse cultures and populations in a global context. The wide distribution of the communication of health information is beneficial to the efficacy of culture and health promotion.

Knowledge-based platforms is of importance to the advancement of the practicality and usage of system capabilities that are beneficial to system-wide performance. The research and development of knowledge-based platforms show the design of computational tools and technologies that are beneficial to the interface of system capabilities that advance the quality of data and information of system capabilities ("telemedicine"). Knowledge-based platforms advance the development of evidence-based databases that facilitate the performance of digital tools and technologies. The design of knowledge-based platforms contributes to the advancement of the functional performance of system capabilities in health and medicine.

The design of knowledge-based platforms advances the quality of care of system capabilities in health and medicine. The development of knowledge-based platforms that interface with medical system capabilities illustrates the use of computational tools and technologies that inform evidence-based approaches. The use of knowledge-based platforms in health and medical settings improves the quality of data and information of evidence-based resources to inform the design and implementation of bioinformatic databases. The design and implementation of bioinformatic databases with the functional capabilities of knowledge-based platforms facilitates the usage and availability of quality of data and information in the health and medical setting. The advancement of the development of knowledge-based platforms contributes to the knowledge-based resources that facilitate system-wide capabilities.

The usage and availability of knowledge-based platforms is beneficial to the prevention and intervention of health and medicine. Knowledge-based platforms contribute to the use and availability of knowledge-based resources that provide information and data as societal resources to promote the strategies of health intervention in medicine. The use and availability of societal resources contributes to the quality of information that is beneficial to the efficacy of health intervention. The design of knowledge-based platforms contributes to the integral use of tools and technologies that facilitate the functional performance of system-wide capabilities. The development of knowledge-based platforms is fundamental to ensure the use and availability of information and data that promote the efficacy of health intervention.

The advancement of knowledge-based platforms is of importance to the societal resources that inform on the prevention and intervention of health and medicine. The development of knowledge-based platforms facilitates the improvement of the quality of data and information to promote health prevention and intervention. The equitable access to societal resources is fundamental to the development of knowledge-based platforms that provide quality of data and information on culture and health equity. The design of knowledge-based platforms that facilitate the functional capabilities of discovery and delivery science is essential to the efficacy of health prevention and intervention. The advancement of knowledge-based platforms is fundamental to the enrichment of societal resources that demonstrate the efficacy of culture and health promotion.

Discussion

Computational tools and technologies are fundamental to the innovation of discovery and delivery science in health and medicine. The research and development of computational tools and technologies advances the understanding of the brain basis of culture. Evidence-based research on computational cultural neuroscience contributes to the scientific discovery of the brain basis of culture. The building of evidence-based resources on computational cultural neuroscience contributes to the innovation of the practical applications of culture and technology. The advancement of the research and development of computational tools and technologies is beneficial to the innovation of discovery and delivery science in health and medicine.

Research and development on culture and technology contributes to the evidence-based practice and policy of health and medicine. The building of evidence-based resources on culture and health informs the innovation of tools and technologies that serve as a foundation of the evidence-based approaches of culture and technology. The building of scientific infrastructure and research capabilities in health and medicine with computational tools and technologies facilitates the innovation of evidence-based resources on culture and health. The research and development on culture and technology benefits from the innovation of computational tools and technologies for the benefit of evidence-based research and its practical applications for practice and policy.

The advancement of culture and technology contributes to the use and availability of digital tools and technologies that promote culture and health. Digital tools and technologies facilitate the wide dissemination of cultural content that promotes the societal and public health awareness on culture. The design and construction of digital tools and technologies advances the functional capabilities of system performance for the promotion of culture and health. The development of the cultural content of digital tools and technologies provides scientific and societal resources that inform on culture and health promotion. The use and availability of digital tools and technologies is beneficial to the promotion of the public health awareness of the benefits of culture.

Conclusion

Translational research on computational cultural neuroscience informs the development of practice and policy. The development of research capabilities and scientific infrastructure on computational cultural neuroscience advances the building of evidence-based resources on culture, technology and health. Evidence-based resources contribute to the knowledge generation on culture and brain function and its practical applications. The development of evidence-based resources is essential to the systematic study of culture and brain function in health and medicine. The systematic study of culture and brain function with computational tools and technologies is beneficial to the building of evidence-based approaches and their implications for practice and policy.

Evidence-based resources on culture and brain function facilitate the development of computational cultural neuroscience. The use of computational tools and technologies for the systematic study of culture and brain function contributes to the understanding of the functional architecture of culture and brain function. The use of multilevel approaches to the study of the functional architecture of culture and brain function demonstrates the modeling of the structural and functional principles of cultural brain function. The design of computational models to investigate the cultural patterns of brain function contributes to the computational discovery of the patterns and regularities of large-scale datasets.

The scientific and technological innovation of computational tools and technologies for the study of culture and brain function is fundamental to the advancement of culture, health and technology. The innovation of computational tools and technologies is fundamental to the advancement of the practical applications of culture and technology in health and medicine. The building of research capacity and scientific infrastructure on culture, health and technology advances the capabilities of scientific and medical discovery for the amelioration of complex disease. The development of prevention and intervention strategies on culture and health is fundamental to culture and health promotion.

Implications

Health

Research and development on culture and technology is beneficial to the promotion of the interests of culture and health. Evidence-based approaches to culture and technology contribute to the understanding of the reliability and effectiveness of computational tools and technologies for the purpose of scientific and medical discovery. The use and availability of computational tools and technologies that are culturally and ethnically appropriate facilitate the advancement of discovery and delivery science in health and medicine. Computational tools and technologies contribute to the precision of computational discovery that informs the strategies of cures, preventions and interventions to complex disease. The research and development on culture and technology is of importance to the promotion of the quality of data and information that informs the interests of culture and health.

Policy-based approaches to culture and technology also are fundamental to culture and health promotion. Knowledge-based resources provide information and data that inform on the impact of technology on culture and health. The building of knowledge-based resources contributes to the use and availability of information and data that is beneficial to the understanding of the impact of technology for the purpose of culture and health promotion. Knowledge-based resources contribute to the use and availability of information and data of diverse cultures and populations. The integral use of knowledge-based resources with technology contributes to the effectiveness of culture and technology in the promotion of health interests.

Medicine

The research and development of culture and technology informs on the innovation of computational tools and technologies that advance the precision of computational discovery in health and medicine. The innovation of computational tools and technologies facilitates the quality of data and information that build from evidence-based approaches in biomedical and biobehavioral research. Computational tools and technologies are fundamental to the innovation of discovery and delivery science that leads to cures, preventions and interventions to complex disease. The precision of computational discovery advances the study of cultural patterns and regularities of large-scale datasets that are fundamental to the building of evidence-based resources on culture and health The building of evidence-based resources with computational techniques contributes to the design of strategies of prevention and intervention that lead to the amelioration of complex disease.

Public Policy

Policy-level approaches to culture and technology are of importance to the development of scientific infrastructure and capacity development of that ensures the functional capabilities of complex systems. Policy development and implementation on culture and technology is essential to the use and availability of tools and technologies that impact societal and public interests. The development of programs and policies on culture and technology facilitates the policy frameworks that inform the innovation of the functional capabilities of complex systems. The programs and policies of culture and technology are fundamental to the building capacity development that

Reference

World Health Organization. (2005). *Global observatory for eHealth*. Geneva.

11

CULTURAL INTELLIGENCE

Introduction

The scholarly and scientific interest in thought and reason explores the importance of rationality as a basis of social and cultural life. The societal importance of rationale discourse as foundation of the socio-political thought that comprises social and cultural life illuminates on the importance of thought and reason in society and culture. The reliance on rational discourse for the discernment of the interests and concerns of social, political and cultural facets of society demonstrates the shared meaning and significance of social influence and social action as mechanisms of social coordination. The development of social and political thought illuminates on mutual interests and understanding of the interests and concerns that impact societal and cultural life.

The discernment of social, political and cultural interests based on rationale discourse provides a foundation for the cultivation of social influence and social action as a foundation of civilization and society. The idealism and materialism of cultural intelligence illustrate the functional role of thought and reason as a fundamental basis of culture and behavior. The emergence of rationality as a foundation of social, political and cultural discourse illustrates the societal significance of inference and reasoning on social, political and cultural life. The sharing of mutual interests and concerns that impact social, political and cultural interests is fundamental to the social coordination and cultural involvement in societal and public interests.

The pontification on cultural thought and reason serves as a foundation of the cultural intelligence of minds and machines. The design of machine

DOI: 10.4324/9781003384236-12

intelligence that is cultural illuminates on the cultural computation of simple to complex devices. The cultural computation of computing machines is essential to the functional performance of cultural computation. The manifestation of cultural machine intelligence in real and artificial systems shows the functionality and adaptability of culture and complex systems. The development of cultural machine intelligence as a functional performance of devices and computers illuminates on the physical instantiation of cultural intelligence as machine computation. Cultural intelligence as machine computation demonstrates the practicality and feasibility of the implementation of cultural machine intelligence as a foundation of cultural machine computation.

The goal of the chapter is to provide a review on the foundations of cultural intelligence. The foundations of cultural intelligence entail the understanding of the fundamentals of cultural thought and reason. The fundamentals of cultural thought and reason comprise the complexity of thought and reason that are of importance to the higher-level function of culture and behavior. The societal discourse on cultural thought and reason illustrates the functional significance of inference and rationality on culture and society. Cultural intelligence illuminates on the complexity of emergent thought of the functional adaptation of biological organisms as a cultural species. Practical applications of cultural intelligence for the research and development on the cultural machine intelligence of simple to complex devices and the cultural computation of real and artificial systems are discussed. Implications of research on cultural intelligence for health, medicine and practical fields are discussed.

Evolutionary Cultural Neuroscience

The scholarly interest in evolutionary cultural neuroscience describes the evolutionary basis of the cultural brain. The study of evolutionary cultural neuroscience provides a foundation for the understanding of the evolutionary processes that contribute to the functional machinery of biological organisms and its functional role in multilevel adaptation. The evolutionary basis of cultural capacities describes the higher-level processing of behavior that characterizes biological organisms as cultural species. The functional adaptation of the cultural brain illustrates the emergence of cultural capacities and its underlying mechanisms that are essential to biological and cultural adaptation. The cultural brain as a functional adaptation is fundamental to the emergence of cultural capacities that are essential to multilevel adaptation.

The study of cultural brain evolution explores the notion of the cultural brain as a functional adaptation of biological organisms in response to evolutionary processes. The structural and functional basis of the cultural brain illustrates the functional role of basic mechanisms in response to the ecological and cultural niche. The structure and function of the cultural brain illustrates the functional architecture of adaptive machinery that is responsive to the cultural environment. The structural and functional basis of the cultural brain entails the processes and mechanisms that are essential to the core capacities of cultural adaptation. The cultural brain as adaptive machinery depicts the functional mechanisms that facilitate the adaptive response of biological organisms to cultural adaptation.

The functional adaptation of the cultural brain facilitates core capacities as the adaptive responsiveness of biological organisms. The core capacities of the cultural brain function demonstrate the functional processes and mechanisms that are fundamental to ecological and cultural niche construction. The core capacities that correspond with cultural brain function consist of the adaptive machinery that is fundamental to the alteration of the ecological and cultural niche. The cultural brain function of biological organisms facilitates the ecological and cultural changes of the environment that are responsive to the adaptation of biological organisms. The ecological and cultural niche construction of biological organisms is based on the adaptive machinery of the cultural brain and its responsiveness to the cultural environment.

The evolution of the cultural brain entails the understanding of the evolutionary foundations of the functional adaptation of cultural capacities. The evolutionary foundations of cultural adaptation illustrate how cultural capacities serve as functional adaptation of biological organisms. The emergence of cultural capacities demonstrates the preparedness and responsiveness of biological organisms as cultural species. The adaptive machinery of biological organisms illustrates the functional architecture of the cultural brain as a mechanistic basis of cultural capacities. The functional architecture of the cultural brain is fundamental to processes and mechanisms that are essential to the information processing mechanisms of cultural capacities.

The emergence of cultural capacities such as cultural intelligence demonstratese adaptive preparedness of biological organisms for cultural adaptation. Cultural intelligence as a functional adaptation details the preparedness and responsiveness of the adaptive machinery of biological organisms in response to the cultural environment. Cultural intelligence as a cultural capacity illustrates the functional basis of cultural thought and the higher-level processes that underlie behavior. The functional processes of cultural thought and reason show the adaptive machinery of complex organisms that are essential to the adaptive preparedness of cultural species.

The functional basis of cultural intelligence illustrates the core capacities of higher-level processes of the cultural brain that are essential to cultural adaptation. The cultural brain consists of the functional architecture that describes the processes and mechanisms of cultural capacities. The functional processes and mechanisms of cultural brain function depict the mechanistic basis of cultural processing. The information processing mechanisms of the cultural brain show the adaptive machinery that is essential to the cultural processing of core capacities. The core capacities of the cultural brain are comprised of the functional processes and mechanisms that are of importance to cultural adaptation.

The core capacities of cultural intelligence demonstrate the higher-level processes of behavior that are fundamental to cultural thought and behavior. The functional processes of cultural thought and behavior encompass the higher-level function that is essential to cultural intelligence. The functional processes of cultural thought and behavior comprise cultural patterns that are of importance to the behavioral and linguistic expression of culture. The behavioral and linguistic expression of culture encompasses cultural patterns that facilitate the higher-level function of cultural thought and behavior. The core capacities of cultural intelligence are fundamental to the functional processes of cultural thought and behavior.

The control and regulation of cultural capacities entails the higher-level function of cultural thought and behavior. The core capacity of cultural thought and behavior illustrates the functional importance of the control and regulation of the behavioral and linguistic expression of culture. The control and regulation of cultural thought and behavior show the importance of regulatory control on the behavioral and linguistic basis of cultural expression. Cultural capacities comprise the behavioral and linguistic patterns of cultural expression that are fundamental to cultural adaptation. The control and regulation of cultural capacities demonstrate the functional significance of regulatory control on the patterns of cultural thought and behavior.

The core capacities of the cultural brain function demonstrate the functional preparedness of biological organisms as cultural species. The functional architecture of the cultural brain depicts the processes and mechanisms that are essential to the adaptive responsiveness of biological organisms to the cultural environment. The functional processes of core capacities show the functional role of cultural information processing mechanisms as a mechanistic basis of functional and cultural adaptation. The core capacities of the cultural brain function detail the functional processes that contribute to the control and regulation of cultural capacities.

The evolutionary basis of the cultural brain entails the functional machinery of cultural adaptation. The evolutionary processes that are fundamental to adaptation and behavior illustrate the importance of biology and culture as mutual influences. Cultural brain evolution shows how the adaptive machinery of the cultural brain reveals the functional adaptation of cultural capacities. The functional adaptation of the cultural brain illustrates the core capacities that are fundamental to cultural capacities. The mutual influences of biology and culture on adaptation and behavior demonstrate the evolutionary basis of the functional processes and mechanisms of multilevel adaptation. The foundations of the evolutionary basis of the cultural brain are essential to the characterization of the functional architecture of cultural adaptation.

The Biology and Culture of Behavior

The study of the biology and culture of behavior consist of the understanding of the functional basis of the biological basis of culture and behavior. The biological basis of culture and behavior entails the functional mechanisms that are of importance to the complexity and higher-level function of behavior. The functional mechanisms of cultural capacities illuminate on the complexity of higher-level function that is of importance to the manifestation of behavior. The biological mechanisms of culture and behavior show how the physical implementation of cultural capacities is emergent. The biological basis of culture and behavior shows the functional significance of basic mechanisms as fundamental to the manifestation of the complexity and higher-level function of behavior.

The functional adaptation of culture and behavior relies on the adaptive machinery of biological organisms that is responsive to the environment. The adaptive machinery of biological organisms shows the functional adaptation of higher-level function that comprises cultural capacities. Cultural capacities consist of the functional processes of thought and reason that are fundamental to complex behavior. The core capacities of cultural adaptation show how the higher-level processes of thought and reason are fundamental to the intention and action of biological organisms as cultural species. The core processes of thought and reason serve as a functional basis of the higher-level function that is essential to the cultural processing of core capacities and multilevel adaptation.

The functional architecture of the brain serves as a mechanistic basis of culture and behavior. The cultural brain as a biological computing machine illuminates on the functional role of the brain as a biological basis of culture and behavior. The functional components of cortical mechanisms detail the core capacities that comprise the complexity of culture and behavior. The functional basis of core capacities shows the processing of core capacities that are fundamental to functional mechanisms. The functional processing of cortical mechanisms demonstrates the functional basis of cultural capacities. The functional components of the cultural brain show the basic mechanisms that are essential to culture and behavior.

The functional adaptation of the cultural brain illustrates the complexity of higher-level function and behavior. Cultural capacities demonstrate the functional processes that are essential to the higher-level function of behavior. The cultural processing of higher-level function encompasses the thought and reason that are essential to the levels of processing of the cultural brain. The functional mechanisms of cultural processing depict the levels of complexity of thought and behavior that comprise higher-level function. The biological mechanisms of the cultural brain detail the functional processes that comprise the intention and action of higher-level function. The functional basis of cultural processing further illuminates on the thought and reason that is essential to complex behavior.

The cultural processing of functional mechanisms facilitates the adaptive responsiveness of biological organisms. The functional basis of thought and reason is of importance as the functional and emergent properties of the cultural brain. The emergence of thought and reason illustrates the functional basis of the core processing of cultural capacities. The core processing of cultural capacities shows the causal–functional significance of the functional and emergent properties of the cultural brain. The functional and emergent properties of the cultural brain illustrate the functional significance of thought and reason as higher-level function.

The biological and cultural adaptation of behavior illuminates on the functional processing and biological mechanisms of cultural capacities. The biological basis of cultural capacities demonstrates the preparedness and adaptiveness of biological organisms as a cultural species. The functional processing and biological mechanisms of cultural capacities depict the core components of multilevel adaptation. The biological mechanisms of cultural processing demonstrate the mechanistic basis of functional and behavioral adaptation. The fundamentals of biological and cultural adaptation illustrate the processes and mechanisms that are central to the core capacities of cultural processing.

The Cultural Brain Hypothesis

The study of the cultural brain encompasses a wide realm of considerations on the origin and nature of cultural brain function. The functional architecture of the cultural brain illustrates the adaptive machinery of biological organisms as cultural species. The cultural brain hypothesis posits on the functional mechanisms of biological organisms as a cultural species. The structural and functional basis of cultural brain function encompasses the adaptive mechanisms that are fundamental to cortical brain organization. The cultural brain hypothesis ascertains on the multilevel mechanisms that are of importance to the functional adaptation of cultural capacities. The functional processes and mechanisms of the cultural brain are fundamental to cultural processing. The cultural brain hypothesis serves as a conceptual foundation for the study of cultural brain function and its link to cultural capacities.

The foundations of cultural brain evolution illustrate the evolutionary processes that serve as a basis for the adaptation of the cultural brain. The cultural brain as a functional adaptation of biological organisms demonstrates the functional specialization and functional integration of cortical brain organization for cultural capacities. The functional basis of the cultural brain entails the adaptive function of core capacities that are essential to cultural processing. The functional capacities of cortical brain organization for the processing of cultural information demonstrate the adaptive preparedness of biological organisms for cultural adaptation. The study of cultural brain evolution encompasses the notion of the evolutionary basis of the cultural brain and its functional significance.

The evolution of the cultural brain entails the functional importance of cortical brain mechanisms on multilevel adaptation. The adaptive preparedness of biological organisms to alter or change the environment demonstrates the functional role of core capacities. The functional adaptation of cortical brain mechanisms facilitates the changes of core capacities that are of importance to the interaction of biological organisms with the cultural environment. The preparedness and responsiveness of biological organisms to the interaction with the cultural environment illustrate the functional role of core capacities. The interaction of biological organisms with the cultural environment contributes to the functional basis of core capacities and their functional significance. The evolution of the cultural brain is fundamental to the functional processes and cortical brain mechanisms of multilevel adaptation.

The evolutionary foundations of cultural capacities illuminate on the mutual influence of biology and culture on the functional adaptation of the cultural brain. The biological evolution of cortical brain organization illustrates the functional specialization and functional integration of cortical mechanisms for core capacities; the adaptive machinery of cortical brain organization comprises the functional basis of information processing mechanisms. The information processing of cortical mechanisms shows the functional basis for the biological transmission of functional characteristics. The biological adaptation of cortical brain organization demonstrates the adaptive preparedness of biological organisms for multilevel adaptation.

The biological mechanisms of cortical brain organization detail the functional role of information processing mechanisms in the multilevel processing of cortical brain function. The structural and functional basis of cortical brain organization depicts the functional architecture of cortical mechanisms that are of importance to core capacities. The biological mechanisms of cortical brain function demonstrate the functional processing that is a fundamental biological adaptation. The functional processing of biological mechanisms illustrates the core capacities that are essential to the multilevel processing of cortical brain function.

The biological adaptation of complex organisms as a cultural species shows the functional processing that is essential to core capacities. The biological mechanisms of functional processes demonstrate the information processing that is essential to cortical brain function. The information processing mechanisms of cortical brain function detail the mechanistic basis of core capacities and higher-level function. The functional processing of core capacities comprises the information processing of cortical brain mechanisms as biological adaptation. The biological adaptation of complex organisms illustrates the functional processes and mechanisms that are fundamental to cultural capacities.

The cultural evolution of functional adaptation demonstrates the adaptive response of the functional mechanisms of cortical brain function to the cultural environment. The cultural capacities of biological organisms show the adaptive preparedness of simple to complex organisms for the functional adaptation of cultural brain evolution. The functional adaptation of cultural processing illustrates the functional role of cortical brain mechanisms. The functional role of cortical brain mechanisms for cultural processing illustrates the functional basis of core capacities. The functional adaptation of cultural capacities shows the cultural patterns of cortical brain function that are essential to information processing. The cultural evolution of the functional adaptation of cortical brain function shows the adaptive preparedness of biological organisms for interaction with the cultural environment.

Cultural mechanisms comprise the functional basis of the causal processes that are essential to information processing and cortical brain function. The cultural mechanisms of cortical brain function illustrate the information processing that is essential to cultural capacities. The causal role of cultural mechanisms depicts the cultural patterns of functional processing that are specific to culture. The functional processing of cultural mechanisms demonstrates the causal relation of cortical brain function and cultural processing. The changes in cortical brain function and cultural processing that are based on cultural mechanisms illustrate the functional role of cultural capacities.

The cultural adaptation of biological organisms as complex species details the cultural processes and mechanisms that are essential to multilevel adaptation. The cultural adaptation of functional processes shows the information processing mechanisms that serve as a mechanistic basis of cultural capacities. The adaptiveness and responsiveness of biological organisms to the cultural environment illustrate the processes and mechanisms that are essential to adaptation. Cultural mechanisms facilitate information processing and functional processes that are essential to cultural capacities. The information processing of functional mechanisms shows the mechanistic basis of cultural capacities that are essential to cultural adaptation.

The cultural brain demonstrates the functional processes and cortical mechanisms of the higher-level function of behavior. Cultural brain function depicts the functional processing of cortical brain mechanisms that are of importance to the higher-level function of core capacities. The culture of thought and rationality demonstrates the levels of complexity that are consistent with higher-level function. The complexity of cultural thought that comprises cultural inference and reasoning facilitates the functional processing of complex organisms. The causal role of cultural inference and reason as a functional basis of complex behavior illustrates the functional importance of higher-level function.

The functional adaptation of the cultural brain is central to the emergence of cultural patterns of thought and behavior. The cultural brain as a functional adaptation demonstrates the functional processes of biological organisms as a cultural species. The functional processes of biological mechanisms depict the emergent patterns of cultural thought and behavior that are essential to adaptation. The functional mechanisms of cultural brain function show the information processing that facilitates cultural patterns of thought that are emergent. The cultural brain as a functional adaptation illustrates the information processing mechanisms of biological organisms for cultural capacities.

The functional mechanisms of the cultural brain illustrate the information processing mechanisms that are fundamental to cultural capacities. The cortical mechanisms of functional architecture show the functional specificity of information processing that is essential to cultural thought and behavior. The cultural processing of cortical brain mechanisms shows the functional role of core capacities on cultural thought and behavior. The functional role of cultural processing depicts the functional mechanisms that facilitate the processing of cultural information. The functional mechanisms of the cultural brain demonstrate how cortical brain function is fundamental to cultural processing.

The cultural brain hypothesis serves as a conceptual foundation for the postulation on the functional architecture of the cultural brain and its causal effects. The postulation on cultural brain function encompasses the characterization of the multilevel mechanisms of cultural processing. The functional basis of cultural processing illuminates on the core capacities of biological organisms for cultural adaptation. The functional processing of cultural brain function depicts the higher-level function that comprises cultural capacities. The cultural brain hypothesis provides a conceptual basis for the formulation of scientific theories and hypothesis testing on the origin and nature of the cultural brain.

Culture as Inference and Reason

Culture as inference and reason describes the manifestation of culture as mental states that are fundamental to the higher-level function of behavior. The manifestation of cultural states as mental states illustrates the functional basis of inference and reason as higher-level function. The core capacities of inference and reason detail the mental states that serve as a functional basis of mentality and thought. The cultural states of mentality and thought are fundamental to the functional basis of inference and reason. The core capacities of higher-level function illustrate the functional processes of inference and reason as adaptation.

Cultural states as mental states illuminate on the functional basis of inference and reason. The core capacities of inference and reason comprise functional processes that reflect higher-level function of importance to behavior. The functional processes of inference and reason serve as a functional basis of the mental states that comprise cultural capacities. Cultural capacities rely on the inferential reasoning of mental states as a fundamental basis of core capacities. The core capacities of inferential reasoning facilitate the understanding of others that are essential to the understanding of others. The functional processes of inferential reasoning show the levels of processing that serve as a functional basis of core capacities. Cultural states as mental states entail the understanding of inferential reasoning as a fundamental basis of cultural capacities.

The core capacities of inference and reason consist of the functional processes of higher-level function and behavior. The functional processes of core capacities comprise the mental states of inference and reason that are cultural. The cultural states of mental state inference imply the levels of processing that are specific to culture. The inference and reason on mental states that are culture-specific encompass cultural states that illustrate the depth of processing that comprises cultural capacities. The cultural capacities of inference and reason show that the functional specificity of cultural processing demonstrates the higher-level function of culture and behavior.

Cultural capacities entail the functional processing of cultural patterns of thought that are based on algorithmic functions. The cultural patterns of thought that comprise higher-level processing illustrate the functional use of algorithmic function as a basis of cultural inference and reason. The higher-level processing of cultural capacities shows the importance of cultural inference and reason on the understanding of cultural causation. The cultural causation of mental inference and reason illustrates the functional significance of higher-level processing on patterns of cultural thought. The cultural capacities of inference and reason depict the functional use of algorithmic function as a higher-level function.

The core processes of cultural inference and reason entail the functional components of cultural capacities. The functional processing of cultural inference and reason encompasses the patterns of thought that are emergent. The cultural inference and reason facilitate the emergence of patterns of thought that are fundamental to cultural capacities. The emergent thought of cultural patterns illustrates the functional role of cultural inference and reason that is based on functional performance. The functional performance of cultural capacities demonstrates the emergent patterns of thought that facilitate cultural inference and reason. The core processes of cultural inference and reason illustrate the functional components of cultural patterns of thought.

Cultural inference and reason demonstrate the complexity of cultural thought that comprise the core capacities of cultural mental function. The cultural inference of functional processes shows the levels of complexity that comprise the causal reasoning on abstract conceptual knowledge and its causal effects. The role of conceptual representation as a functional basis of causal reasoning demonstrates a causal mechanism of cultural inference. The causal reasoning on conceptual knowledge illustrates the levels of complexity that comprise the prediction and inference on the causation of cultural inference. The core capacities of cultural thought depict the patterns of causal reasoning that are fundamental to the prediction and causation on cultural inference.

The functional processes of inference and reason illustrate the intrinsic properties that are of importance to culture. The cultural processing of functional capacities details the extent to which the mental states comprise an inherent cultural function. Cultural states as mental states depicts the intrinsic properties that are based on the inherent naturalism of cultural function. The cultural states of mental function demonstrate the functional processing that encompasses the mental states of inherent cultural function. The cultural processing of functional capacities shows the mental states that are of importance to inference and reason of cultural capacities.

The inference and reason on culture encompasses a wide realm of functional roles that facilitate the causal–functional significance of cultural capacities. Cultural inference and reason facilitates the cultural patterns of thought that are specific to culture. Cultural patterns of thought entail the functional and emergent properties that are essential to cultural mental causation and its functional significance. The cultural patterns of thought and its causal effects illustrate the internal content of cultural inference and reason that hold. The internal representation of cultural patterns of thought facilitate the representation and maintenance of cultural inference and reason as internal states that are fundamental to the cultural causation of minds and machines.

Cultural Intelligence

The study of cultural intelligence consists of the mentality and thought that underlie the inference and reason on culture. Cultural intelligence encompasses the thought and rationality that comprise cultural mental life. Cultural intelligence as the mental states of cultural thought and reason illustrates the function and adaptation of culture and behavior. Cultural intelligence consists of the complexity of thought and reason that is of importance to the higher-level function of behavior. The scholarly interest in cultural intelligence illustrates the functional significance of the inference and reason on the cultural intelligence of minds and machines in the world.

The mental states of thought and rationality that comprise cultural intelligence are fundamental to cultural mental life. The thought and rationality on cultural life demonstrate the causal role of mental states as a functional basis of cultural capacities. The functional role of mental states illustrates how inference and reason are fundamental elements of the cultural processing of core capacities. The functional basis of core processes details the patterns of inference and reasoning that show the levels of processing of cultural capacities. The cultural patterns of inference and reason depict the core processes that are essential to cultural capacities.

Cultural patterns of thought are fundamental to the understanding of the inference and reason of cultural capacities. The cultural capacities of functional mechanisms illustrate the levels of processing that are essential to cultural and cortical processing. The cultural patterns of thought illustrate the cultural inference and reason that are of importance as core capacities of functional mechanisms. The functional basis of cultural capacities demonstrates how cultural inference and reason illustrate levels of processing that are fundamental as cultural patterns of thought. Cultural inference and reason show the cultural patterns of thought that are essential to functional mechanisms of culture and behavior.

Cultural patterns of thought contribute to the understanding of the functional basis of cortical and cultural processing. The cultural processing of brain function facilitates the levels of complexity. The cultural patterns of the brain function demonstrate the functional processing of inference and reason that characterize the levels of complexity of cultural thought. Cultural patterns of thought are essential for the cultural inference and reason that is specific to culture and complexity. The functional processes of the cultural brain function show the cultural patterns of thought that comprise the functional basis of levels of complexity of cultural thought.

The patterns of cultural thought that comprise cultural intelligence are fundamental to adaptation and behavior. Cultural patterns of thought demonstrate the functional specificity of core processes that are essential to the function and adaptation of the cultural mind and its physical instantiation. The functional specificity of core processes shows the cultural patterns of thought that illustrate the preparedness and adaptiveness of the cultural mind. The core processes of cultural thought demonstrate the functional basis of the adaptation of cultural capacities and their functional significance. The patterns of cultural thought that are of importance to cultural intelligence show the functional specificity of core processes that are fundamental to functional and cultural adaptation.

Cultural intelligence illustrates the cultural patterns of thought that have functional significance. The functional role of cultural intelligence as inference and reason shows how higher-level processes are essential to the understanding of the causal–functional significance of cultural mental states. The cultural patterns of thought that comprise higher-level processes demonstrate the states of cultural inference and reason that are fundamental to cultural mental states and their causal effects. Cultural inference and reason show the functional significance of cultural mental states as a causal impetus of higher-level processes that are specific to culture. The patterns of cultural thought illustrate the cultural inference and reason that are essential to cultural intelligence.

Conceptual Foundations of Cultural Intelligence

The foundations of cultural intelligence consist of the understanding of the thought and rationality of cultural mental life that comprise the inference and reason on culture. The conceptual basis of cultural intelligence comprises the specific constructs that characterize the multifaceted dimensions of cultural intelligence. The fundamentals of cultural intelligence constitute the thought and rationality of mental function that is inherently cultural. The conceptual basis of cultural intelligence illuminates on the complexity of culture as thought and reason of minds and machines in the natural world. The complexity of thought illustrates the levels of processing that are characteristic of cultural intelligence.

Scientific concepts and language of cultural intelligence entail the conceptual foundations of the functional operationalization of cultural intelligence. The functional operationalization of cultural intelligence is comprised of the specific constructs that are fundamental to the systematic study of cultural intelligence. The scientific concepts of cultural intelligence consist of a broad range of conceptual language that encompasses the functional significance of cultural intelligence. Cultural intelligence is essential to the understanding of the inference and reasoning of diverse cultures. The conceptual language of cultural intelligence is fundamental to the understanding of the functional significance of the complexity of the thought and rationality of culture.

Theory building on cultural intelligence consists of the scientific theories that abound with the hypotheses on cultural intelligence and its functional significance. Theory building of scientific concepts and conceptual language entails the characterization of the functional basis of cultural intelligence. The scientific theories on cultural intelligence entail the understanding of the fundamentals of the thought and rationality that comprise the inference and reason on culture. Theory building on cultural intelligence investigates the processes and mechanisms of cultural intelligence and its functional significance. Scientific theories aim to characterize the fundamental principles of cultural intelligence and its functional role as a causal mechanism of cultural thought and reason. The theory building on cultural intelligence elaborates on the causal–functional role of cultural thought and rationale on the understanding of inference and reason on culture.

Conceptual development of scientific theories and empiricism on cultural intelligence facilitate the development of scientific concepts and conceptual language of cultural intelligence. The development of scientific concepts seeks to characterize the fundamental elements of cultural intelligence ("cultural inference"). The conceptual language of cultural intelligence entails the broadening of the scientific terminology that encompasses the cultural patterns of thought and reason that comprise the complexity of mental function that is inherently cultural. The cultural patterns of thought and reason comprise the cultural inference that is of importance to cultural capacities. The notion of cultural inference encompasses the cultural patterns of thought and reason that are fundamental to cultural capacities.

Conceptual frameworks on cultural intelligence explore the core processes that comprise the thought and reason on culture. The functional basis of cultural intelligence elaborates on the specific core processes that consist of the cultural patterns of thought and reason that are fundamental to cultural intelligence. The core processes of thought and reason facilitate the cultural patterns that are essential to cultural inference. The functional processes of thought and reason describe the cultural patterns that comprise the inferential thought that is cultural. The tacit and automatic thought on culture details the formulation of cultural thought and reason that demonstrates the inherent cultural function of the mind.

Conceptual models on cultural intelligence consist of the fundamental elements of cultural intelligence and its causal–functional significance. Conceptual models on cultural intelligence aim to characterize the impact of cultural intelligence on functional and behavioral outcomes. Conceptual models on cultural intelligence elaborate on the core processes that comprise the patterning of thought and reason that are essential to cultural intelligence. The causation of core processes illustrates the cultural patterns of thought and reason that comprise cultural intelligence. The development of conceptual models on cultural intelligence is fundamental to the understanding of cultural intelligence as a precursor to functional performance.

Other conceptual models on cultural intelligence seek to understand the role of cultural intelligence as a functional basis of cultural performance. The conceptualization of models of cultural intelligence considers the impact of cultural intelligence on the functional performance of cultural tasks. Conceptual models of cultural intelligence further seek to characterize the causal role of cultural intelligence as a functional performance that is specific to culture. The functional role of cultural intelligence is fundamental to the understanding of the cultural patterns of thought and reason that are essential to the functional performance of cultural tasks.

Conceptual models further consider how the functional performance of cultural tasks as routine builds the emergent patterns of cultural thought and reason that broaden the higher-level function of cultural intelligence. Cultural intelligence consists of the emergent patterns of thought and reason that are specific to the cultural inference that holds. Cultural patterns of thought and reason encompass the functional role of cultural inference and its causal effects. The functional performance of cultural tasks as routine serves as a causal mechanism for the emergence of cultural patterns that are fundamental to cultural inference and its causal–functional significance. The systematic study of cultural intelligence is fundamental to the understanding of the causal relation of cultural intelligence as a higher-level function.

The conceptual foundation of cultural intelligence is fundamental to the characterization of the processes and mechanisms of cultural intelligence and its functional significance. The characterization of the core processes and mechanisms of cultural intelligence illuminates on the functional significance of the cultural patterns of thought and reason that comprise cultural inference. The cultural patterns of thought and reason serve as a functional basis of the cultural inference that is essential to higher-level function. The systematic study of cultural intelligence is essential to the understanding of the fundamental elements of cultural inference and reason as a basis of the thought and rationality of cultural mental life.

Methodological Foundations of Cultural Intelligence

Methodological approaches to the study of cultural intelligence include multilevel approaches to the observation and measurement of cultural intelligence and its functional significance. The functional operationalization of cultural intelligence with multilevel methods contributes to the systematic investigation of cultural intelligence and its causal effects. The multilevel approach to the study of cultural intelligence includes behavioral and neuroscience methods to systematically investigate the functional processes that are fundamental to cultural intelligence. Multilevel methods are fundamental to the characterization of the functional processes of cultural intelligence and its causation. The development of methodological approaches contributes to the understanding of cultural intelligence and its physical implementation.

Behavioral methods are fundamental to the observation and measurement of the scientific constructs that are foundational to cultural intelligence. The use of behavioral methods demonstrates the functional basis of behavioral measures as a functional operationalization of cultural intelligence. The functional operationalization of cultural intelligence facilitates the use of behavioral surveys and functional tasks as observational methods of behavioral performance. The design of behavioral surveys and functional tasks is fundamental to the functional operationalization of cultural intelligence. Behavioral methods show how core processes of mental function are central to the observational measurement of cultural intelligence and its functional and behavioral outcomes.

Behavioral surveys are fundamental to the observation and measurement of behavioral responses based on self-report. The use of behavioral surveys to study cultural intelligence is of importance to assess the validity and reliability of the functional operationalization of specific constructs. The functional operationalization of specific constructs facilitates the characterization of the functional components of cultural intelligence and its causal effects. The reliance on behavioral surveys shows the importance of self-report on the assessment of the functional components of cultural intelligence.

Empirical paradigms with functional tasks further illustrate a methodological approach to the functional operationalization of cultural intelligence. Cultural intelligence as a construct may be assessed as the functional performance on cultural tasks. The design of functional tasks facilitates the functional use of cultural task activity as the functional basis of mental operation. The functional performance on cultural tasks shows a methodological approach to assess the levels of processing to infer on the functional components of cultural intelligence. The functional use of cultural tasks is also beneficial to determine the levels of complexity of mental operations that have functional significance for cultural intelligence.

Neuroscience methods are fundamental to the characterization of the functional mechanisms that are of importance to cultural intelligence. The use of neuroscience methods contributes to the understanding of the functional mechanisms that are fundamental to cultural patterns of thought and reason that are central to cultural intelligence. Neuroscience methods facilitate the understanding of the physical instantiation of cultural patterns of thought as a functional basis of cultural intelligence and its functional significance. The methodological approaches of neuroscience contribute to the characterization of the functional relations of cultural patterns of thought and cultural intelligence indices. The development of neuroscience methods facilitates the systematic study of the functional relations of cultural patterns of thought based on functional task performance.

Neuroscience methods such as functional magnetic resonance imaging provide an indirect means for the observation and measurement of cultural patterns of thought and behavior. The use of functional neuroimaging methods contributes to the methodological approaches that are fundamental to cultural patterns of thought and its underlying neural mechanisms. The methods of functional neuroimaging show the functional brain activity of core brain regions based on cultural task performance and its causal relation to functional and behavioral outcomes. The development of functional neuroimaging paradigms is essential to the identification of core brain regions that comprise the brain basis of cultural intelligence.

The development of functional neuroimaging paradigms is also beneficial to the characterization of the functional significance of cultural intelligence. The characterization of the causal relations of core brain regions as a predictor of functional and behavioral outcomes is of importance to the understanding of the causal–functional significance of cultural intelligence. The systematic study of the cultural patterns of functional brain activity contributes to the understanding of the functional role of cortical brain mechanisms as a predictor of functional and behavioral outcomes. The causal relations of functional brain activity based on cultural task activity and functional and behavioral outcomes are fundamental to the understanding of the cultural predictors of cultural brain activity.

The design of functional neuroimaging paradigms is also beneficial to the systematic study of the intrinsic properties of cultural brain activity and its functional significance. The functional processes of cultural brain activity based on the resting-state function show the intrinsic properties of functional brain activation. The functional processing of the cultural brain illustrates the intrinsic activity of cortical brain mechanisms that are of importance to inherent cultural function. The study of the intrinsic processing of the cultural brain contributes to the understanding of the inherent cultural function of functional brain activity and its causal effects.

The study of the brain basis of cultural intelligence is fundamental to the understanding of the processes and mechanisms of cultural capacities. The cultural patterns of functional brain activity show the functional role of cortical mechanisms as a mechanistic basis of cultural processing. The cultural processing of cortical mechanisms depicts the cultural patterns of functional brain activity that are linked to mental function and behavior. The identification of core brain regions is of importance to the functional processes of cultural intelligence. The characterization of the causal relations of functional brain activity and cultural intelligence is of further importance to the understanding of the causal–functional role of cultural brain function and its functional and behavioral outcomes.

Empirical Foundations of Cultural Intelligence

Empirical approaches on cultural intelligence systematically investigate the cultural patterns of thought and reason that comprise cultural mental life. The cultural patterns of thought and reason demonstrate the functional processes that are of importance to cultural capacities. The systematic study of the brain basis of cultural intelligence illustrates the functional processes and mechanisms that characterize cultural capacities. The identification of core processes that are fundamental to cultural intelligence illustrates the cultural patterns of thought and reason that are essential to the higher-level function of behavior. Empirical approaches on cultural intelligence show how programs of research are fundamental to the quality of evidence and data on culture and health.

Empirical research illustrates systematic programs of research that are fundamental to the functional processes and mechanisms that characterize cultural capacities. The core processes and functional mechanisms of cultural intelligence show the cultural patterns of thought that are fundamental to cultural inference and reason. The systematic investigation of the core processes and functional mechanisms of cultural intelligence demonstrates the functional architecture that is essential to cultural processing. The cultural processing of functional mechanisms shows the causal–functional relation of cultural capacities. The empirical research on cultural intelligence furthers the understanding of the functional significance of cultural intelligence and its causal effects.

The study of the brain basis of cultural intelligence is fundamental to the characterization of functional processes and mechanisms that comprise the inference and reason of cultural capacities. The identification of the core brain regions of cultural processing illustrates the functional basis of cultural intelligence. The characterization of the functional mechanisms of cultural capacities is essential to cultural brain function. Cultural inference and reason consist of the cultural patterns of thought that are essential to the understanding of inferential reasoning that is cultural. The study of the cultural patterns of functional brain activity that correspond with cultural thought illustrates the functional relation of the cortical and cultural processing of mental function. The study of the brain basis of cultural intelligence is essential to the understanding of the cultural processing of cortical brain function.

Cultural inference and reason comprise the emergent patterns of thought that illustrate the higher-level function of behavior. The study of cultural inference and reason facilitates the understanding of the functional and emergent properties that comprise higher-level processes. The cultural patterns of thought are fundamental to the cultural inference and reason of higher-level function and behavior. The emergent patterns of cultural thought show the functional role of cultural inference and reason as a functional basis of cultural causation. The cultural causation of inference and reason is fundamental to the cultural patterns of thought that comprise higher-level function and behavior.

Evidence-based approaches are fundamental to the building of quality of data and evidence on culture and health. The systematic study on cultural intelligence builds evidence-based resources that contribute to the scientific and societal enrichment of culture and health promotion. The promotion of culture and health benefits from the understanding of the benefits of cultural intelligence and its impact on cultural life patterns. The evidence-based approach to cultural intelligence facilitates the quality of data and evidence that is beneficial to the broadening of understanding on the scientific and societal enrichment of cultural and health promotion.

Evidence-based resources on cultural intelligence provide information and data that inform on the practical benefits of culture and health promotion. Evidence-based research on cultural intelligence demonstrates the systematic approaches to the study of cultural brain function. The quality of data and evidence on cultural brain function is essential to the development of evidence-based resources on culture and health. The building of scientific infrastructure and research capacity for the study on cultural intelligence facilitates the understanding of the functional architecture of the cultural mind and brain that is fundamental to cultural capacities.

Policy-based approaches to cultural intelligence focus on the scientific and societal benefits of culture and health promotion. The scientific and societal enrichment on cultural intelligence promotes the programs and policies that advance the understanding of the benefits of culture and health promotion. The programs and policies of culture and health promotion contribute to the design of policy-based programs that broaden awareness of the practical benefits of cultural intelligence as a scientific and societal resource. The development of programs and policies contributes to the policy-based resources on cultural intelligence that contributes to the information and data on culture and health.

The Brain Basis of Cultural Intelligence

The study of the brain basis of cultural intelligence provides a characterization of the core brain regions that serve as a functional basis of cultural inference and reason. The systematic investigation of the brain basis of cultural intelligence shows the functional processes and mechanisms that are central to cultural capacities. The characterization of the functional mechanisms of cultural brain function is essential to the understanding of the cultural patterns of cortical processing. The cultural patterns of cortical brain function show the levels of processing that correspond with cultural inference and reason. The systematic investigation of cortical brain function during the resting state illustrates the functional processes that comprise inherent cultural function. The study of the brain basis of cultural intelligence is fundamental to the understanding of the mechanistic basis of cultural inference and reason.

The identification of the core brain regions of cultural brain function is essential to the characterization of the functional mechanisms of cultural processing. The core brain regions of cultural brain function demonstrate the mechanistic basis of cortical and cultural processing. Core brain regions located within frontotemporal, parietal lobules and interconnected brain regions show cultural patterns of functional brain activity that are linked to the functional processing of abstract conceptual knowledge. The cultural patterns of functional brain activity located within core brain regions illustrate the functional role of cortical mechanisms in cultural processing. The functional role of core brain regions as a mechanistic basis of cultural brain function depicts the cortical basis of cultural processing. The core brain regions of cultural brain function serve as a functional basis of cultural processing.

Core brain regions responsible for cultural brain function demonstrate the functional processes that are essential to cultural capacities. Core brain regions such as the frontotemporal and parietal lobes comprise interconnected brain areas that are responsible for abstract conceptual knowledge. The functional basis of abstract conceptual knowledge contributes to the higher-level function of cultural processing. The functional processes of the abstraction of cultural inference and reasoning are essential to the emergent patterns of cultural thought that comprise cultural capacities. The functional capacities for the control and regulation of cultural processing contribute to the abstract conceptual knowledge of cultural capacities. The core brain regions of cultural brain function demonstrate the causal–functional role of functional machinery for cultural processing.

The characterization of the functional mechanisms of cultural brain function demonstrates the functional role of core brain regions and its causal–functional relation on culture and behavior. Multilevel mechanisms of cultural brain function demonstrate the functional correspondence of cortical and cultural processing. The functional mechanisms of cultural brain function depict the cultural patterns of cortical brain activity that coincide with cultural task activity. The functional relation of cortical and cultural processing shows how cultural brain function is essential to the functional processing of cultural capacities. The multilevel mechanisms of cultural brain function illustrate the functional correspondence of cultural processing and its underlying cortical mechanisms.

The functional correspondence of cultural brain function and functional outcomes illustrates the functional significance of cortical and cultural processing. The cultural patterns of cortical brain function show the functional processing of task activity that is specific to culture. The functional correspondence of cultural brain function and functional outcomes depicts the cultural patterns of cortical brain activity that predict functional outcomes. The functional processing of cultural task activity elicits cortical processing that is specific to the functional performance of culture. The cultural patterns of cortical brain function detail the functional processing that is essential to the functional performance of cultural tasks.

The cultural processing of brain function illustrates functional patterns of cortical brain activity that correspond with the levels of complexity of cultural thought. The functional patterns of cortical brain activity correspond with the cultural mental function that is illustrative of culture and complexity. The spontaneous fluctuations of cortical brain activity and cultural task activity depict the cultural patterns of brain function that are concordant with the complexity of cultural thought. The coincidence of spontaneous fluctuations of physiological and mental activity that is cultural illustrates the importance of cortical and cultural processing as a functional basis of the culture and complexity of thought.

The study of the brain basis of cultural intelligence entails the systematic investigation of the intrinsic properties of the cultural brain function and their functional significance. The intrinsic properties of the cultural brain function illuminate on the functional processing that is linked to inherent cultural function. The cultural brain function of intrinsic processing shows the patterns of functional brain activity that are based on the inherent mental function that is cultural. The intrinsic properties of the cultural brain function demonstrate the cultural patterns that are essential to the inherent cultural function and its causal effects.

Cultural Machine Intelligence

Research on cultural machine intelligence investigates the functional capabilities of computing machines for the functional performance of culture. From simple to complex devices, the computation of machines illustrates the levels of capabilities that comprise the computational performance of cultural processing. The cultural brain as a biological computing machine illustrates the feasibility of the physical implementation of a mechanistic engine that is capable of cultural performance. The cultural computer as a digital computing machine demonstrates the plausibility of the computer as a physical realizer of culture that is capable of the functional capabilities of cultural computation. Research on cultural machine intelligence explores the physical implementation of cultural computation of minds and machines and its functional equivalence. The functional significance of cultural computation as a mechanistic basis of cultural performance entails the functionality and adaptability of minds and machines that are capable of cultural machine intelligence.

Cultural machine intelligence encompasses the functional capabilities of computing machines that show the functional performance of cultural computation. The functional capabilities of computing machines illustrate the internal architecture of machine computation that demonstrates the functional performance of cultural processing. The internal components of cultural computation detail the functional performance of cultural computing. The internal states of functional components entail the functional operations that are essential to cultural performance. The internal states of computation depict the physical state and physical state transitions that are fundamental to cultural computation.

The internal architecture of cultural machines entails the sets of physical states that comprise cultural performance. The physical states of cultural machines illustrate the automated production of cultural states based on functional operations or algorithmic function that is cultural. The functional use of cultural algorithms as the internal operations of physical states and physical state transitions serves as the mechanical states of automated cultural performance. The mechanical states of automated cultural performance show the cultural states that comprise the algorithmic function of functional operation. The mechanical states of cultural machines serve as a functional operation of the automated production of cultural performance.

Cultural machine intelligence is comprised of the functional capabilities that comprise the internal states of cultural performance. The functional capabilities of internal components detail the functional states that are fundamental to the transitions of cultural input–output relations. The functional operations of initial to final cultural states depict the mechanical states and their transitions. The functional use of internal components as a mechanical basis of functional operations describes the computational basis of cultural performance. The functional capabilities of internal components detail the functional relations of cultural states that are of importance to the functional states of cultural performance.

The internal states of functional operations describe the cultural representation of mechanical states and their transitions. The functional operations of internal components consist of the cultural content of internal representations of mechanical states. The cultural representation of mechanical states comprises the cultural states that perform functional operations such as algorithmic computation. The cultural representation of internal states details the functional operations of internal components that facilitate the computation of algorithmic function. The functional operations of internal components illustrate the functional role of cultural representation as an internal state of machine computation that is cultural.

The functional operation of algorithmic function that is cultural serves as a basis of cultural machine computation. Cultural heuristics describe the algorithmic function of functional states that is cultural. The use of cultural heuristics encompasses the functional states that are of importance to the problem-solving of cultural tasks. The functional basis of cultural heuristics entails the use of cultural-based algorithms as the functional states of problem-solving. The internal states of cultural heuristics consist of the internal operations of algorithmic function that comprise the functional states of problem-solving. Cultural heuristics illustrate the differential use of functional processing as a computational basis of algorithmic function. The use of cultural heuristics underscores the functional importance of cultural algorithms as the internal states of functional operations.

The integral use of cultural heuristics serves as a functional basis of cultural machine intelligence. The design of computing machines that perform functional operations based on cultural heuristics illustrates a high-level functional performance of cultural computation. The functional operations of cultural heuristics encompass a high-level performance that illustrates the levels of capabilities of cultural machines. The use of cultural heuristics demonstrates the high-level performance and real-world advantages of internal operations that are inherently cultural. The functional performance of cultural computation encompasses cultural states that comprise functional operations that depict the high-level performance of problem-solving. The reliance on cultural heuristics as algorithmic function serves as a functional basis of cultural machine intelligence.

The use of cultural heuristics as the functional basis of high-level performance demonstrates the real-world advantages of cultural computation. Cultural heuristics describe the use of algorithmic function that is cultural with a high-level performance of problem-solving. The performance of cultural computation with cultural heuristics shows how functional capabilities facilitate a high level of efficiency in cultural computing. The demonstration of the functional performance of cultural machines and devices with cultural heuristics describes the functional compatibility of internal components that show the efficiency and reliability of high-level computing performance. The integral use of cultural heuristics as the algorithmic function of high-level performance facilitates the demonstration of the real-world advantages of cultural computation.

Cultural machine intelligence also relies on the functional capabilities of computing machines for internal programming. The functional capabilities of computing machines describe the internal operations of functional components that demonstrate the capabilities of computing machines. The capabilities of computing machines for the high-level performance of cultural computation show how internal components demonstrate the functional states of computation that are autonomous. The functional capabilities of computing machines for internal programming show the functional purpose of the internal operations of functional components. The autonomous computation of the functional states of computing machines illustrates the levels of capabilities of cultural machine intelligence.

The internal programming of computing machines shows the internal operations that comprise cultural machine intelligence. The functional operations of cultural computation encompass the cultural states of internal components that perform internal programming. The functional operations of cultural states as functional states illustrate computing states of internal programming that facilitates the recursion of internal function. The computation of cultural states as functional states and other states demonstrates the recursion of internal operation that comprises the high-level performance of cultural computing. The capabilities of computing machines to perform internal operations that entail cultural states as functional states illustrate the high-level performance of internal programming. The internal programming of cultural states of computing machines shows the functional basis of computation that is autonomous with real-world advantages. The internal programming of machine computation illustrates the functional use of internal operations as a functional basis of cultural machine intelligence.

The computing states of internal programming rely on functional operations to describe the high-level performance of cultural computation. The higher-level processing of functional operations depicts the internal states that encompass the complexity of functional capabilities. The complexity of functional operations illustrates the internal states that facilitate high-level computing performance. The functional operations of internal states are fundamental to the complexity of functional capabilities that are cultural. The higher-level processing of functional operations illustrates the functional capabilities of computing machines that are essential to cultural computation.

Culture and Complex Systems

Culture and complex systems describe the levels of functional performance of multilevel systems. Cultural systems consist of the functional characteristics that entail the cultural processes and mechanisms of fundamental dimensions; complex systems are comprised of fundamental components that facilitate the functionality and adaptability of performance. Culture and complex systems abound with the fundamental components of high-level processes that are fundamental to multilevel systems. The understanding of the causation of culture and complex systems is fundamental to the characterization of cultural computation.

Cultural thought describes the cultural patterns of inference and reason that are of importance to cultural computation. The cultural patterns of inference and reason are fundamental to the complexity of cultural thought. Cultural inference and reason as functional processes describe the levels of complexity that serve as a functional basis of cultural thought. The complexity of cultural thought encompasses the cultural patterns of inference and reason that show the levels of processing that comprise high-level function. Cultural thought consists of the inference and reason that are fundamental to the high-level function of culture and behavior.

The levels of complexity of cultural thought describe the cultural inference and reason that are of importance to cultural capacities. Cultural thought consists of cultural patterns that demonstrate the complexity of cultural inference and reason. Cultural patterns of inference and reason comprise the higher-level processing that is fundamental to the complexity of cultural thought. The levels of complexity of cultural patterns of thought detail the patterns of inference and reason that are cultural. The functional role of cultural thought as a causal impetus of the complexity of cultural patterns implies the causal significance of cultural complexity as a functional adaptation. The complexity of cultural thought illuminates on the causal–functional role of the adaptation of cultural capacities.

Cultural Computation of Real and Artificial Systems

The cultural computation of real and artificial systems illuminates on the multiple physical realizability of the functional performance of culture. The characterization of cultural patterns and regularities as fundamental laws and principles illustrates the importance of culture in the causal structure of the natural world. From biological organisms to complex devices, the cultural computation of computing machines shows the physical realization of cultural states as sets of physical states that encompass the functional performance of culture and its causal effects. The functional capabilities of computing machines demonstrate the physical states and physical state transitions that comprise the computational performance of culture. Computing machines that perform the computation of cultural states show the functional capabilities that facilitate the physical implementation of cultural computation. The cultural computation of computing machines entails multiple physical realizers that demonstrate the functional equivalence of cultural performance of minds and machines.

The cultural computation of biological organisms entails the functional machinery that comprises cultural capacities. The cultural computation of biological organisms details the inherent cultural processing that demonstrates the preparedness and adaptiveness of biological organisms for cultural adaptation. The inherent cultural function of adaptive machinery illustrates the causal mechanisms that serve as a functional basis of cultural processes. The causal mechanisms of cultural computation show the functional processes that comprise the functional performance of culture. The computational basis of cultural capacities demonstrates the cultural patterns and regularities of functional mechanisms and their causal effects. The functional machinery of cultural capacities comprises the biological mechanisms of cultural computation.

The characterization of the biological mechanisms of cultural computation illustrates the physical instantiation of cultural states as brain states. The functional processing of cultural capacities shows the mechanistic basis of cultural brain function and its causal–functional role. The brain basis of cultural capacities details the functional mechanisms that have causal–functional significance on functional performance. The cultural computation of cortical brain function depicts the cultural processing that is essential to functional adaptation. The cultural patterns of cortical brain function depict the functional mechanisms that demonstrate cultural processing. The biological mechanisms of cultural computation entail the functional states of cultural brain function.

The causal interaction of cultural brain function depicts the multitude of functional relations of cultural states as brain states. The physical instantiation of cultural states as brain states illustrates the functional role of cultural processing and its causal effects. The causal interaction of cultural and brain states arises functional states that contribute to functional and behavioral performance. The functional role of cultural processing as a causal impetus for functional and behavioral performance illustrates the functional significance of cultural states. The interaction of cultural and brain states facilitates the causal influence of functional states and their relation to cultural performance. The causal interaction of cultural brain function is fundamental to the functional relations of cultural and brain states.

The physical realization of cultural and brain states demonstrates the inherent naturalism of the cultural brain as a biological computing machine. The mechanistic basis of cultural brain function illustrates the inherent cultural function of the cultural brain as a biological computing machine. The naturalistic realism of the cultural brain as a biological computing machine shows the functional processes that enact causal effects on cultural computation. The computation of cultural brain function details the physical states that serve as a mechanistic basis of cultural capacities. The cultural computation of the biological computing machine shows the inherent cultural function of biological mechanisms that are essential to cultural adaptation. The physical realization of functional states of the cultural brain shows the feasibility of cultural computation as a naturalistic mode of cultural brain function.

The cultural computation of computing machines illustrates the functional performance of machine computation that is cultural. The design of simple to complex devices as computing machines demonstrates the functional capabilities of cultural machine computation. The development of devices and computers that have functional capabilities for cultural performance shows the internal components of computing machines. The internal architecture of computing machines consists of the internal components that demonstrate the functional capabilities of importance to cultural computation. The internal states of computing machines that comprise cultural states depict the functional role of cultural input–output relations that is fundamental to cultural computation. The functional performance of computing machines consists of the internal states of cultural computation that are fundamental to the functional capabilities of cultural machines. The functional performance of computing machines shows the plausibility of devices and computers that perform cultural computation.

The functional capabilities of computing machines that demonstrate cultural computation show the functional equivalence of the cultural performance of minds and machines. The levels of capabilities of computing machines demonstrate the functionality and adaptability of cultural machines. The design and construction of devices and computers that show cultural performance are essential to the demonstration of the computation of cultural machines. Cultural computers as digital computing machines show the plausibility of the physical implementation of cultural computation. Cultural performance entails the levels of capabilities that comprise the programming of internal states and algorithmic function that is fundamental to cultural computation. The cultural performance of machines illustrates the functional capabilities that facilitate the automated production of the computation of cultural devices and machines. Cultural machine computation is of importance to the functional equivalence of the cultural computation of minds and machines in the world.

The design of cultural devices and computers with functional capabilities entails the automated production of cultural performance. Cultural devices and computers – cultural machines – show the levels of capabilities of internal components that facilitate the automated production of cultural content. The cultural content of machine computation as internal states of the computing machine shows the functional operations that comprise the functional performance of cultural computation. The cultural content of internal states entails the cultural states that are in causal relation with other functional states. The functional performance of cultural computation shows the levels of capabilities that are essential to the automated production of cultural content.

The internal components of cultural machines facilitate the functional performance that shows the programming function of cultural devices and computers. The internal components of cultural devices and computers illustrate the levels of capabilities of programming function that is essential to the functional operations of cultural computers. The functional capabilities for internal programming show the functional components that are fundamental to cultural performance. The demonstration of levels of capabilities of internal programming illustrates the functional performance of higher-level processes. Cultural computers that show functional capabilities of internal programming manifest a functional performance of cultural computation that entails the prediction and performance of algorithmic function that is cultural. The functional performance of cultural devices and computers shows the functional capabilities of cultural performance.

The functional equivalence of the cultural computation of minds and machines illustrates the functional performance of computing machines that is comparable across multiple physical realizers. The functional capabilities of computing machines for cultural performance that is comparable demonstrate the internal architecture and functional operations that are consistent based on the physical implementation of cultural states. The physical implementation of cultural machines built on the internal architecture of computing machines shows the importance of the functional capabilities of machine computation. The functional performance of cultural machine computation shows that the functional equivalence of minds and machines is essential to the plausibility of the multiple physical realizers of culture.

The cultural computation of real and artificial systems is fundamental to the understanding of the computation of cultural minds and cultural machines. From the cultural brain to cultural computer, the cultural computation of the computing machine demonstrates the functional capabilities of mechanistic engines for the functional performance of culture. The cultural brain as a biological computing machine illustrates the natural realism of cultural performance that is essential to cultural life patterns. The cultural computer depicts the computation of cultural states that manifest the functional operations of cultural performance. The cultural computation of real and artificial systems shows the levels of capabilities of complex systems for the manifestation of culture.

Practical Applications

The practical application of scientific research on cultural intelligence is foundational to the design of computational tools and technologies that have functional capabilities for cultural performance. The design of computational tools and technologies is beneficial to the physical implementation of simple to complex devices that show the functional capabilities of computing machines for cultural computation. From simple to complex devices, computing machines demonstrate the internal components and functional capabilities of cultural performance. The scientific and technological innovation of computational tools and technologies illustrates the practical application of research and development on the cultural intelligence of minds and machines for the benefit of the functional capabilities of scientific and technological innovation.

Computational tools and technologies are beneficial to the study of the core brain regions of cultural brain function. The development of computational techniques for the study of cultural brain function contributes to the methodological approaches for the study of the functional mechanisms of cultural processing. Computational techniques are of importance to the prediction and measurement of the causal interaction of core brain regions that comprise cultural brain function. The use of computational techniques for the study of cultural brain function is beneficial to the characterization of the multilevel mechanisms of cultural processing.

The design of devices and computers that show the functional capabilities of cultural machine intelligence illustrates the internal components that are essential to the functional performance of cultural machine computation. The development of internal components of devices and computers that show the levels of capabilities of cultural machine intelligence demonstrates the functional performance to perform the computational operations of internal components that are compatible with cultural machine intelligence. The integration of the internal components of cultural intelligence ("cultural algorithm") with functional capabilities demonstrates the levels of capabilities of devices and computers that perform cultural computation. From heuristic searches to knowledge-based platforms, the design of devices and computers that show the functional capabilities of cultural intelligence demonstrates the functionality and adaptability of computational tools and technologies for cultural machine computation.

Practical applications of computational tools and technologies demonstrate the feasibility of the design of devices and computers that show the functional capabilities of cultural computation. The functional capabilities of devices and computers that perform cultural tasks illustrate the level of capabilities that are fundamental to the computation of cultural performance. The scientific and technological innovation of computational tools and technologies details the functional performance of cultural intelligence. The development of devices and computers that perform cultural computing is fundamental to the demonstration of the functional capabilities of cultural intelligence.

Discussion

The scholarly and scientific interest in cultural intelligence encompasses a wide realm of interests on the inference and rationality of culture. The functional role of thought and inference as a rational basis of culture illuminates on the function and adaptation of culture and behavior. The cultural patterns of thought and rationality comprise a functional basis of the inference and reason on culture. The notion of cultural inference connotates othe functional processes and mechanisms of the cultural mind that facilitate cultural capacities. The functional basis of the culture of minds and machines is fundamental to cultural performance. The inference and rationality of culture are essential to the inferential reasoning on the mental states of culture.

The systematic study of cultural intelligence is essential to the understanding of the fundamental elements of cultural intelligence and its functional significance. The fundamental elements of cultural intelligence illuminate on the causal–functional role of cultural inference as a functional basis of thought and reason of diverse cultures. The manifestation of cultural inference as mental states illustrates the functional significance of the thought and reason of cultural capacities. Cultural inference demonstrates the core processes that comprise the thought and reason of cultural capacities. The thought and reason on cultural patterns are central to the cultural causation of mental inference.

The exploration of cultural intelligence illustrates the fundamentals of the thought and reason of diverse cultures. The cultural patterns of thought and reason that hold illuminate on the functional significance of cultural inference as a higher-level function. Cultural thought and reason show the cultural patterns of thought that are essential to higher-level processes. The cultural patterns of thought facilitate the functional role of inference and reason as a functional basis of adaptation and behavior. The higher-level processes of cultural inference and reason illustrate the functional significance of cultural patterns of thought as functional components of cultural capacities.

The humanistic and scientific understanding of cultural intelligence is fundamental to the social and educational enrichment on the societal significance of cultural mental life. The core capacities of cultural intelligence enrich cultural mental life with moral benefaction and recognition of the causal power of cultural causation on mental life. The social and educational enrichment on the fundamentals of cultural intelligence emboldens the understanding of the potency of culture as a causal impetus of the mental causation of culture. The building of humanistic and scientific understanding of cultural intelligence contributes to the broadening of understanding of the importance of cultural inference as a fundamental basis of moral benefaction.

Conclusion

The study of cultural intelligence encompasses the cultural thought and rationality of cultural capacities. Research on cultural intelligence investigates the functional processes and mechanisms that underlie cultural capacities. The study of cultural intelligence is of importance to cultural inference and reason that characterizes the higher-level function of culture and behavior. The systematic investigation of the brain basis of cultural intelligence contributes to the understanding of cultural brain function. The study of cultural intelligence is essential to the understanding of the complexity of cultural thought and reason. The systematic investigation of cultural intelligence provides a foundation for the understanding of the inherent cultural function of the cultural brain.

The brain basis of cultural intelligence illustrates the function and adaptation of cortical brain function for cultural inference. The functional adaptation of the cultural brain demonstrates the preparedness and adaptiveness of cortical brain function for cultural inference. The functional processes of cultural brain function show the functional and emergent properties that comprise cultural inference and reason. The functional and emergent properties of the cortical brain illustrate the levels of processing that characterize cultural brain function. The characterization of the core brain regions of cultural intelligence shows the functional machinery that underlies the cortical brain function of cultural inference and reason.

Research on cultural intelligence illuminates the functional basis of cultural inference and reason. The systematic study of cultural intelligence broadens scientific and societal enrichment on the benefits of cultural intelligence and its societal impact. The building of evidence-based resources on cultural intelligence contributes to the scientific and societal impact of culture and health promotion. Evidence-based resources on cultural intelligence are essential to the information and data that inform on the practical benefits of culture and health promotion. The study of cultural intelligence illuminates on the practical impact of scientific and societal enrichment for the benefit of culture and health promotion.

Implications

The study of cultural intelligence has practical implications for health, medicine, public policy and related fields of study. Evidence-based research on cultural intelligence contributes to scientific and societal resources that inform on the practice and policy of culture and health promotion. The evidence-based resources on cultural intelligence facilitate the understanding of the promotion of culture and health. The building of evidence-based research serves as a foundation of the quality of data and information on the scientific and societal benefits of cultural intelligence and its practical applications to cultural and health promotion. The development of scientific infrastructure and research capacity building for evidence-based research on culture and health promotion is beneficial to equitable societal and health outcomes.

The evidence-based resources on cultural intelligence facilitate the innovation of culture and health promotion in health and medicine. Evidence-based research on cultural intelligence generates novel knowledge of the scientific and societal benefits of cultural and health. The building of support for programs and policies of cultural and health promotion facilitates the innovation of strategies for the building of awareness of the importance of culture and health. The promotion of programs and policies that promote culture and health is fundamental to the societal benefits of cultural enrichment. Evidence-based resources on cultural intelligence are fundamental to culture and health promotion.

Policy-based research on cultural intelligence is foundational to the understanding of the policy impact of programs and policies on culture and health promotion. The policy-based resources on culture and health provide information that supports the building of awareness on the equitable standards that promote culture and health equity. Policy-based research is essential to the discernment of the policy impact of cultural and health promotion programs that build awareness of the scientific and social enrichment on culture and health. The building of public awareness on culture and health broadens the societal impact of cultural intelligence and its benefits for culture and health equity.

12
COMPUTATIONAL CULTURAL NEUROSCIENCE AND PUBLIC POLICY

Introduction

Culture and health promotion are foundational to the promotion of issues and stances that impact the interests and concerns of culture, health and society. The development of strategies for culture and health promotion is beneficial to the consideration of the interests and concerns of the impact of scientific and technological innovation on societal and health outcomes. The understanding of a wide range of thematic issues on culture and health promotion is of importance to the advocacy and public health action on culture and health. The building of public health action on culture and health promotion advances the public and societal understanding of the importance of equitable access to resources that impact societal and health outcomes. The equitable access to informational resources on culture and health promotion contributes to the attainment of health equity.

Translational research on computational cultural neuroscience informs on the practice and policy of culture and health promotion. Evidence-based research on computational cultural neuroscience builds scientific knowledge that leads to the development of scientific and educational resources on culture and health. Integrative approaches to research on computational cultural neuroscience broaden societal and public understanding of the tools and technologies that benefit culture and health promotion. The building of scientific knowledge on computational cultural neuroscience leads to the advancement of the evidence base of knowledge for the benefit of the promotion of culture and health.

DOI: 10.4324/9781003384236-13

Policy development and implementation on culture, technology and health is fundamental to the advancement of culture and health promotion. Policy development contributes to the considerations of scientific and technological innovation for the advancement of culture and health promotion and its societal and public impact. Policy considerations on culture and health promotion detail the understanding of the societal and public health impact of scientific and technological innovation. Scientific and technological innovation advances the knowledge base that informs on the societal and public understanding of culture and health promotion. The implementation of policies and programs on culture, technology and health is of importance to the societal resources that inform on the promotion of culture and health. The development and implementation of public policy is beneficial to the broadening of societal and public understanding of culture, technology and health for the benefit of the promotion of culture and health.

The goal of the chapter on computational cultural neuroscience and public policy is to provide a comprehensive review on translational research and its impact on the advocacy and policymaking on culture, technology and health. Translational research on computational cultural neuroscience advances integrative approaches to the understanding of the computational foundations of the cultural brain. The evidence-based research on computational cultural neuroscience informs the practice and policy of culture and health. The building of evidence-based resources broadens scientific and societal resources that benefit the advocacy and policymaking for the promotion of culture and health. The policy development and implementation on culture, technology and health contribute to the policy-based approaches to the scientific and educational enrichment of culture and health promotion.

Translational Research on Computational Cultural Neuroscience

Research on computational cultural neuroscience investigates the computational foundations of cultural neuroscience. The study of computational cultural neuroscience is foundational to the understanding of the neurocomputational study of culture and behavior. The systematic investigation of computational cultural neuroscience contributes to the development of practical applications such as computational tools and technologies that are foundational to the attainment of societal quality and health equity. The computational modeling of the cultural brain contributes to the understanding of the causal mechanisms that affect the functional machinery of behavioral and cultural adaptation. The understanding of the computational machinery of the cultural brain is essential to the scientific and educational enrichment that leads to the promotion of culture and health.

The neurocomputational study of culture and behavior builds the evidence base of knowledge on the causal mechanisms that affect health outcomes. Research on the neurocomputational study of culture and behavior contributes to the evidence base of information and data that are fundamental to the understanding of the causal mechanisms of health outcomes. The study of the understanding of the causal mechanisms of health outcomes contributes to the prevention and treatment of complex disease. The systematic study of cultural neurocomputation contributes to the building of the evidence base of information and data that are beneficial to the understanding of the processes and mechanisms of culture that lead to the improvement of health outcomes.

The development of computational tools and technologies is of importance to the availability and use of technological innovation that advances the improvement of societal and health outcomes. Computational tools and technologies contribute to the research and system capabilities that lead to the advancement of benefits of technology in culture and health. The use and availability of computational tools and technologies that are culturally and ethnically appropriate are beneficial to the development of research and system capabilities in health and medicine. The scientific and societal enrichment on computational tools and technologies is of importance to the promotion of culture and health. The availability of research and system capabilities in health and medicine contributes to the planning and programs that advance the resources for scientific and societal enrichment on culture, technology and health.

Research and system capabilities in health and medicine provide a wide range of resources that are beneficial to the prevention and promotion of culture and health. The research capabilities of health and medicine provide opportunities to build scientific infrastructure that is appropriate to meet the public health needs of the population. The development of scientific infrastructure advances the opportunities for research and training on computational cultural neuroscience and its practical applications on culture, technology and health. The building of scientific infrastructure contributes to the development of the societal resources that improve the use and availability of the information and data that are essential to culture and health promotion.

The development of computational tools and technologies is foundational to the availability and use of resources that lead to the improvement of societal and health outcomes. The use and availability of computational tools and technologies (e.g., knowledge platforms, digital technologies) contribute to the efficiency and efficacy of the strategies of prevention and intervention in health and medicine. The use of computational tools and technologies with research and system capabilities contribute to the advancement of the quality of data that improves health outcomes. The development of computational tools and technologies that show the capabilities for research with diverse cultures, populations and settings illustrates the adaptability and complexity of system capabilities of health and medicine. The use and availability of computational tools and technologies that are culturally and ethnically appropriate is essential to societal and health outcomes.

Computational tools and technologies that advance the quality of data and information to inform research and system capabilities are beneficial to diverse cultures, populations and settings. The building of scientific infrastructure for the discovery and delivery science that are culturally and ethnically appropriate improves the research and system capabilities for culture and health promotion. The discovery science of computational cultural neuroscience advances evidence-based approaches to the systematic study of the computational level of the cultural brain. The understanding of the computational foundations of the cultural brain is beneficial to the research and development of its practical applications. The understanding of the functional and mechanistic basis of the cultural brain is essential to discovery and delivery science in health and medicine.

The development of scientific and educational resources on computational cultural neuroscience is beneficial to the improvement of societal and health outcomes. Scientific and educational resources on computational cultural neuroscience contribute to the understanding of the societal and public impact of the discovery and delivery science in health and medicine. The public understanding of computational cultural neuroscience broadens the awareness and advocacy of culture, technology and health for the promotion of culture and health. The building of the scientific and educational resources to inform the public awareness and advocacy of culture, technology and health is fundamental to the broadening of the societal and public impact of culture and health promotion and its benefits.

The building of public awareness and advocacy of culture and health promotion is bolstered with the building of scientific and educational resources. The scientific and educational resources on computational cultural neuroscience to inform on the societal and public impact of computational tools and technologies that are culturally and ethnically appropriate are beneficial to societal equality and health equity. Computational informatics and data broaden the wide reach of information and data that are available to build public awareness on culture, technology and health. Computational tools and technologies facilitate the research capabilities of scientific and technological innovation for the use and availability of scientific and societal resources that are culturally and ethnically appropriate.

Research capabilities in health and medicine advance the research and development of computational tools and technologies that show effectiveness in the improvement of the societal and health outcomes that meet public health needs of the population. The integration of computational tools and technologies in system-wide approaches and empirical studies of its public health impact advances the understanding of computational informatics and data to inform on the strategies of prevention and intervention in culture and health promotion. Computational tools and technologies show the performance and capabilities of technology for the promotion of culture and health. Computational tools and technologies that are designed to show performance and adaptability as capabilities of technology contribute to the advancement of scientific and technological innovation.

Policy Framework on Computational Cultural Neuroscience

Research on the policy framework of computational cultural neuroscience advances the planning of considerations regarding the stances and issues that inform on culture, technology and health and its societal and public impact. The development of the policy framework on computational cultural neuroscience provides a systematic approach to the consideration of ideas that inform the development of policy stances across a myriad of policy areas on culture, technology and health for the societal and public health benefit of culture and health promotion (Figure 12.1). Scientific and

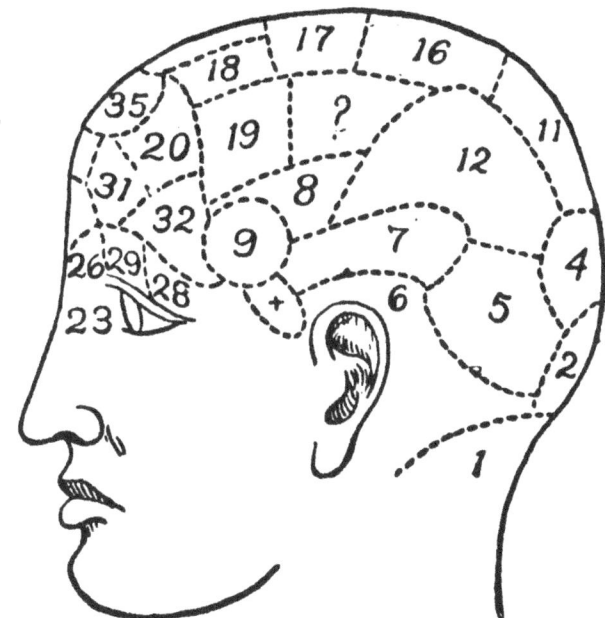

FIGURE 12.1   Policy Framework on Computational Cultural Neuroscience: (A) Translational Research on Computational Cultural Neuroscience; (B) Planning and Programming on Culture, Technology and Health; (C) Public Health Awareness and Advocacy on Culture, Technology and Health; (D) Policies and Public Health Action on Culture and Health Promotion

technological innovation is a central issue that is essential to the improvement of research and system capabilities. The advancement of scientific and technological innovation promotes research and development of its practical applications for the benefit of the discovery and delivery of research and system capabilities in health, medicine and related areas. The promotion of culture and health relies on the advancement of scientific and technological innovation for the improvement of societal and health outcomes.

The considerations regarding the societal and public health impact of culture, technology and health detail a myriad of stances and issues that affect the scientific and technological innovation on computational cultural neuroscience. The promotion of the planning and policies on culture, technology and health contributes to the building of system capabilities that advance the delivery of equitable societal and health outcomes for the promotion of culture and health across nations. The international guidance on the planning and policies of culture, technology and health advance the development of policies and programs that contribute to the promotion on culture and health. The international experts on computational cultural neuroscience are essential to advance the issues and stances of national and local legislation that impact the efficacy of the scientific and technological innovation on computational cultural neuroscience.

The international development of policy stances on culture, technology and health is beneficial to the planning and policies that are of importance to the promotion of culture and health. The international guidance on policy stances on culture, technology and health includes the cultural and societal considerations in the response to the issues of public health and its societal impact. Cultural and societal considerations consist of a myriad of issues that affect the quality of response to the improvement of societal and health outcomes of the population. Cultural and societal considerations (e.g., language, cultural diversity) comprise the specific issues that impact the efficacy of policy development and implementation on culture and health promotion. The understanding of cultural and societal considerations is fundamental to the development of policy stances and issues that improve societal outcomes.

Cultural and societal considerations affect the understanding of the development of programs and policies that have societal impact and public health outcomes. The considerations regarding the perception of the programs and policies of culture and health promotion and their societal impact are formidable. The development of scientific and informational resources is beneficial to inform on the efficacy of programs and policies that are beneficial to culture and health promotion. The use of scientific and informational resources to inform on the cultural and societal considerations of culture and health promotion is of importance to ensure the use and availability of policies and programs on culture, technology and health.

The design of prevention and intervention strategies on culture and health promotion benefits from the use of technological innovation in culture and health. The promotion of scientific and technological innovation in computational cultural neuroscience is beneficial to the design of prevention and intervention strategies and their societal and public health impact. The consideration of the strategies of prevention and intervention in health and medicine is essential to the improvement of societal and health outcomes across cultures, populations and settings. The understanding of the cultural and societal considerations on strategies of prevention and intervention is beneficial to ensure the ethical standards in health and medicine.

The consideration of the values, principles and vision that are of importance to policy development and implementation is essential. The development of the core values, principles and vision of programs and policies is foundational to the understanding of the priorities and plans that lead to the implementation of programs and policies. The priorities and plans of culture and health promotion are essential to the understanding of the vision of national policies and its goals of development. The development of plans and goals that benchmark the attainment of key priorities of policy development facilitates the implementation of the programs and policies of culture and health promotion that are efficacious. The determination of the core values, principles and vision of programs and policies on culture and health promotion contributes to the design of policies and programs that lead to societal equality and health equity.

The development of national, state and local legislation on culture and health promotion is fundamental to the implementation of programs and policies to the population. The consideration of the societal and public impact of program and policies on culture and health promotion affects the public support and societal influence of information and data that pertains to the public programs on culture and health. The building of public support and societal influence for programs on culture and health leads to the development of legislation that implements societal and informational resources for the improvement of societal and health outcomes. The understanding of the importance of the perception of public support and societal impact of policy development and planning is essential to the building of public planning and programs that lead to goal attainment.

The development and implementation of programs and policies on culture and health promotion benefit from the global policy environment. The global policy environment provides the societal resources that are foundational to the promotion of culture and global health. The programs and policies of culture and global health policy provide information and data that are beneficial to the prevention and intervention of health promotion of diverse cultures, populations and settings in global context. The strategies of prevention and intervention of culture and health promotion are beneficial to the understanding of the priorities and planning that are essential to the advocacy and policymaking of culture and health.

The advocacy and policymaking of culture and health promotion in the global policy environment benefit from the building of evidence-based resources that are representative of diverse cultures, populations and settings. Evidence-based resources on computational cultural neuroscience inform the advocacy of culture and health promotion in the global context. The building of evidence-based approaches on computational cultural neuroscience contributes to the quality of information and data to inform the advocacy of culture and health promotion. The scientific and societal resources on computational cultural neuroscience contribute to the building of public awareness of the research capabilities of computational tools and technologies in health and medicine. The public advocacy of culture and health promotion builds public support and societal influence that are beneficial to policymaking on culture, technology and health.

The development of policy stances on culture, technology and health is beneficial to policy implementation. The development of policy stances that promote culture and health is beneficial to the programs and policies that raise public awareness and advocacy of health equity. The public advocacy of policy stances that support culture and health promotion shows the societal and public benefits of culture, technology and health as well as their practical applications. The public support on culture, technology and health shows improvement in societal and health outcomes based on policy-level intervention on culture, technology and health. The main priority of policy stances in support of culture, technology and health is that the use and availability of programs and policies on culture, technology and health are efficacious to the amelioration of disease and the improvement societal and health outcomes.

The policy development and implementation on culture, technology and health lead to the control and regulation of the use and availability of technology for the goals of culture and health promotion. The development of policies and programs on culture and health promotion benefits from the priority areas of culture, technology and health. The innovation of science and technology in culture and health promotion ensures the equitable use and availability of scientific and societal resources to ameliorate public health concerns. The building of scientific and societal resources in the area of culture, technology and health contributes to the advocacy and policymaking of culture and health promotion that is foundational to health equity.

Computational Cultural Neuroscience and Public Policy

The study of computational cultural neuroscience contributes to the development and implementation of public policy. Research on computational cultural neuroscience informs on the evidence-based approaches that advance policy development and implementation. The building of research on computational cultural neuroscience contributes to the development of policy stances that are foundational to the development of policies and programs. The evidence base on computational cultural neuroscience is comprised of the informatics and data that inform societal influence and public interest at the intersection of culture, technology and health. The development and implementation of public policy benefit from the advancement of evidence-based approaches on culture and health.

The empirical study of computational cultural neuroscience builds evidence-based knowledge on culture, technology and health and their impact on society. Research on computational cultural neuroscience contributes to the evidence-based knowledge that is of importance to the innovation of science and technology on culture and health and its societal and public impact. The scientific and technological innovation on culture and health broadens evidence-based understanding of the computational approaches to culture and health. The development of computational tools and technologies for the empirical investigation on culture and health contributes to the capabilities of health and medicine for the development of cures, prevention and intervention of complex disease. The development of scientific and technological innovation on computational cultural neuroscience is foundational to the efficacy of prevention and intervention in health and medicine.

Research on computational cultural neuroscience builds the evidence base of information and resources that inform on computational approaches of culture and health. Evidence-based approaches on culture and health show the building of evidence base and scientific infrastructure in the development of scientific and technological innovation. The building of evidence-based resources and scientific infrastructure is fundamental to the development of computational tools and technologies of culture and health. The development of an evidence base of scientific and technical knowledge is fundamental to the understanding of the role of computation on the processes and mechanisms of culture and health. The development of scientific infrastructure broadens the wide reach of evidence-based resources that inform on culture and health.

The innovation of science and technology for the benefit of culture and health illustrates the wide realm of benefits of computational approaches to culture and health. The development of computational approaches on culture and health shows the importance of the innovation of the capabilities of health and medicine for scientific and public health impact. The innovation of medical system capabilities to provide computational tools and technologies that support the discovery of novel tools and techniques contributes to the development of computational technologies in health and medicine.

The development of computational tools and technologies benefits from scientific infrastructure that broadens the wide range of capabilities of the medical and health systems. The scientific infrastructure that contributes to the building of evidence-based approaches of computational cultural neuroscience is beneficial to ensure the development of scientific and technological innovation that is effective in its societal and public health impact. Understanding the scientific and public health impact of scientific and technological development is essential to the development and implementation of policy in the areas of health and medicine. The discovery of the scientific and public impact of scientific and technological innovation contributes to the building of medical and health system capabilities.

The scientific and technological innovation of computational tools and technologies contributes to the research capabilities of scientific infrastructures in health and medicine. The research capabilities of scientific infrastructures benefit from the development of computational tools and technologies that lead to the advancement of cures, preventions and interventions in health and medicine. The design of cures, preventions and interventions is of importance to the innovation of science and technology that is effective in the prevention and treatment of complex disease. The innovation of computational tools and technologies builds capabilities in health and medicine to advance the prevention and treatment of complex disease.

Policy-based approaches to computational cultural neuroscience detail the considerations on the use of policies and programs for the promotion of culture and health. The development of policy-based approaches considers the role of policies and programs on culture, technology and health that inform on culture and health promotion. Policy-based approaches on culture, technology and health are beneficial to inform on the strategies of prevention and intervention that are efficacious for the improvement of societal and health outcomes. The design of strategies of prevention and intervention contributes to the wide range of resources that are of importance to inform on the programs that advance culture and health promotion.

The policies and programs of culture and health promotion benefit from the consideration of the policy stances on culture, technology and health. Evidence-based research on computational cultural neuroscience shows that computational tools and technologies are a fundamental source of the innovation of scientific and technological advancement in the discovery and delivery of research capabilities in health and medicine. The computational informatics on cultural neuroscience advance the development of innovative tools and technologies that advance the quality of care and treatment of complex disease. The development of computational tools and technologies that are efficacious across culturally and ethnically diverse populations contributes to the equitable conditions that are fundamental to societal equality and health equity. Computational tools and technologies are beneficial to the scientific and technological innovation of culture and health.

Policy considerations on the areas of culture and health promotion are formidable. The policy areas of culture, technology and health are comprised of the issues with regard to their societal and public health impact. The capabilities in health and medicine to provide discovery and delivery on culture and health promotion are strengthened with the use of computational tools and technologies that advance the availability and use of research and medical capabilities for culturally and ethnically diverse populations. The use of policy frameworks on computational cultural neuroscience in the development of legislation on culture and health promotion is beneficial for the planning and implementation of policies and programs. Research on policy frameworks of computational cultural neuroscience is beneficial to the breadth of considerations that inform on the policy and programs on culture and health promotion.

The development of the scientific infrastructure on computational cultural neuroscience promotes the innovation of science and technology that is beneficial to the advancement of the research capabilities in health and medicine. The building of scientific resources that contribute to the design of the strategies of prevention and intervention is beneficial to inform on development of computational tools and technologies that are efficacious. The advancement of computational informatics contributes to the development of prevention and intervention strategies for the promotion of culture and health. The development of information and data of computational informatics ensures that scientific and societal resources are available to provide equitable societal and health outcomes.

Evidence-Based Approaches to Computational Cultural Neuroscience

Evidence-based approaches to computational cultural neuroscience contribute to the use and availability of the evidence base to inform on the advocacy and policymaking of culture and health promotion. Evidence-based approaches are comprised of the empirical data and information from scientific resources that show an evidence base to inform on the scientific knowledge that guides the priority setting on culture, technology and health in global context. The building of scientific resources on computational cultural neuroscience contributes to the systematic investigation of the computational approaches to cultural neuroscience. The scientific and technological innovation on computational cultural neuroscience advances the use and availability of computational tools and technologies for the promotion of culture and health.

The development of evidence-based resources on computational cultural neuroscience entails the use of computational tools and technologies that are culturally and ethnically appropriate. The design of computational tools and technologies that are appropriate for diverse cultures, populations and settings is foundational to the equitable conditions of health and medicine. Empirical studies with computational tools and technologies are of importance to ensure the understanding of its causal effects on societal and health outcomes. Systematic investigation of the public and societal impact of computational tools and technologies on societal outcomes is essential to the evidence-based approaches to the development of policies and programs of culture and health promotion. Computational tools and technologies that show effectiveness with culturally and ethnically diverse populations demonstrate the adaptability and complexity of system-wide approaches.

Evidence-based research on computational cultural neuroscience contributes to the development of an evidence base on computational approaches to cultural neuroscience that informs policy development on culture, technology and health. Empirical research on computational cultural neuroscience shows the prediction and explanation of the functional pathways of cognition and brain function that are essential to multilevel adaptation. The use of computational tools and technologies in the scientific discovery of the cultural brain in health and medicine contributes to the broader understanding of the causal mechanisms of cognition and brain function in cultural context.

The contribution of evidence-based research on computational cultural neuroscience to policy development and implementation illustrates the role of the evidence base to inform the wide range of policy areas that benefit from culture, technology and health. The policy development on culture, technology and health details the thematic issues that show improvement in societal and health outcomes based on the implementation of planning and programs. Thematic issues on culture, technology and health include (a) the development of computational tools and technologies that are culturally and ethnically appropriate; (b) the design of prevention and intervention strategies that are responsive to diverse cultures, populations and settings; (c) the development of system-wide approaches that integrate computational informatics and technological platforms in the discovery and delivery science of complex disease. The development of evidence-based research advances the knowledge base to inform on policy development to ensure societal and health resources that bolster health equity.

Evidence-based research on the development of computational tools and technologies entails the building of scientific infrastructure that broadens the evidence base of knowledge to inform on the discovery and delivery science of culture and health. The building of scientific infrastructure contributes to scientific research and development that improves the understanding of the efficacy of computational tools and technologies in cultural neuroscience. Computational tools and technologies facilitate the quality of data that is available to inform the discovery and delivery science of culture and health. The availability of quality of data ensures a breadth of approaches to the design of prevention and intervention that is culturally- and ethnically-appropriate. Computational informatics are beneficial to the depth of understanding and breadth of evidential approaches that are sufficient in the prevention and intervention of complex disease.

The building of scientific infrastructure ensures that an evidence base of knowledge that shows cultural congruence. The manifestation of public health needs arises from considerations of ecological fit and cultural congruence. The development of research to build an evidence base of knowledge is essential to ensure the local ownership of scientific research and production of discovery science that is responsive to the public health needs of the population. The cultural congruence of the evidence base with the local population facilitates the public and scientific understanding of thematic issues to inform on the prevention and intervention strategies on culture and health. The readiness of public and scientific understanding in the development of research and innovation is essential to ensure that the highest quality of data and evidence in the production of the scientific infrastructure.

Computational tools and technologies facilitate the technological innovation of systematic approaches to the discovery and delivery science of prevention and intervention on culture and health. The technological innovation of computational tools and technologies that improve the efficiency and accuracy of the quality of data ensures the highest quality of data and evidence in the production of the evidence base of data and informatics. The computational tools and technologies that are available for scientific research and development on key priorities contribute to the technological innovation that is essential to the design of prevention and intervention strategies. The computational approaches to the understanding and prediction of data and informatics in large-scale data sets are foundational to the scientific understanding of the patterns and regularities on culture and health.

Computational tools and technologies are essential to the scientific understanding of large-scale data sets. The scientific investigation of the causal mechanisms of complex disease builds on a wide range of studies of population health that advance the availability of large-scale datasets. From single studies to epidemiological studies, the systematic approaches to the study of complex disease relies on the technological innovation of computational tools and technologies that advance the efficiency and accuracy of quality of data and informatics. The observational tools of computational tools and technologies allow for the understanding of the causal mechanisms that are of importance in the understanding of the patterns and regularities on culture and health. The identification of the cultural patterns and regularities that show causal influence on biological mechanisms and health contribute to the prevention and intervention of complex disease.

The design of prevention and intervention strategies on culture and health benefits from the integration of computational tools and technologies in the systematic investigation of discovery and delivery science. Computational tools and technologies facilitate the prevention and promotion on culture and health. The computational informatics and data on causal mechanisms of complex disease help to inform the prevention and promotion on culture and health. The public health understanding of prevention and intervention on culture and health benefits from the integration of computational informatics and data to inform scientific knowledge and public awareness on disease prevention and health promotion.

The myriad of scientific and societal resources to inform scientific knowledge and public awareness on disease prevention and health promotion is beneficial to the efficacy of health prevention and promotion. The use of computational informatics and data that are culturally and ethnically appropriate facilitates the breadth of scientific knowledge and public awareness on disease prevention and health promotion. The development of knowledge databases on culturally and ethnically diverse populations is essential to ensure the availability of scientific and educational resources that are effective in health prevention and promotion. The building of scientific knowledge and public awareness on health prevention and promotion is foundational to ensure the availability of resources to inform on the scientific and societal advocacy on culture and health.

The development of knowledge-based resources is of importance to the design of prevention and intervention strategies that are integrative with system-wide approaches. Knowledge-based resources (e.g., "knowledge-based platforms", "digital technologies") ensure the use and availability of digital technology in system-wide approaches of health and medicine. The availability of knowledge-based resources provides large databases of informatics and data that contribute to the design of strategies of prevention and intervention strategies on culture and health. Knowledge-based platforms with large databases facilitate the discovery and delivery science in health and medicine. The knowledge-based platforms of system-wide approaches are essential to ensure the quality of data and evidence that are fundamental to scientific and public understanding in health and medicine.

Knowledge-based platforms are scientific resources that demonstrate the feasibility and effectiveness of computational informatics and data in health and medicine. The design of knowledge-based platforms benefits from the computational informatics and data of diverse cultures, populations and settings. The development of large databases that consist of the population-level health data and informatics for specific complex disease ensures the breadth of an evidential base that is essential to the prevention and intervention strategies of health and medicine. The presence of large databases with knowledge-based platforms for prevention and intervention strategies ensures the use and availability of computational informatics and data of diverse cultures, populations and settings. The availability of computational informatics and data of culturally and ethnically representative populations is essential to ensure the scientific resources and technological platforms that contribute to the feasibility and effectiveness of health prevention and promotion.

The knowledge-based resources of technological platforms benefit from the computational informatics and data of culturally and ethnically representative populations. The use and availability of technological platforms that show readiness and feasibility for use with culturally and ethnically diverse populations is essential to the public and scientific understanding of health and disease. Technological platforms and knowledge-based resources are informative of the digital technologies that show the efficacy of system-wide approaches in health and medicine. The development of digital technologies benefits from the use and availability of technological platforms and knowledge-based resources that are efficacious with the diverse cultures, populations and settings. The presence of digital technologies in system-wide approaches advances the efficiency and effectiveness of prevention and intervention strategies in health and medicine.

Digital technologies in system-wide approaches for health and medicine show substantial benefits in the capabilities to provide scientific and informational resources that contribute to the efficacy of health prevention and promotion. Digital technologies demonstrate the capabilities of technology and health for the benefit of health prevention and promotion. Digital technologies broaden scientific and public awareness of disease prevention and health promotion. Digital technologies show the adaptability and complexity of the research and development capabilities in health and medicine. The adaptability and complexity of digital technologies show how technological innovation advances the precision of the highest quality of data to inform on health prevention and promotion. The use and availability of digital technologies as an equitable societal resource advances the societal conditions of health equity.

The equitable access to digital technologies is essential to the efficacy of system-wide approaches in health and medicine. Digital technologies show the adaptability and complexity of capabilities of the knowledge base of diverse populations, cultures and settings and the feasibility of technological innovation to advance the quality of data and evidence that is essential to cures, preventions and interventions in health and medicine. From vaccines to cures, the capabilities of digital technologies ensure the quality of data and evidence that contributes to the advancement of preventions and interventions that show the preparedness and responsiveness of disease prevention and health promotion. Digital technologies provide the capabilities of system-level approaches in the discovery and delivery science of cures, preventions and interventions.

The integration of digital technologies with system-wide approaches shows the adaptability and complexity of system-wide capabilities in health and medicine. The knowledge base of technological platforms and digital technologies ensures a wide range of informatics and data to ensure the use and availability of prevention and intervention strategies that are efficacious. The system-wide capabilities in health and medicine that are responsive to culturally and ethnically diverse populations demonstrate the equitable access to scientific and societal resources that ensure health equity. The equitable access to scientific and societal resources is essential to the attainment of health equity of diverse cultures, populations and settings.

Evidence-Based Research on Computational Cultural Neuroscience

Evidence-based research on computational cultural neuroscience is of importance to ensure the development of scientific knowledge to inform on health prevention and promotion. Evidence-based research generates novel scientific knowledge that is beneficial to inform on the scientific and technological innovation of culture, technology and health. The generation of novel scientific knowledge of computational cultural neurosciences shows the research and development of computational tools and technologies to systematically investigate culture, brain and behavior. The use of computational tools and technologies provides observational methods for the systematic investigation of the cultural patterns and regularities of the cognition and higher-level function of brain and behavior. The scientific understanding of the brain–behavior relationships of cognition and brain function of diverse cultures, populations and settings is essential to the broader practical application of computational cultural neuroscience in health and medicine.

The generation of scientific knowledge on computational cultural neuroscience benefits from the use of large-scale data sets of diverse populations. Large-scale data sets on brain function and behavior of culturally and ethnically diverse populations contribute to the use and availability of knowledge-based resources that inform the prevention and intervention of health and disease. Large-scale representative samples of brain function and behavior that are culturally and ethnically diverse illustrate the research capabilities of health and medicine for the systematic investigation of brain–behavior relationships of diverse cultures and populations. The development of standards in the research and methods of neuroscience with diverse cultures and populations is essential to ensure the equitable access to knowledge-based resources to inform on the design of prevention and intervention in health and medicine.

The building of the evidence base of computational cultural neuroscience provides novel scientific knowledge that shows the highest levels of quality of data and evidence on culture and health. The scientific knowledge on the study of brain–behavior relationships of cognition and brain function across cultures is comprised of the large-scale data sets that include comparative studies on culture and health. Large-scale comparative studies that comprise cross-cultural research methods contribute to the understanding of the brain–behavior relationships of cognition and brain function of culturally and ethnically diverse populations. The systematic investigation of brain–behavior relationships benefits from the large-scale data sets of computational informatics and data. Computational tools and technologies provide large-scale data sets on the brain–behavior relationships of diverse cultures, populations and settings.

The use of cross-cultural methods in neuroscience is beneficial to the scientific and public understanding of the causal mechanisms that underlie the biological pathways of health of diverse cultures and populations. The systematic study of contextual and environmental influences contributes to the understanding of the biological pathways of health in context. The identification of specific factors that facilitate the social selection and causation of functional mechanisms that contribute to the understanding of contextual and environmental influences in health. The systematic study of the causal mechanisms that affect biological pathways of health in context contributes to the understanding of contextual and environmental influences. Contextual and environmental influences affect the interaction of biological pathways in cultural context.

The development of computational tools and technologies that show effectiveness as a research method for the observational study of the brain function and behavior of diverse cultures and populations is beneficial to scientific and technological innovation. The observational study of brain function and behavior benefits from the development of the methodological standards of computational tools and technologies to ensure the use and availability of knowledge-based resources. The empirical demonstration of the feasibility and appropriateness of methodological standards of computational tools and technologies for use and availability with diverse cultures and populations is of importance to ensure the development of knowledge-based resources.

The building of the scientific evidence base with computational tools and technologies enhances the quality of data on culture, brain and behavior. Novel scientific knowledge with computational tools and technologies shows the breadth of empirical approaches for the observational study of the brain function and behavior of diverse cultures and populations. Computational tools and technologies facilitate the generation of novel scientific knowledge and the efficiency and accuracy of observational study. The use and availability of computational tools and technologies in cultural neuroscience is essential to demonstrate the equitable access to scientific and educational resources for the benefit of the design of prevention and intervention in health and medicine. The equitable access to computational tools and technologies for the study of brain function and behavior of culturally and ethnically diverse populations is fundamental to the societal resources that ensure health equity.

Computational tools and technologies as an evidential approach in computational cultural neuroscience provide a means to ensure the highest level of quality in data and evidence in neuroscience. The development of technological innovation ensures that computational approaches are in use and available to provide quality data and evidence that is beneficial to health prevention and intervention. The benefits of computational tools and technologies across diverse populations are substantive and evidential. Computational tools and technologies harness the potency of technological innovation to ensure the precision in accuracy and performance of knowledge-based platforms and digital technologies. With computational approaches, the systematic investigation of cultural neuroscience advances the public and scientific understanding of the importance of contextual and environmental factors in health and medicine.

Computational approaches in cultural neuroscience are essential to ensure the generation of novel scientific knowledge that demonstrates the benefits of large-scale data. The feasibility and accuracy of computational approaches to large-scale data show the importance of computational tools and technologies on the prediction and explanation of the patterns and regularities of large-scale datasets. The complexity of large-scale data sets details the importance of computational tools for the data extraction of patterns and regularities that are beneficial to the understanding of culture, brain and behavior. The feasibility and accuracy of computational tools to simplify the computational complexity of the data analytics of cultural patterns and regularities facilitates the generation of scientific knowledge that is beneficial to public and scientific understanding.

The research capabilities of computational tools and technologies for the systematic investigation of causal mechanisms are foundational to evidence-based research on computational cultural neuroscience. The computational modeling of the functional pathways of the brain and behavior of diverse cultures and populations shows the feasibility and efficiency of the analytic approaches to the understanding of the causal pathways of functional mechanisms. The computational modeling of the cultural brain stands as a feat in the understanding of the causation of functional mechanisms of brain function and behavior. The prediction and explanation of the causation of functional mechanisms that comprise the biological pathways of brain and behavior are formidable as an evidence-based approach to the scientific understanding of the cultural brain with computational approaches.

Computational approaches are essential to the data mining of large-scale data sets. The use of biomedical technologies such as functional magnetic resonance imaging and other brain imaging tools is foundational to the observational study of the brain function and behavior of diverse cultures. The large-scale data sets of biomedical technologies show the computational complexity of bioinformatics that is essential to the understanding of complex data patterns. Cultural patterns and regularities as a complexity of bioinformatics show how computational approaches are beneficial to the data mining of causal relations that comprise large-scale data sets. Computational approaches are beneficial to the understanding of large-scale data sets and the data extraction of complex data patterns.

Evidence-based research on computational cultural neuroscience is fundamental to the understanding of the computational approaches to the cultural brain. The systematic investigation of the computational principles of the cultural brain contribute to the scientific and public understanding of the structural and functional principles that explain the cultural patterns and regularities of the brain. The identification of the cultural factors that affect the biological pathways of behavior contributes to the understanding of contextual and environmental influences. The use and availability of computational tools and technologies for the empirical study of computational cultural neuroscience are fundamental to ensure the research capabilities of health and medicine for the prevention and intervention in health and medicine of diverse cultures, populations and settings.

Evidence-Based Resources on Computational Cultural Neuroscience

Evidence-based resources on computational cultural neuroscience provide systematic approaches to the development of the evidence base to inform on thematic issues of culture and health. The development of evidence-based resources demonstrate the knowledge generation that informs on the computational informatics and data on the cultural brain. The scientific knowledge on the computational informatics and data of the cultural brain is essential to ensure the comprehensive and systematic approaches to the equitable access to the highest quality of prevention and treatment in health and medicine. The knowledge generation on the cultural brain details the computational approaches that show the technological innovation of importance to the understanding of the cultural brain and its practical applications on culture and health.

The development of evidence-based resources shows the importance of quality of data on the efficacy of prevention and intervention in health and medicine. Evidence-based resources on computational cultural neuroscience show the scope of approaches to the building of knowledge-based platforms that show the practical application of empirical research. Empirical research and development facilitate the development of research capabilities of large-scale data sets and methodological standards of biomedical technologies for the prevention and intervention in health and medicine. The development of research capabilities of large-scale data sets and methodological standards shows the importance of evidence-based resources in health prevention and promotion.

The quality of data and information of evidence-based resources provides reliability and feasibility in the standards of methods that are available for the prevention and promotion of health. The highest standards of quality of data and information contribute to the advancement of the tools and technologies for health prevention and promotion. The evidence-based resources on computational cultural neuroscience demonstrate the quality of data and information that facilitates the development of knowledge-based platforms and digital technologies. The development of knowledge-based platforms and digital technologies is essential to ensure the equitable access to system capabilities of health and medicine that are beneficial to diverse cultures, populations and settings.

Knowledge-based platforms contribute to the use and availability of digital resources that are informative and beneficial to the advancement of quality of information and care in health and medicine. Digital resources such as knowledge-based platforms provide digital technologies and system-wide approaches to advance the highest standards of disease prevention and health promotion in health and medicine. Knowledge-based platforms show the breadth of informatics and data on culture and health that are beneficial to the discovery and delivery science of cures, prevention and intervention. The knowledge-based platforms of digital technologies are essential to the research capabilities that inform on discovery and delivery science of health prevention and promotion.

The use and availability of knowledge-based platforms facilitate the research infrastructure for the production of societal and educational resources that are responsive to public health needs. The research development of knowledge-based platforms contributes to the production of societal and educational resources that demonstrate the use and availability of quality data to inform on the design of prevention and intervention strategies. Health prevention and intervention benefit from the quality of data of computational informatics that informs knowledge-based platforms and digital technologies. The design of knowledge-based platforms that show compatibility with diverse cultures, populations and settings illustrates the availability of evidence-based resources to inform the design of health prevention and intervention strategies.

Evidence-based resources on computational cultural neuroscience advance the discovery and delivery science in health and medicine. The design of knowledge-based platforms facilitate the ease and availability of evidence-based resources to inform on the prevention and promotion of health of diverse cultures and populations. Knowledge-based platforms depict the practical applications that are based on the technological innovation on culture, technology and health. The adaptability and complexity of knowledge-based platforms ensure the feasibility and availability of information and data that informs on the strategies of prevention and intervention. Knowledge-based platforms are beneficial to the demonstration of the adaptability and complexity of research capabilities that are comprehensive as an evidence-based resource.

Evidence-based resources show the wide range of benefits to empirical approaches to the systematic study of culture and health. Empirical approaches generate novel scientific knowledge that broadens the quality and availability of evidence-based resources for the advancement of quality of information and data in health and medicine. Empirical approaches advance the scientific and public understanding of the etiology of complex disease and the myriad of approaches to cures, prevention and intervention to complex disease. Evidence-based resources are essential to the scientific and public advocacy and policymaking on culture and health.

Scientific and educational enrichment on culture and health benefits from evidence-based resources on computational cultural neuroscience. The development of scientific and educational resources on computational cultural neuroscience broadens scientific and public awareness of the importance of computational tools and technologies for the discovery and delivery science of health and medicine. The promotion of scientific and public awareness on computational cultural neuroscience builds on the benefits of scientific and technological innovation as scientific and educational enrichment. The scientific and public appreciation of the importance of scientific and technological innovation is fundamental to the effectiveness of scientific and educational programs that enrich society and the public.

The building of scientific and public awareness on computational cultural neuroscience enhances the scientific and public understanding of comprehensive care in health and medicine. The scientific and public support for computational tools and technologies of cultural neuroscience in health and medicine is essential to ensure the prevention and intervention strategies that are effective to meet public health needs. The computational tools and technologies of cultural neuroscience build understanding of the scientific and educational resources that are of importance to the equitable access to strategies that are efficacious in the prevention and intervention of health and medicine.

Policy-Based Approaches to Computational Cultural Neuroscience

Policy-based approaches to computational cultural neuroscience consist of the development and implementation of policy on culture, technology and health. Policy-based approaches provide a breadth of consideration of the policy mechanisms that are foundational to the policies and programs that are responsive to the public interests. The planning and programs of culture, technology and health are beneficial to the understanding of the policy-based approaches that are beneficial to the preparedness and responsiveness of the population to public interests and needs. The interests and concerns of policies and programs on culture, technology and health ensure the policy-based approaches that facilitate the development of priorities and policy areas that contribute to national policies and planning on culture and health.

Policy-based approaches facilitate the policy development on culture, technology and health. Policy development consists of a multitude of approaches to provide information and data that contribute to societal and public understanding of the main issues that impact health and society. The considerations of culture, technology and health abound with concerns and interests of key priorities to ensure societal equality and health equity. The specific concerns and interests of culture, technology and health include (a) the administration and support infrastructure to provide planning and programming on culture, technology and health; (b) the equitable access to societal and educational resources on culture, technology and health to promote societal equality and health equity; (c) the provision of information and data of system-wide capabilities and resources to ensure the availability of policies and programs on culture, technology and health that are representative of diverse culture, populations and settings. The concerns and interests on culture, technology and health detail the specific policy issues that are of importance to the preparedness and readiness of advocates, policymakers and government leaders to support and promote culture, technology and health in society and the public.

The key priorities on culture, technology and health show international support for the development of policies and planning on culture and health and its scientific and technological innovation. International support is beneficial to the identification of key priorities that inform on the advocacy and policymaking of culture, technology and health. The development of the quality of data and evidence of international experts is essential to ensure that the local scientific infrastructure shows appropriateness and effectiveness in the responsiveness to societal and public concerns. The development of international experts on computational cultural neuroscience informs on the setting of key priorities on culture, technology and health. The building of international standards on computational cultural neuroscience and its practical implications informs the advocacy and policymaking on culture, technology and health.

International standards are essential to the development of guidance that informs advocacy and policymaking on culture, technology and health at all levels. The development of international standards on computational tools and technologies is beneficial to ensure the highest standards of quality of data and evidence for prevention and intervention in health and medicine. The international cooperation for the establishment of standards of scientific and technological innovation of computational cultural neuroscience is fundamental to the reliability and effectiveness of international guidance and support. International expertise in computational cultural neuroscience is beneficial to the demonstration of the efficacy of the international standards on culture, technology and health. The international cooperation on computational cultural neuroscience is foundational to the building of international guidance and support on the highest level of standards in the scientific and technological innovation on culture, technology and health.

The development of international standards and guidance on computational cultural neuroscience provides a scientific infrastructure that is a foundation for international development in culture, technology and health. Scientific and technological innovation is beneficial to ensure the advancement of the standards of production that are beneficial to society and the public. The innovation of science and technology on computational cultural neuroscience shows the feasibility and availability of technological resources for the development of practical applications that show the advancement of the key priorities of culture, technology and health. The international development of key priorities on culture, technology and health advances the policy cycle to ensure the responsiveness of advocacy and policymaking on culture and health.

International development that contributes to the key priorities on culture, technology and health is foundational to the advocacy and policymaking on culture and health. The advocacy and policymaking on culture, technology and health entails a policy cycle that is designed to ensure that translational research informs on the integrative approaches to the practical applications of basic research on computational cultural neuroscience. The generation of scientific knowledge on computational cultural neuroscience contributes to the evidence base that informs the advocacy and policymaking on culture, technology and health. The advocacy and policymaking on key priorities advance with the scientific base to inform advocates, policy-makers and government leaders on the importance of culture, technology and health for the benefit of society and the public. The advocacy and poli-cymaking on culture, technology and policy benefits from the translational research on computational cultural neuroscience to inform practice and policy.

The sustained investment in scientific infrastructure shows the political will of decision-makers that contributes to the societal and public support for the advocacy and policy development of culture, technology and health. The development of scientific infrastructure provides an educational and public resource for the generation and dissemination of scientific knowl-edge that is responsive to the public interests and concerns of the local population. The local ownership of the production of scientific knowledge is essential to ensure that the scientific production is comprehensive and effective in preparedness and responsiveness to public interests. The for-mulation of societal and public issues on culture, technology and health benefits from the sustained investment of scientific infrastructure on com-putational cultural neuroscience and its related fields of study.

The political will of decision-makers on culture, technology and health is foundational to the building of societal influence and public support on policy stances. The scientific and educational resources on computational cultural neuroscience demonstrate the feasibility and reliability of scientific and technological innovation that leads to advancement of research and system capabilities in health and medicine. The public and societal under-standing on the benefits of culture and technology in health and medicine is beneficial to ensure the public support on national and local legislation that supports programs and policies on culture, technology and health. The building of societal influence with scientific and educational resources shows the importance of quality of data and evidence in the building of public and societal awareness of the benefits of culture, technology and health. The scientific and educational enrichment in computational cultural neurosci-ence builds public support for the advocacy and policymaking on key issues.

The building of societal influence and public support on culture, tech-nology and health benefits from the depth of public interest on scientific and technological innovation and its broad range of societal influence. Scientific and technological innovation on computational cultural neuro-science broadens the wide range of computational tools and technologies that show feasibility and availability in the design of prevention and inter-vention of culture and health promotion. The promotion of culture and health with technology is central to the advancement of the highest stand-ards of quality of data and information for scientific and public advocacy and policymaking. Culture and health promotion that highlights the ben-efits of technology for society and the public advances public and scientific awareness of the efficiency, accuracy and reliability of the use of technology in culture and health promotion. The scientific and societal resources that support scientific and technological innovation of local scientific infrastructures and that meet the public health needs of local populations is beneficial to improve societal outcomes.

The public health campaigns on culture, technology and health benefit from the broad societal and educational enrichment on the use of technology in culture and health promotion. The public advocacy of the societal benefits of technology for the promotion of health interests with diverse cultures and populations is of importance to the building of societal influence and public support. The widespread interest and enthusiasm for scientific and educational resources on computational informatics and data on the cultural brain are beneficial to the show of societal influence and public support on culture, technology and health. The public interest in the use of technology for culture and health promotion shows the feasibility and availability of the societal resources that are necessary for effectiveness with public health campaigns.

The involvement and leadership of government and the public are beneficial to the public advocacy on culture, technology and health. The public support of government and leadership for policies and programs on culture, technology and health is foundational to the building of public advocacy and policymaking on key stances. The involvement and leadership of government and the public in the support of key stances for the promotion of culture, technology and health demonstrate the political will and commitment of key decision-makers to advance the public agenda and key priorities on such issues. The advancement of key agenda items of government leadership with public support is crucial to the efficacy of advocacy and policy development on culture, technology and health.

The political commitment of key decision-makers on culture, technology and health demonstrates the sustained investment and support for programs and policies that impact public health needs and societal outcomes. The programs and policies of culture, technology and health contribute to the equitable access to societal and educational resources that enrich public and societal understanding. The sustained investment for programs and policies on culture, technology and health is fundamental to ensure the use and availability of resources that lead to the attainment of health equity. The policy development and implementation on key priorities advance with the political commitment of key decision-makers to show public and societal support for planning and policies that implement the equitable access to public programs on culture, technology and health.

Policy development and implementation on key priorities and issues advance the use of policy mechanisms on culture technology and health.

From national legislation to policy reform, the wide range of policy mechanisms for the policy development and implementation on culture and health promotion demonstrate the importance of the preparedness and readiness of policymakers to advance the key agenda items with efficiency and effectiveness to meet concerns that are in the public interest. Policy mechanisms advance the implementation of planning and policies that contribute to the programming on culture, technology and health. The use of policy mechanisms shows the effectiveness of the involvement and leadership of government and policymakers to advance the interests and concerns on culture, technology and health of the state and the public. The building of state capabilities in the implementation of culture and health promotion shows the adaptability and complexity of system-wide approaches to support tthe wide distribution of societal resources that impact public health awareness and needs.

The considerations of policy implementation show a range of policy mechanisms that support the building of state capabilities in health and medicine. The monitoring and evaluation of the use of policy mechanisms to meet public health needs require accountability and reliability of government and policymakers. The development and use of key indicators for the monitoring and evaluation of system performance are essential to ensure the improvement of societal outcomes that benefit the public. The use of information and data of key indicators for monitoring and evaluation of the performance of state capabilities is beneficial to the adaptability of system-wide approaches that lead to improvement of societal outcomes. The development of policy mechanisms that ensure the building of state capabilities in health and medicine is of importance to ensure that the equitable access to resources that are beneficial to the public.

Policy implementation on culture, technology and health benefits from the monitoring and evaluation of system performance in health and medicine. The use of key indicators in the monitoring and evaluation of system-wide approaches contributes to the use and availability of information and data for the building of state capabilities. The monitoring of key indicators on culture, technology and health (Appendix A) are foundational to the evaluation of system performance in health and medicine. The monitoring and evaluation with key indicators facilitate the information and data to ensure the reliability and accountability of system performance for the benefit of public interests. The overall effectiveness of system performance is essential to the building of system capabilities in health and medicine that meets public health needs.

The demonstration of public health needs is of importance to considerations of policy-level intervention in culture, technology and health. The development and implementation of policy-level intervention contribute to a wide range of policy-based approaches to the provision of societal and educational resources that meet public health needs. The development of policy-level intervention considers a wide range of key issues regarding the public health concern and the policy-based approaches that may lead to effectiveness to meet public interest. Policy-level intervention in culture, technology and health facilitates the use and availability of policies and programs that are specific to public health needs. Policy-level intervention is comprised of the specific policies and programs on culture, technology and health that are tailored to provide societal resources that show effectiveness in the improvement of societal and health outcomes of the local population.

Policy-level intervention in culture, technology and health shows the responsiveness of government leaders and policymakers to meet public health concerns. The preparedness and readiness of policymakers to support policy-level intervention show the effectiveness of policymakers in response to the interests and concerns of public health. The policy-level intervention in culture, technology and health that respond to specific public health needs of the local population shows the preparedness and readiness of government and public health experts at all levels. The efficacy of policy-level intervention demonstrates the capabilities of policy mechanisms for the improvement of societal and health outcomes.

Policy-based approaches to culture, technology and health are of importance to the public and societal understanding on culture and health promotion. The policy-based approaches to the promotion of culture and health advances the societal and educational resources that are beneficial to meet public health needs. The building of societal and educational resources contributes to policy development and implementation on culture, technology and health. The policy-based approaches on culture, technology and health facilitate the wide range of policy mechanisms that show preparedness and readiness of advocates, policymakers and government leaders to show public support for the policies and programs that lead to the improvement of societal and health outcomes of the population.

Policy-Based Research on Computational Cultural Neuroscience

Policy-based research on computational cultural neuroscience is of importance to the understanding of the wide range of approaches to policy development and implementation on culture, technology and health. The wide realm of policy mechanisms describes the production of planning and policies on key issues that impact society and the public. Policy-based research is fundamental to the translation of scientific research to inform public policy. Translational research on computational cultural neuroscience is of importance to inform on the policy development and implementation on key issues of culture, technology and health. The policy-based research on computational cultural neuroscience advances the understanding of the wide range of approaches to scientific and technological innovation on culture and health promotion.

The policy development and implementation of culture, technology and health consist of the understanding of the policy mechanisms that impact the planning and programs on culture and health promotion. Policy development requires planning of policy-based approaches that are fundamental to the implementation of policies and programs. The policy development on culture, technology and health relies on the identification of specific policy areas that require policy response. The wide range of policy areas on culture, technology and health are comprised of the concerns and interests of policy development and implementation (Table 12.1). The development of policy areas on culture, technology and health facilitates the development of policy-based approaches that are effective in policy response.

Policy areas on culture, technology and health consider the interests and concerns of scientific and technological innovation for culture and health promotion. Specific policy areas on culture, technology and health consist of thematic issues regarding topics such as (a) the impact of technology on culture and health promotion; (b) the equitable access to the resources of technology and health of culturally and ethnically diverse populations; (c) the provision of scientific, societal and educational resources on culture, technology and health in the promotion of culture and health equity. The consideration of specific policy areas facilitates the formulation of policy

TABLE 12.1 Policy Areas on Culture, Technology and Health

Policy Area		Example
Policy Area A	Policy-Based Research on Culture, Technology and Health	National Policy Center on Culture, Technology and Health
Policy Area B	Policy-Based Resources on Culture, Technology and Health	National Informatics and Databases on Computational Cultural Neuroscience
Policy Area C	Policy-Level Intervention on Culture, Technology and Health	Programming on Culture, Technology and Health to specific population or target group
Policy Area D	Public Health Advocacy and Policymaking on Culture, Technology and Health	Public Health Campaigns on Culture, Technology and Health Awareness
Policy Area E	Public Health Action on Culture, Technology and Health	National Programs on Culture, Technology and Health

stances from which to policy development ensures the policy response that is effective in meeting public health interests and concerns.

Policy-based research on policy areas on culture, technology and health is fundamental to the building of societal and educational resources to inform advocacy and policymaking on culture and health promotion. The consideration of interests regarding the impact of technology on culture and health promotion is one of the foremost issues that require the information and data of policy-based research. First, international and scientific experts are a source of information and data to inform the consideration of interests in the impact of technology on culture and health promotion. The policy-based research on the impact of technology on culture and health promotion provides information and data that contribute to the understanding of the policy mechanisms that are beneficial to societal and health outcomes.

Policy-based research on the impact of technology on culture and health promotion provides a wide range of information on the benefits of policies and programs for society and the public. Policy-based research investigates the efficacy of the impact of technology on the programs on culture and health promotion. The development of policy-based resources that provide information and data on key indicators is of importance to the evaluation of planning and programs. The identification of key indicators is essential to the formulation of conceptual and empirical approaches for the implementation of policy-based research. The use of multilevel approaches in policy-based research for the study and analysis of information and data on key indicators is beneficial to the implementation of research.

Policy-based resources are important to ascertain the information and data that are of importance to the evaluation of policy goals and outcomes. The development of policy-based resources (e.g., population data, study data) on key indicators contributes to the feasibility and reliability of information and data for study and analysis on the effectiveness of policy measures. The study and analysis of the effectiveness of policy measures consider a range of key indicators that pertain to the change of policy outcomes based on policy measures. Policy-based resources contain multilevel methods for the study and analysis of information and data to ascertain the change of policy outcomes that are of relevance to policy response.

The development of policy-based resources contributes to the evaluation of the policy outcomes based on specific policy response that centers on the technology of culture and health promotion. Planning and policies that facilitate the wide dissemination of scientific and educational resources on culture, technology and health to inform the public and societal understanding on culture and health promotion provide a policy response that aims to improve societal and health outcomes. The evaluation of the effectiveness of planning and policies on the public advocacy of culture, technology and health contributes to the scientific and educational enrichment and at the same time provides a policy response that leads to the attainment of priority goals. Policy-based resources are beneficial to determine the overall effectiveness of the public advocacy of technology in culture and health promotion.

Policy-based resources on computational cultural neuroscience show the impact of the innovation of science and technology on culture and health promotion. Planning and policies that support the building of scientific infrastructure that is responsive to the public health concerns of the local population contribute to the use and availability of policy-based resources for the evaluation of the efficacy of the impact of technology on culture and health promotion. The building of scientific infrastructure for the use of computational tools and technologies in the study of cultural neuroscience requires the adaptability and complexity of research capabilities in health and medicine. The local ownership of scientific knowledge and the building of scientific infrastructure are elements of key indicators that are of importance to the evaluation of policy measures and responses. Policy-based research that evaluates the effectiveness of scientific and educational resources to promote culture and health contributes to the policy-based resources that inform on the impact of technology on culture and health promotion.

Another consideration of policy-based research on computational cultural neuroscience concerns the use and availability of technology among culturally and ethnically diverse populations. The equitable access to technology that is amenable to the study of culturally and ethnically diverse populations is of importance to the understanding of the overall effectiveness of technology in health and medicine. Planning and policies are beneficial to ensure the equitable access to technology for the promotion of culture and health. Planning and policies that provide culture and health promotion programs contribute to the scientific and societal resources that support the equitable access to technology in health and medicine that is available to diverse cultures and populations. Policy-based research that evaluates the improvement of societal and health outcomes based on the planning and policies that promote culture, technology and health contributes to the understanding of the overall effectiveness of policy response.

Policy-based research on computational cultural neuroscience contributes to the understanding of the policy response that advances equitable access to culture and technology in health and medicine. Planning and policies that promote culture, technology and health facilitate the equitable access to scientific and societal resources for the benefit of societal and health outcomes. Computational tools and technologies provide research capabilities that are beneficial to the strategies of cures, prevention and intervention in health and medicine. The advancement of research capabilities in health and medicine leads to the quality of care in health and medicine. The evaluation of societal and health outcomes based on the highest standards of technological innovation contributes to the understanding of the effectiveness of the equitable access to culture and technology on the improvement of societal and health outcomes of the population.

Further considerations of policy-based research include the evaluation of the effectiveness of equitable access to culture and technology on population health. The implementation of policies and programs that provide equitable access to culture and technology is of importance to the understanding of the overall improvement of societal and health outcomes of diverse cultures and populations. The equitable access to culture and technology in health and medicine builds the research and delivery capabilities of health systems. The planning and programs that support equitable access to culture and health contribute to the enhancement of system-wide approaches in health and medicine. The enhancement of research and scientific capabilities facilitates the policy implementation of culture and technology that is beneficial to system-wide approaches. Policy-based research that ensures the equitable access to culture and technology contributes to the improvement of societal and health outcomes of the population.

Policy-based research on computational cultural neuroscience is beneficial to the understanding of the impact of scientific and educational enrichment on culture, technology and health for societal and health outcomes. Scientific and educational enrichment broadens public awareness and advocacy on culture, technology and health for the benefit of society and the public. The provision of scientific, societal and educational resources on computational cultural neuroscience builds scientific and public understanding of the impact of technological innovation on culture and health. Scientific and educational resources broaden public awareness of the impact of technological innovation on the societal and health outcomes of diverse cultures and populations. The wide dissemination of scientific and educational resources provides information and data that informs the public and societal understanding on the impact of technology on culture and health.

The policy-based approaches of empirical research on computational cultural neuroscience broaden the societal influence and public support on the policies and programs on culture, technology and health. Policy-based research provides information and data to inform the impact of societal influence and public support on the effectiveness of policies and programs on culture, technology and health. Culture and health promotion programs that include public health campaigns that broaden societal influence and public support for the planning and programming on culture, technology and health benefit from a base of public support that leads to the improvement of societal and health outcomes of the population. Policy-based research contributes to the scientific resources that are beneficial to culture and health equity.

The translational research on computational cultural neuroscience informs the development of societal and educational resources on culture, technology and health. The planning and programming of societal and educational resources advances the public and societal advocacy and policymaking on culture, technology and health. Societal and educational resources facilitate the development of public health campaigns that broaden public and societal awareness of key issues that impact the interests and concerns of public health. The building of public and societal awareness on culture, technology and health highlights the policy issues that are of importance to meet public health concerns. The tailoring of key thematic issues that are central to the health prevention and promotion of diverse cultures and populations is foundational to the practice and policy of culture and health.

The planning and programming of culture and health promotion are central to the development of key themes and issues of public health campaigns on culture, technology and health. The broadening of societal and public understanding of the importance of key ideas on disease prevention and health promotion is central to the promotion of culture and health. Key themes on disease prevention and health promotion of public health campaigns with diverse cultures and populations include (a) the building of public awareness on the benefits of disease prevention and health promotion that is culturally and ethnically diverse is necessary to the amelioration of disease and to the protection of public health interests; (b) societal and public support is essential to the prevention and promotion on culture and health for the benefit of diverse cultures and populations; (c) disease prevention and health promotion are a public interest that has ethical, societal and public health impact on the efficacy of culture and health promotion. The building of public advocacy on culture, technology and health leads to consideration of the design of public health campaigns that are efficacious in the building of societal and public support for policy issues and stances.

Public health campaigns show the effectiveness of the planning and programming on the advocacy of key themes of culture and health promotion. The tailoring of key messages on public health interests is essential to ensure the effectiveness of the health communication of public health campaigns. The building of key messages that are central to the public health interests on culture, technology and health facilitates the development of health communication that is efficacious to the broadening of public and societal understanding. Health advocacy that is central to the building of public awareness on culture, technology and health includes key themes on the role of technological innovation in culture and health promotion. The thematic content on technological innovation in culture and health promotion provides health communication messages that are accurate and reliable on the efficacy of the culture and health promotion that is based on technology in health and medicine that is culturally and ethnically appropriate. The development of public health campaigns on culture, technology and health is essential to the public health advocacy of culture and health promotion.

Policy-based research is beneficial to the understanding of the effectiveness of the health communication on the public health campaigns of diverse cultures and populations. Policy-based research on culture, technology and health considers themes and topics such as (a) the effectiveness of health communication on the public health advocacy on culture, technology and health; (b) the impact of technological innovation on culture and health promotion on the effectiveness of public health advocacy; (c) the impact of scientific, societal and educational resources on the efficacy of health communication on culture, technology and health and their impact on culture and health promotion. The policy-based research on culture, technology and health contributes to the public and societal understanding of the impact of health communication of public health campaigns on culture and health promotion.

Policy-based research on computational cultural neuroscience contributes to the information and data on the effectiveness of policy mechanisms on culture and health promotion. Policy-based research provides the information and data to inform the implementation of policy mechanisms that show the effectiveness of policy response. The formulation of policies and programs on culture, technology and health across levels benefits from the quality of information and data that inform the development of policy mechanisms. The policy development mechanisms demonstrate the formulation of legislation and policies that support the implementation of policy response. Policy development mechanisms illustrate the readiness and preparedness of national, state and local governments for planning and policies on culture, technology and health in a global policy environment that is responsive to public health needs.

Policy-based research on the effectiveness of policy response is foundational to the public and societal understanding of the public health impact of culture, technology and health. The understanding of the public health impact on the provision of scientific, societal and educational resources is essential to the discernment of the effectiveness of advocacy and policymaking on culture and health promotion. The public health advocacy and policymaking on culture and health promotion provide a wide range of policy-based approaches in the implementation of policy response. The policy-based research on computational cultural neuroscience contributes to the building of public and societal awareness of the societal and health benefits of culture and health promotion.

Discussion

The considerations on culture, technology and health in the public interest benefit from the evidence-based resources on computational cultural neuroscience. Evidence-based resources provide quality of information and data that inform on the policy development and implementation on culture, technology and health for the purpose of culture and health promotion. The building of evidence-based approaches on computational cultural neuroscience advances the scientific and technological innovation that informs the practice and policy of culture and health. The development of evidence-based approaches to scientific and technological innovation that is culturally and ethnically appropriate is essential to the strengthening of research capabilities in health and medicine. The building of evidence-based resources on computational cultural neuroscience is beneficial to the advancement of tools and technologies in health and medicine that contribute to the improvement of societal and health outcomes of diverse cultures and populations.

Policy-based approaches on computational cultural neuroscience further consideration of the interests and concerns on culture, technology and health that is beneficial to culture and health promotion. The considration of interests and concerns on culture, technology and health enhance the societal and public understanding of the policy areas and stances that matter in the building of societal influence and public support for culture and health promotion. The involvement and leadership of stakeholders in the global policy environment advances the planning and programming on culture, technology and health for the benefit of society and the public. The understanding of policy-based approaches is essential to the broadening of scientific and societal awareness of the importance of culture, technology and health for public health benefit.

The public health advocacy and policymaking on culture and health promotion advances with the considerations of interests on culture, technology and health. Public health advocacy requires the planning and programming of evidence-based approaches to provide information and data that build scientific and educational enrichment on culture, technology and health. The building of public health awareness is crucial to the broadening of public support that leads to policy development and implementation. Evidence-based approaches to computational cultural neuroscience provide the scientific, societal and educational resources that inform the scientific and educational enrichment on culture, technology and health. The public and societal understanding of the central issues on culture, technology and health is fundamental to the effectiveness of public health advocacy and action.

Conclusion

Translational research on computational cultural neuroscience informs on the practice and policy of culture and health promotion. The integrative research on computational cultural neuroscience builds the evidence base of knowledge on culture, technology and health to inform the planning and policies on culture and health promotion. The building of the evidence base of scientific knowledge on culture, technology and health advances the use and availability of knowledge-based resources to inform the promotion of culture and health. The evidence-based approaches to the advocacy and policymaking on culture, technology and health contribute to the scientific and educational enrichment that is beneficial to health and society.

Evidence-based approaches to the advocacy and policymaking on culture, technology and health advance the scientific infrastructure and the societal and educational resources that inform the advocacy and policymaking on culture, technology and health. Societal and educational resources on culture, technology and health broaden the quality of information and data to inform scientific and educational enrichment on culture and health promotion. The development of societal and educational resources advances the public and societal understanding of the wide scope of benefits of the promotion of culture and health. Evidence-based approaches to the advocacy and policymaking on culture, technology and health are beneficial to the building of scientific and societal resources on culture and health promotion.

Policy-based approaches on computational cultural neuroscience facilitate the understanding of the impact of advocacy and policymaking on culture, technology and health and its societal and public health benefits. Policy mechanisms provide a wide realm of policy-based approaches of the global policy environment to support the planning and policies of culture and health promotion programming. The policy development and implementation on culture, technology and health ensure the equitable access to technological innovation in health and medicine that lead to the improvement of societal and health outcomes of diverse cultures and populations. The policy development on culture, technology and health contributes to the planning and programming on culture and health promotion that meets public health interests and concerns. Policy implementation facilitates the public health advocacy and policymaking on culture, technology and health and the public health action on culture and health promotion.

Implications

Public Health

Translational research on computational cultural neuroscience is beneficial to inform the practice and policy of culture and health. Evidence-based research on computational cultural neuroscience informs on the computational tools and technologies that are of importance to the study of culture and the brain. The scientific and technological innovation on computational cultural neuroscience broadens the practical applications of basic biomedical and biobehavioral research in health and medicine. The advancement of translational research in the field of study is beneficial to ensure the building of scientific infrastructure that is essential to the advocacy and policymaking on culture and health promotion.

Evidence-based resources on computational cultural neuroscience inform the design of cures, prevention and intervention on culture and health. The design of prevention and intervention strategies relies on an evidence base to inform on the practical applications of evidence-based research. Evidence-based research is beneficial to ascertain the feasibility, accountability and effectiveness of prevention and intervention strategies that are based on culture, technology and health. The development of cures, prevention and intervention strategies that build on scientific and technological innovation benefit from the design of tools and technologies that show the efficacy of culture, technology and health with diverse cultures and populations.

Policy-based resources on computational cultural neuroscience are beneficial to the public health advocacy on culture, technology and health in the promotion of culture and health promotion. Public health advocacy on culture, technology and health builds public and societal awareness on the prevention and intervention of culture and health promotion. The broadening of public and societal awareness builds the public support for the advocacy and public health action on culture and health. The public health advocacy and action on culture and health is beneficial to the attainment of societal equality and health equity.

Medicine

Basic biomedical and biobehavioral research on computational cultural neuroscience generates scientific knowledge that informs the research capabilities of system-wide approaches in health and medicine. Research capabilities in health and medicine are essential to ensure the highest standards of quality of care and treatment of complex disease. The development of cures, preventions and interventions on culture and health benefit from knowledge platforms and digital technologies that show efficacy with diverse cultures, populations and settings. The adaptability and complexity of research capabilities of the system-wide approaches in health and medicine are fundamental to the advancement of medical innovation that leads to effectiveness in the prevention and treatment of complex disease.

The development of computational tools and technologies is beneficial to the wide use and accessibility of the highest standards of culture, technology and health in medicine. The medical innovation of computational tools and technologies ensures the wide use and accessibility of culture and technology that has an impact on societal and health outcomes. Computational tools and technologies contribute to the effectiveness of prevention and intervention strategies in health and medicine. The use and accessibility of computational tools and technologies that are culturally and ethnically appropriate as system capabilities advance the highest standards of quality of care and treatment in medicine.

The medical discovery of cures, preventions and interventions to complex disease relies on the scientific and technological innovation of research and medical capabilities in culture, technology and health. The development of medical innovation for the purpose of disease prevention and health promotion demonstrates the importance of scientific and technological innovation on the amelioration of complex disease. Computational tools and technologies that are culturally and ethnically appropriate are beneficial to the design of effective care and treatment to complex disease. Medical discovery advances with the integration of computational tools and technologies that enhance medical capabilities for the discovery of cures, preventions and interventions to complex disease.

Public Policy

The considerations on culture, technology and health benefit from policy-based resources on the interests and concerns of technology on culture and health promotion. The public interest on culture, technology and health aims to lead in the advancement of the impact of scientific and technological innovation for the improvement of societal and health outcomes. The development of policy-based resources facilitates the enrichment of societal and public understanding on the benefits of culture, technology and health. The building of public awareness on the impact of scientific and technological innovation on culture, technology and health for the improvement of health outcomes contributes to the development of policies and programs on culture and health promotion.

Policy development and implementation on culture, technology and health are essential to the formulation of key policy areas that are of importance in the building of public health advocacy and action on culture and health promotion. From policy-based research to public health action, the policy development and implementation on culture, technology and health informs on the policy mechanisms that are beneficial to the improvement of societal and health outcomes. The policy-based approaches to culture, technology and health contribute to the advancement of policies and programs that meet the public health concerns of diverse cultures and populations.

CONCLUSION

The book *Computational Cultural Neuroscience* is an introduction to the study of the computational foundations of cultural neuroscience. The study of computational cultural neuroscience is fundamental for understanding the computational principles of cultural neuroscience. The systematic study of computational cultural neuroscience contributes to the understanding of the computational approaches to the study of the cultural brain. The structural and functional organization of the cultural brain contributes to the understanding of the computational principles that contribute to the cultural processes and mechanisms of cortical brain organization. The structural and functional principles of cortical brain organization contribute to the characterization of the multilevel mechanisms of the cultural brain and the casual–functional role of cultural processes and mechanisms on information processing.

Conceptual foundations of computational cultural neuroscience as a field of study encompass a wide range of conceptual approaches to the understanding of the functional machinery of culture and the brain. The philosophical notion of the mind as a computing machine illustrates the importance of mental causation as a conduit for the computational performance of the cultural brain. The pontification on the cultural causation of the mind in the world illuminates on the functional performance of minds and machines as a demonstration of culture as a causal power. The study of the cultural processes and mechanisms of the brain entails the

DOI: 10.4324/9781003384236-14

understanding of the computational foundations of the culture in the mind and in the world.

Philosophical notions of the cultural mind as a computing machine entail the understanding of the functional machinery of biological organisms as a cultural species. The postulation of the biological organism as a computational machine elaborates the notion of the functional mechanisms that have a causal–functional role. The computational machinery of biological organisms demonstrates the causal mechanisms that show the functional performance of cultural adaptation. The demonstration of the functional processes of biological organisms as a causal mechanism illustrates a mechanistic basis for the functional performance of culture and adaptation. The functional machinery of biological organisms that show cultural capacities illustrate the feasibility of the cultural mind as a computing machine.

The scholarly inquiry on the cultural mind as a computing machine details the understanding of the cultural brain as the functional machinery of biological organisms. The cultural brain as a mechanistic basis of functional and cultural adaptation illustrates the causal–functional role of cortical brain organization on cultural processes and mechanisms. The delineation of the causal–functional role of cortical brain mechanisms and their functional significance is essential to the mechanistic understanding of cultural capacities. The characterization of the multilevel mechanisms of the cultural brain contributes to the understanding of the functional mechanisms that underlie the cultural brain as the functional machinery of biological organisms as a cultural species.

Methodological foundations on computational cultural neuroscience comprise the design of computational tools and technologies for the observation and study of the cultural brain. Computational approaches are essential to the understanding of the causal modeling of the interaction of cortical brain regions and their functional correspondence with culture and behavior. Computational tools and technologies provide a means for the computational modeling of cortical brain dynamics and their causal role in culture and behavior. The computational modeling of the dynamical organization of the cultural brain facilitates the prediction and inference of the causal interaction of cortical brain regions that are based on cultural processing. Computational approaches to cultural neuroscience are fundamental to the empirical study of the cultural brain and its causal interaction.

Empirical foundations on computational cultural neuroscience comprise the empirical study of the computational principles of the cultural brain. Systematic investigation of computational cultural neuroscience provides a foundation for the understanding of the computational basis of cultural processing of the brain. From neurons to networks of neurons, the study of cultural neurocomputation describes the functional components of cultural processing and their physical instantiation in the functional mechanisms of cortical brain organization. Neurocomputational study of cultural processing depicts the functional mechanisms such as distinct types of neuronal responses that encode sensory data of the cultural environment. Empirical study of cultural neural networks consists of the systematic investigation of the structural and functional connectivity of the cultural brain and the role of interconnection in understanding of the cultural brain. The interconnectivity of the cultural brain provides a functional basis for the understanding of the computational principles of culture and functional brain activity.

Research on computational cultural neuroscience is fundamental to the empirical study of the computational foundations of the cultural brain. The development of scientific concepts and language in computational cultural neuroscience facilitates empirical approaches to the study of culture and the brain. Empirical study as a field of study entails the design of computational tools and technologies for the study of structural and functional principles of cultural processes and mechanisms and their functional correspondence to cortical brain organization. The innovation of computational tools and technologies facilitates the understanding of the structural and functional principles that are of importance to the characterization of the functional mechanisms of the cultural brain.

Computational approaches to the study of the cultural brain abound with the computational tools and technologies for the observation and study of the causal–functional relations of culture and cortical brain organization. The causal–functional role of cortical brain mechanisms is of importance to the processes and mechanisms of culture and behavior. The systematic study of the causal–functional relations of culture and cortical brain mechanisms contributes to the understanding of the causal mechanisms of cortical brain organization. The characterization of the causal–functional role of cortical brain mechanisms illustrates the observation and study of the functional correspondence of culture, brain and behavior with computational tools and technologies.

Computational modeling of the cultural brain provides a means for the prediction and inference of the causal dynamics of cortical brain organization and its functional correspondence with culture and behavior. The use of computational tools to investigate the causal dynamics of the functional brain activity of the cultural brain illustrates the role of causal modeling in the prediction and inference of the cultural brain. The study of causal dynamics of culture and cortical brain organization facilitates the understanding of the causal interaction of cortical brain regions and the information processing of culture. The computational modeling on the causal dynamics of functional brain activity of the cultural brain provides a means for the prediction and inference on the functional correspondence of causal interaction of culture, brain and behavior.

The systematic investigation of the structural and functional connectivity of the cultural brain facilitates the understanding of the interconnection of cortical brain regions and their functional significance. The structural and functional connectivity of cortical brain organization shows the fundamental principles that describe the functional processes and mechanisms of brain function and behavior. The structural and functional principles of the cortical brain detail the computational principles that underlie the functional machinery of cultural information processing mechanisms. The interconnection of cortical brain organization provides a mechanistic basis for the understanding of the causal relations of cortical brain regions and their functional correspondence with cultural capacities.

The study of the causal dynamics of the cultural brain contributes to the characterization of the computational principles that underlie cultural brain dynamics. The dynamics of cultural brain activity illustrate the fluctuations of cortical brain function that are of importance to cultural processing. The characterization of the dynamical processes of cultural brain activity describes the changes in spatiotemporal processing that correspond to cultural brain function. The dynamics of cultural brain activity illustrate the autonomous and self-organizing principles of culture and cortical brain function. The bidirectional information processing of cultural brain mechanisms demonstrates the information flow and brain dynamics of cultural computation. The interconnection of the cultural brain shows the functional characteristics of the dynamics of cultural brain activity that contribute to the complexity of cultural and neural systems. The interconnectivity of cortical brain function demonstrates the core structure of structural and dynamical basis of culture and cortical brain function.

The evidence-based research on computational cultural neuroscience builds scientific and societal resources that inform practice and policy. Translational research on computational cultural neuroscience provides the quality of data and information that contributes to the scientific and societal understanding of culture and technology in health and medicine. The systematic study of cultural brain function facilitates the building of scientific and societal enrichment on the cultural brain and its societal impact. The building of evidence-based resources on computational cultural neuroscience contributes to the advocacy and policymaking of culture and health promotion. Evidence-based research on computational cultural neuroscience is fundamental to the building of scientific and societal resources that inform public health advocacy and policymaking for the promotion of culture and health.

In three parts, the book provides a comprehensive review of the conceptual, methodological and empirical foundations of computational cultural neuroscience. Part 1 of the book consists of an introduction to the fundamentals of computational cultural neuroscience. Thematic content on computational cultural neuroscience includes cultural computation, cultural mental computation, cultural machine computation and cultural neurocomputation. Topics of thematic content illustrate a wide range of conceptual approaches to understand the interrelation of culture and computation of minds, brains and machines.

In Part 1, the first five chapters provide a comprehensive review of the fundamentals of computational cultural neuroscience. Chapters 1–5 describe the conceptual foundations of computational cultural neuroscience.

Chapter 1 on Computational Cultural Neuroscience provides a comprehensive introduction to the field of study. The chapter explores the computational foundations of the study of culture and the brain. The chapter introduces a review of computational principles that describe the structural and functional organization of the cortical brain and its relation to cultural adaptation.

Chapter 2 on Cultural Computation posits on the philosophical foundations of cultural computation. The chapter provides a philosophical review on culture and computation of minds and machines. Philosophical inquiry on cultural computation provides insight on a wide range of topics that are of importance to the understanding of culture, mind and machines.

Chapter 3 on Cultural Mental Computation describes the foundations of cultural mental computation. The foundations of cultural mental computation consist of the understanding of the functional components of the cultural mind. The characterization of the functional components of cultural computation of the mind contributes to the understanding of the cultural mind as a computing machine.

Chapter 4 on Cultural Machine Computation details the fundamentals of cultural machine computation. The philosophical interest on cultural machine computation provides a conceptual foundation for the postulation on culture in mind and machine. Cultural machine computation describes the cultural computation of machines. The notion of culture as machine computation illustrates the role of machines as a computational basis of culture in the world. The cultural computation of machines demonstrate the functional significance of computing machines on the production of culture in the mind and in the world.

Chapter 5 on Cultural Neurocomputation provides a review on cultural neurocomputation. Cultural neurocomputation consists of the neurocomputational study of culture and behavior. The neurocomputational study of cultural processing details the encoding of cultural information across cortical layers of brain organization. The study of cultural neurocomputation describes the functional machinery of biological organisms as a cultural species.

In Part 2, the book explores the fundamentals of the cultural brain and its structure and function. Part 2 provides a review on the computational principles of the cultural brain. Chapters 6–9 discuss the functional architecture of the cultural brain and its structural and functional principles.

Chapter 6 on Cultural Neural Networks provides a review on neural networks that are dedicated to cultural processing. The cultural patterns of functional brain activity show the functional role of neural networks on the cultural information processing of functional mechanisms. The study of the patterning of cultural brain activity and its causal–functional role is essential to the understanding of the structural and functional organization of the cortical brain.

Chapter 7 on Cultural Brain Dynamics details the foundations of cultural brain dynamics. Cultural brain dynamics describe the dynamical principles that are fundamental to the understanding of the causal interaction of the cultural brain. The computational modeling of causal dynamics of the cultural brain contributes to the understanding of the causal interaction of cortical brain regions and their causal–functional role in cultural processing.

Chapter 8 on Cultural Machine Learning describes the conceptual approaches to the computational modeling of cultural brain function and the functional role of learning on cultural computation. The systematic study of cultural machine learning is foundational to the characterization of cultural patterns of mental and machine computation.

Chapter 9 on Cultural Connectome discusses the structural connectivity of the cultural brain and its functional significance. The cultural connectome reviews the structural and dynamical principles of interconnectivity as a computational basis of cultural brain dynamics.

In Part 3, the book investigates the translational research on computational cultural neuroscience and its practical applications for health, medicine and public policy. Chapters 10–12 review the scientific and technological innovation on computational cultural neuroscience and its public impact.

Chapter 10 on Culture and Technology investigates the fundamentals of culture and technology. The chapter reviews scientific and technological innovation on computational cultural neuroscience and its practical applications for culture and technology.

Chapter 11 on Cultural Intelligence explores the foundations of cultural intelligence and its practical applications for scientific and technological innovation. Research on cultural intelligence investigate the conceptual foundations of cultural inference and reason as a functional basis of cultural intelligence. Theoretical foundations of cultural intelligence posit on the fundamentals of cultural intelligence.

Chapter 12 on Computational Cultural Neuroscience and Public Policy describes the importance of translational research to inform practice and policy on culture and health. Policy-based approaches benefit from translational research on computational cultural neuroscience. Consideration of the scientific and societal benefits of translational research for policy development and implementation on culture and health promotion are discussed.

The scholarly and scientific interest on computational cultural neuroscience abounds with a wide realm of considerations at the intersection of culture, technology and health. The connotation of culture as minds and machines broadens humanistic and scientific understanding of the functional and mechanistic basis of culture in the world. From biological organisms to computing machines, the multiple realizability of culture as computation illuminates on the breadth of physical realizers that demonstrate the causal potency of culture of the mind and the world. The characterization of the cultural computation of brain function broadens the imagination of the core capacities of biological organisms as cultural species. The functional basis of biological organisms for cultural capacities illustrates the complexity of the brain as a conduit of cultural adaptation.

The systematic study of computational cultural neuroscience allows for the characterization of the laws and principles of culture in the mind and in the world. The scholarly study of the lawlike principles of cultural patterns in the natural world broadens the understanding of culture as a causal mechanism. The philosophical inquiry of culture as a causal system benefits from the pontification on the scientific naturalism of the lawlike principles of culture. The postulation on the breadth of culture as a causal mechanism elaborates on the interaction of culture in the world and its effects. The philosophical and scientific quandary on culture as computation illuminates the scholarly and scientific benefits of interdisciplinary approaches to fundamental questions.

APPENDIX A

Key Indicators on Culture, Technology and Health

Key indicators

1. Public Involvement
   (e.g., How important is public involvement for programs on culture, technology and health?)
2. Public Support
   (e.g., How important is public support for programs on culture, technology and health?)
3. Societal Perception
   (e.g., What is the societal perception of the effectiveness and reliability of programs on culture, technology and health?)
4. Efficacy, Reliability, Accountability
   (e.g., What is the improvement in societal and health outcomes based on the programs on culture, technology and health?)
5. Effectiveness
   (e.g., What is the overall effectiveness of programs on culture, technology and health?)

INDEX